IEE ELECTROMAGNETIC WAVES SERIES 40

Series Editors: Professor P. J. B. Clarricoats
Professor Y. Rahmat-Samii
Professor J. R. Wait

UNIFORM STATIONARY PHASE METHOD

Other volumes in this series:

Volume 1 **Geometrical theory of diffraction for electromagnetic waves** G. L. James
Volume 2 **Electromagnetic waves and curved structures** L. Lewin, D. C. Chang and E. F. Kuester
Volume 3 **Microwave homodyne systems** R. J. King
Volume 4 **Radio direction-finding** P. J. D. Gething
Volume 5 **ELF communications antennas** M. L. Burrows
Volume 6 **Waveguide tapers, transitions and couplers** F. Sporleder and H. G. Unger
Volume 7 **Reflector antenna analysis and design** P. J. Wood
Volume 8 **Effects of the troposphere on radio communications** M. P. M. Hall
Volume 9 **Schumann resonances in the earth-ionosphere cavity** P. V. Bliokh, A. P. Nikolaenko and Y. F. Flippov
Volume 10 **Aperture antennas and diffraction theory** E. V. Jull
Volume 11 **Adaptive array principles** J. E. Hudson
Volume 12 **Microstrip antenna theory and design** J. R. James, P. S. Hall and C. Wood
Volume 13 **Energy in electromagnetism** H. G. Booker
Volume 14 **Leaky feeders and subsurface radio communications** P. Delogne
Volume 15 **The handbook of antenna design, Volume 1** A. W. Rudge, K. Milne, A. D. Olver, P. Knight (Editors)
Volume 16 **The handbook of antenna design, Volume 2** A. W. Rudge, K. Milne, A. D. Olver, P. Knight (Editors)
Volume 17 **Surveillance radar performance prediction** P. Rohan
Volume 18 **Corrugated horns for microwave antennas** P. J. B. Clarricoats and A. D. Olver
Volume 19 **Microwave antenna theory and design** S. Silver (Editor)
Volume 20 **Advances in radar techniques** J. Clarke (Editor)
Volume 21 **Waveguide handbook** N. Marcuvitz
Volume 22 **Target adaptive matched illumination radar** D. T. Gjessing
Volume 23 **Ferrites at microwave frequencies** A. J. Baden Fuller
Volume 24 **Propagation of short radio waves** D. E. Kerr (Editor)
Volume 25 **Principles of microwave circuits** C. G. Montgomery, R. H. Dicke, E. M. Purcell (Editors)
Volume 26 **Spherical near-field antenna measurements** J. E. Hansen (Editor)
Volume 27 **Electromagnetic radiation from cylindrical structures** J. R. Wait
Volume 28 **Handbook of microstrip antennas** J. R. James and P. S. Hall (Editors)
Volume 29 **Satellite-to-ground radiowave propagation** J. E. Allnutt
Volume 30 **Radiowave propagation** M. P. M. Hall and L. W. Barclay (Editors)
Volume 31 **Ionospheric radio** K. Davies
Volume 32 **Electromagnetic waveguides: theory and application** S. F. Mahmoud
Volume 33 **Radio direction finding and superresolution** P. J. D. Gething
Volume 34 **Electrodynamic theory of superconductors** S.-A. Zhou
Volume 35 **VHF and UHF antennas** R. A. Burberry
Volume 36 **Propagation, scattering and dissipation of electromagnetic waves** A. S. Ilyinski, G. Ya. Slepyan and A. Ya. Slepyan
Volume 37 **Geometrical theory of diffraction** V. A. Borovikov and B. Ye. Kinber
Volume 38 **Analysis of metallic antennas and scatterers** B. D. Popović and B. M. Kolundžija
Volume 39 **Microwave horns and feeds** A. D. Olver, P. J. B. Clarricoats, A. Kishk and L. Shafai

UNIFORM STATIONARY PHASE METHOD

V. A. BOROVIKOV

The Institution of Electrical Engineers

Published by: The Institution of Electrical Engineers, London,
United Kingdom

The Institution of Electrical Engineers,
Michael Faraday House,
Six Hills Way, Stevenage,
Herts. SG1 2AY, United Kingdom

British Library Cataloguing in Publication Data

A CIP catalogue record for this book
is available from the British Library

ISBN 0 85296 812 4

Printed in England by Hobbs the Printers, Southampton

Contents

Preface

The book is devoted to generalisations of the stationary phase method, widely used to evaluate integrals of rapidly oscillating functions (IROF), i.e. integrals of the form

$$I(\lambda, \boldsymbol{\alpha}) = \int_C f(\boldsymbol{p}, \boldsymbol{\alpha}) \exp\,[i\lambda\varphi(\boldsymbol{p}, \boldsymbol{\alpha})]\; \boldsymbol{dp}$$

where λ is the large parameter, $\lambda \gg 1$. Integration is over an interval in the case of a one-dimensional integral, over a domain on a plane (p_1, p_2) or in a space $(p_1, \ldots, p_m)$ in the cases of two- and multidimensional integrals. A phase function φ, nonexponential factor f and a boundary of the integration domain C are assumed to depend on a set of parameters $(\alpha_1, \ldots, \alpha_n)$. For example, if in the problems of wave propagation or diffraction the wavefield is written in the form of the integral expression shown, one can consider the co-ordinates of a source or observation point to be such parameters.

The stationary phase method of evaluating asymptotics of the IROF in the one-dimensional case as well as its generalisations to two- and multidimensional integrals are treated in many books, for example, References 1–3. The algorithm to find the asymptotics is the following one. Critical points are introduced, in particular:

(i) Stationary (extremum and saddle) points of the phase function $\varphi(\boldsymbol{p}, \boldsymbol{\alpha})$ in a domain C, i.e. the points for which $\text{grad}_p\,[\varphi(\boldsymbol{p}, \boldsymbol{\alpha})] = 0$
(ii) Singular points (poles, branch points) of the nonexponential factor $f(\boldsymbol{p}, \boldsymbol{\alpha})$
(iii) Critical points on the boundary of the integration domain C, i.e. the points at which $\text{grad}_p\,[\varphi(\boldsymbol{p}, \boldsymbol{\alpha})]$ is normal to the boundary, corner points of the boundary, etc.

Under natural restrictions on functions φ, f and the boundary of the domain C, the asymptotic of $I(\lambda, \boldsymbol{\alpha})$ as $\lambda \to \infty$ proves to be a sum of contributions u_j of each critical point $\boldsymbol{p}_j$. In the leading term of the asymptotic

$$u_j = \lambda^{-q}\Psi(\boldsymbol{p}_j, \boldsymbol{\alpha}) \exp\,[i\lambda\varphi(\boldsymbol{p}_j, \boldsymbol{\alpha})]\, f(\boldsymbol{p}_j, \boldsymbol{\alpha})$$

where the function Ψ is expressed in terms of the phase function derivatives at the point $\boldsymbol{p}_j$ and is determined (along with an exponent q) by the critical point type. The subsequent terms of the asymptotic u also depend on the nonexponential factor derivatives at the point $\boldsymbol{p}_j$.

Let, for example, $u(k, Q)$ be a solution of the shortwave diffraction problem or the problem of wave propagation. Here, $k \gg 1$ is a wave number, Q denotes an observation point. Assume that the solution is written in integral form shown with the phase function $k\varphi(\boldsymbol{p}, Q)$. Then each critical point $\boldsymbol{p}_j$ corresponds to a ray of one of the ray fields which form the field u, the ray being passed through the point Q. Here $k\varphi(\boldsymbol{p}_j, Q)$ is a phase of the appropriate wave and $k^{-q}\Psi(\boldsymbol{p}_j)f(\boldsymbol{p}_j, Q)$ is its amplitude. In other words, the asymptotic obtained of the field u corresponds to the field representation as a sum of a few fields of geometrical optics.

However, the algorithm described appears to fail when critical points approach each other, merge, vanish or arise again as $\boldsymbol{\alpha}$ changes. For example, if an observation point approaches a caustic (ray envelope), the critical points $\boldsymbol{p}_1$ and $\boldsymbol{p}_2$ (corresponding to a ray that is not yet tangent to the caustic and to a ray that has already passed the caustic) approach each other. The components u_1 and u_2 (computed according to algorithms of the stationary phase method) corresponding to these stationary points grow infinitely when approaching the caustic. It is clear that such asymptotics could not be valid. Similar situations arise in all cases when the observation point is in a transient zone, i.e. the zone in which the number of rays passing through observation point is changed.

One can say that the asymptotic obtained according to the algorithm of the stationary phase method is nonuniform. The natural problem arises in proceeding from the nonuniform asymptotic expansions to the uniform ones applicable for any distances between critical points and describing their merging and vanishing. The price of a wider region of uniform asymptotic applicability is a more complicated form of the asymptotics; they are expressed in terms of special functions (the Airy function, the Fresnel and Piercey integrals, etc.). The choice of special function depends on the type and the number of critical points.

In textbooks devoted to asymptotic methods only two examples of uniform asymptotic are usually considered: the case of two close stationary points (the Airy function) and the case of a stationary point close to the integration domain boundary or to a pole of the nonexponential factor (the Fresnel integral). Both these and more complicated situations of critical-point merging are considered.

To obtain uniform asymptotics one reduces the IROF under consideration to model integrals, i.e. to the simplest integrals with similar configurations of critical points. Some model integrals coincide immediately with integral representations of the known special functions; the rest of them are new special functions. It is important to note that solutions of a wide variety of particular problems are reduced to a relatively small number of special functions to compute for which one can easily find appropriate standard programs in the literature or write such programs (see Chapter 6 and the Appendix).

The first Chapter describes the classical stationary phase method for one-dimensional integrals and contributions of various isolated critical points into the integral asymptotics are computed. In the second and third Chapters the cases of two (Chapter 2) or more (Chapter 3) critical-point merging are considered. In the fourth Chapter the uniform asymptotics of two-dimensional integrals are obtained.

The fifth Chapter is devoted to examples of how to apply the algorithms considered and treats a number of problems of diffraction theory and wave propagation, the short-wave asymptotics of both exact and approximate (physical optics approximation) solutions are obtained and analysed.

The sixth Chapter is of a reference nature; definitions and descriptions of special function main properties are included. Only those data are used which are useful for computation of these functions when applying algorithms for the IROF. Finally, the tables of special function expansions in Chebyshev polynomials are given in an Appendix. Using these tables as well as standard programs to evaluate Chebyshev polynomials one can easily write short and, at the same time, rapidly operating programs to calculate the appropriate special functions.

When writing I bear in mind physicists and radio engineers who face IROF and need efficient methods to evaluate them. This explains both the choice of the material and the style of its presentation. The reader is required to have the minimum of knowledge in calculus and the theory of functions of complex variables (contour integrals and the Cauchy theorem). When reading the book and using algorithms the reader needs no additional literature; all necessary information (definitions of the asymptotic expansion and initial formulas of the stationary phase method) is given.

When outlining the material the main attention is paid to the ideas and algorithms. In many cases the proofs given are not strict. For simplicity, asymptotic equalities are differentiated with respect to parameters without substantiating this operation, which is incorrect, strictly speaking, from a mathematical point of view. At the same time, I endeavoured to account for accumulated practical experience (of my colleagues and my own) concerned with the asymptotic evaluation of the IROF. Some recommendations grounded on this experience have no strict mathematical substantiation at all. For each case considered in what follows the results of evaluation of the uniform asymptotics are formulated as theorems which include all necessary formulas. This makes it possible to use the book as a reference and to use the results, main properties and tables of the relevant special functions (Chapter 6 and Appendix) without studying the book in detail.

Let me clarify why the book, which in essence is a book on applied mathematics, appears in a series of monographs on electromagnetic wave theory published by the IEE. The reason is in the deep intrinsic relations between the asymptotic methods of evaluating the IROF and the problem of diffraction and short-wave propagation.

In all cases, when the solution (both exact and approximate) of a short-wave problem is written as an integral over a real domain this integral is of form shown i.e. the IROF. Often, to evaluate them numerically appears to be impossible even using modern computers. For example, when calculating a radiation pattern for a two-mirror antenna according to the physical optics approximation one should compute a fourfold integral. If the size of the large mirror is of the order of 300λ (λ is the wavelength) and the small one is of the order of 100λ, the corresponding integral sum includes about 10^9 summands. At the same time, computations according to the algorithms of the stationary phase method and their uniform modifications are reduced to calculations according to explicit (though sometimes cumbersome) equations expressed in terms of elementary and a few special functions. The problem of calculating rapidly oscillating fields is reduced to that of calculating slowly varying functions that are amplitude and phase factors and arguments of the special functions used. The expressions obtained can be interpreted from the physical point of view using the terminology of geometrical optics and geometrical diffraction theory. This form of solution often makes it possible to describe the qualitative properties of the sought-for field and its radiation pattern without calculation. It is essential that the applicability ranges for numerical and asymptotic methods to evaluate the IROF turn out to be mutually complementary; the greater the parameter λ in the integral is, the more complicated numerical computations and the more exact the asymptotic equations are. The role of the stationary phase methods in solving problems of short-wave propagation and diffraction is so important that in many monographs and textbooks on these problems there are special sections outlining the methods, though formally they are not related to the main text (for example, References 5–7).

The opposite case is also true: most methods of evaluating IROF asymptotics are, in essence, the generalisations and formalisation of the approaches developed (starting from the classical works by Fresnel) to evaluate integrals in the wave theory of light. For example, the Huygens–Fresnel principle of light propagation leads to an integral of the type

$$U(P) = \frac{\text{const. exp}\,(ikr_0)}{r_0} \iint_S \frac{\exp\,(iks) K(\chi)\, dP_s}{s}$$

where the integration is performed over the sphere S of the radius r_0, with the source P_s as the centre. The distance between P_s and an observation point P is denoted by s, $K(\chi)$ is a diagram of the Huygens source at P_s, and χ is a diffraction angle, i.e. the angle between the normal to S and the direction from P_s to P. Finally, k is the wave number. To show how rectilinear light propagation follows from the Huygens principle one should find the asymptotic of the integral for kr_0, and $ks \gg 1$. The asymptotic evaluation performed by Fresnel by summing contributions of the Fresnel zones (see, for example, Reference 5, Section 8.1) coincides practically with the integral evaluation by the stationary phase method (see also Reference 8).

Further, almost all the special functions used in what follows to obtain the uniform asymptotic of the IROF are introduced for the first time to describe wavefields in the transient zones where the standard geometrical optics approximation fails. The list of such functions starts from the Fresnel integral and the Airy functions introduced by Fresnel [9] and Airy [10] to describe wavefields in the vicinity of geometrical shadow and near a caustic, and completed by the incomplete Airy function introduced by Felsen [11] to describe wavefields when the caustic overlaps the boundary of the geometrical shadow zone and by the Airy–Fresnel integral introduced by Orlov [12] to describe the field when the caustic of edge-wave overlaps this boundary.

This relation between asymptotic methods of IROF evaluation and the short-wave theory of electromagnetic wave diffraction and propagation justifies the book as part of the series of monographs on electromagnetic wave theory.

Introduction

Asymptotic expansions

Asymptotic expansions, asymptotic series and the not-so-definite concept of asymptotics are the main mathematical concepts used in this book.

The concepts of asymptotic expansions and asymptotic series were introduced by H. Poincare [13] and Th. Stieltjes [14], though the particular cases of asymptotic expansion were obtained as early as in the eighteenth century (for example, References 15–18).

To clarify the main difference between the representation of the function under consideration in the form of an asymptotic series and its conventional representation in the form of a converging series consider a particular problem. Let it be required to evaluate the integral

$$I(x)=\int_0^x \exp\,(it^2)\,dt=\sqrt{(\pi/2)}[C_1(x)+iS_1(x)]$$

for large x. The functions $C_1(x)$ and $S_1(x)$ are the Fresnel integrals (Section 6.1)

$$C_1(x)=\sqrt{(2/\pi)}\int_0^x \cos t^2\,dt$$

$$S_1(x)=\sqrt{(2/\pi)}\int_0^x \sin t^2\,dt$$

Since

$$I(\infty)=\int_0^\infty \exp\,(it^2)\,dt=\sqrt{(\pi/2)}[C_1(\infty)+iS_1(\infty)]=\sqrt{(\pi i)}/2$$

(see eqn. 1.6),

$$I(x)=\left[\int_0^\infty-\int_x^\infty\right]\exp\,(it^2)\,dt=\sqrt{(\pi i)}/2-\Phi(x)$$

where

$$\Phi(x)=\int_x^\infty \exp\,(it^2)\,dt \qquad (0.1)$$

and the problem is reduced to the evaluation of $\Phi(x)$ for $x\gg 1$. To this end, integrate $\Phi(x)$ by parts

$$\Phi(x)=\int_x^\infty \frac{d\exp\,(it^2)}{2it}=\frac{i\exp\,(ix^2)}{2x}+\frac{1}{2i}\int_x^\infty \frac{\exp\,(it^2)\,dt}{t^2}$$

and obtain for $\Phi(x)$

$$\Phi(x)=\frac{i\exp(ix^2)}{2x}+\delta_1 \tag{0.2}$$

where the error is

$$\delta_1=\frac{1}{2i}\int_x^\infty\frac{\exp(it^2)\,dt}{t^2}$$

It is estimated by

$$|\delta_1|\leq\frac{1}{2}\int_x^\infty\frac{dt}{t^2}=\frac{1}{2x}$$

This estimation is unsatisfactory since it coincides (in magnitude) with the first term in eqn. 0.2. However, repeating the integration by parts gives

$$\delta_1=\left[\frac{1}{2i}\right]^2\int_x^\infty\frac{d\exp(it^2)}{t^3}$$

$$=\frac{-\exp(ix^2)}{(2i)^2x^3}+\frac{3}{(2i)^2}\int_x^\infty\frac{\exp(it^2)\,dt}{t^4}$$

and obtains

$$\Phi(x)=-\frac{\exp(ix^2)}{(2i)x}-\frac{\exp(ix^2)}{(2i)^2x^3}+\frac{3}{(2i)^2}\int_x^\infty\frac{\exp(it^2)\,dt}{t^4}$$

Repeating this procedure N times, gives

$$\Phi(x)=\frac{i}{2x}\exp(ix^2)\sum_{n=0}^{N}C_n(x)+\delta_N(x) \tag{0.3}$$

where

$$C_0(x)=1,\qquad C_n(x)=\frac{1\cdot 3,\ldots,(2n-1)}{(2ix^2)^n}\qquad(n>1) \tag{0.4}$$

and the remainder

$$\delta_N(x)=\frac{3\cdot 5,\ldots,(2N+1)}{(2i)^{N+1}}\int_x^\infty\frac{\exp(it^2)\,dt}{t^{2N+1}}$$

$$=\frac{3\cdot 5,\ldots,(2N+1)}{(2i)^{N+2}}\left[-\frac{\exp(it)}{x^{2N+3}}+(2N+3)\int_x^\infty\frac{\exp(it^2)\,dt}{t^{2N+4}}\right]$$

admits the estimation

$$|\delta_N(x)|\leq\Delta_N=\frac{3\cdot 5,\ldots,(2N+1)}{2^{N+1}}\left[\frac{1}{2x^{2N+3}}+\frac{2N+3}{2}\int_x^\infty\frac{dt}{t^{2N+4}}\right]$$

$$=\frac{3\cdot 5,\ldots,(2N+1)}{2^{N+1}x^{2N+3}} \tag{0.5}$$

Since

$$\frac{\Delta_N(x)}{\Delta_{N-1}(x)}=\frac{2N+1}{2x^2}$$

the estimation $\Delta_N(x)$ decreases when N increases for $N<x^2$ and increases for $N>x^2$, its minimum value $\Delta_{\min}(x)$ is reached for $N=x^2-1$. Estimating $\Delta_{\min}(x)$ according to eqn. 0.5 and using the approximate Stirling formula for $3\cdot 5,\ldots,(2N+1)!!$, gives

$$\Delta_{\min}(x)\simeq\sqrt{2}\exp(-x^2)/x$$

For $x=2$ one should take four terms in eqn. 0.3 to reach maximum accuracy. In this case the error is $\Delta_{\min}(2)\simeq 0.013$. For $x=3$ one should take nine terms and $\Delta_{\min}(3)\simeq 6\cdot 10^{-5}$, for $x=4$ $\Delta_{\min}(4)\simeq 4\cdot 10^{-8}$, while for $x=5$ $\Delta_{\min}(5)\simeq 4\cdot 10^{-12}$. Thus, eqn. 0.3 makes it possible to evaluate $\Phi(x)$, though with limited but sufficiently high accuracy even for not too large x, the accuracy being increased when x increases.

At the same time, if $N=\infty$ in eqn. 0.3 one obtains the series

$$\begin{aligned}\Phi(x)&=\frac{i}{2x}\exp(ix^2)\sum_{n=0}^{\infty}C_n(x)\\&=\frac{i}{2x}\exp(ix^2)\sum_{n=0}^{\infty}\frac{\Gamma(n+0.5)x^{-2n}}{\Gamma(0.5)i^n}\end{aligned}\tag{0.6}$$

(here $\Gamma(z)$ is the gamma function) diverging for any x since for $n>x^2$,

$$\left|\frac{C_{n+1}(x)}{C_n(x)}\right|=\left|\frac{2n+1}{2ix^2}\right|>1$$

Thus, performing multiple integration by parts in the integral of eqn. 0.3 obtains the formal representation of the function $\Phi(x)$ as a diverging series (eqn. 0.6). At first, the terms of the series decrease when n increases, then reach their minimum for some $n=n(x)$ and then increase. Nevertheless, a partial sum of the series truncated at that n for which $C_n(x)$ is a minimum gives a good approximation which is the more accurate, the greater x is.

These properties of the eqn. 0.6 series are typical for asymptotic series. Let us give exact formulations.

Let two functions, $f(x)$ and $\varphi(x)$ defined in a domain S_x be given. Denote a limiting point of S_x by α. Introduce the symbols $\sim$, O, o by means of which the relations between the functions $f(x)$ and $\varphi(x)$ are described as $x\to\alpha$. Namely, write

$$f(x)\sim\varphi(x)\qquad \text{as } x\to\alpha,\text{ if }\lim_{x\to\alpha}\frac{f(x)}{\varphi(x)}=1\tag{0.7}$$

$$f(x)=o[\varphi(x)]\qquad \text{as } x\to\alpha,\text{ if }\lim_{x\to\alpha}\frac{f(x)}{\varphi(x)}=0\tag{0.8}$$

$f(x) = O[\varphi(x)]$ as $x \to \alpha$, if there is a vicinity S_α of point α in which the inequality

$$\left|\frac{f(x)}{\varphi(x)}\right| < C \qquad \text{(where } C \text{ is a constant)} \tag{0.9}$$

holds. Usually the domain S_α is a part of the real axis or the complex plane and α is $x=0$ or $x=\infty$.

Let me give some examples of relations 0.7–0.9. If $\lambda > \mu$, then $x^\lambda = o(x^\mu)$ as $x \to 0$ and $x^\mu = O(x^\lambda)$ as $x \to \infty$. For any λ, $\exp(-x) = o(x^{-\lambda})$ as $x \to \infty$. Besides, $\sin(x) \simeq x$ and $1 - \cos(x) = O(x^2)$, as $x \to 0$.

$$\Phi(x) \sim \frac{i \exp(ix^2)}{2x} \qquad \text{as } x \to \infty$$

where $\Phi(x)$ is the integral of eqn. 0.1.

Note that the condition $f(x) \sim \varphi(x)$ (as $x \to \alpha$) is equivalent to the condition

$$f(x) - \varphi(x) = o[\varphi(x)] \text{ as } x \to \alpha$$

If the functions $f(x)$ and $\varphi(x)$ depend on a few parameters $q_1, \ldots, q_m$ varying in a region Q, the concept of the uniformity of the symbols $\sim$, o, O with respect to these parameters is introduced. For example, the relation $f(x) = o[\varphi(x)]$ is uniform with respect to $q_1, \ldots, q_m$, if for each ε there is a vicinity S_ε of the point α such that

$$\left|\frac{f(x)}{\varphi(x)}\right| < \varepsilon$$

for each $x \in S_\varepsilon$ and $(q_1, \ldots, q_m) \in Q$. Similarly, the relation $f(x) = O[\varphi(x)]$ is uniform with respect to $q_1, \ldots, q_m$ as $x \to \alpha$, if there are such a constant C and a vicinity S_α of the point α that

$$\left|\frac{f(x)}{\varphi(x)}\right| < C$$

for each $x \in S_\alpha$ and $(q_1, \ldots, q_m) \in Q$.

A sequence of functions $\varphi_1(x), \varphi_2(x), \ldots, \varphi_n(x), \ldots$ is called the asymptotic sequence as $x \to \alpha$, if

$$\varphi_{n+1}(x) = o[\varphi_n(x)]$$

For example, as $x \to \infty$ the sequence $\varphi_n(x) = x^{-n}$ is the asymptotic one. One can say that the function $f(x)$ is expanded into the asymptotic series

$$f(x) \simeq \sum_{n=0}^{\infty} C_n \varphi_n(x) \qquad (x \to \alpha) \tag{0.10}$$

(where $\varphi_n(x)$ is the asymptotic sequence as $x \to \alpha$) if for any N the difference between $f(x)$ and a finite sum of the series tends to zero more rapidly than $\varphi_N(x)$

$$\delta_N(x) = f(x) - \sum_{n=0}^{N} C_n \varphi_n(x) = o[\varphi_N(x)], \qquad x \to \alpha \tag{0.11}$$

If $f(x)$ depends on the parameters $(q_1, \ldots, q_m)$ and if the estimation of eqn. 0.11 is uniform with respect to these parameters for any N, then the expansion of $f(x)$ into an asymptotic series is uniform with respect to $q_1, \ldots, q_m$. The representation of $f(x)$ as one or a few summands of the form

$$\Psi(x)S(x) = \Psi(x) \sum_{n=0}^{\infty} C_n\varphi_n(x) \qquad (x \to \infty) \tag{0.12}$$

(where $\Psi(x)$ is a certain function and $S(x) = \sum_{n=0}^{\infty} C_n\varphi_n(x)$ is the asymptotic series) is called the asymptotic expansion (or asymptotic) of the function $f(x)$. For example, the integral of eqn. 0.1 has the asymptotic expansion of eqn. 0.6 where

$$\Psi(x) = i \exp(ix^2)/(2x), \qquad \varphi_n(x) = x^{-2n}$$

$$C_n = (-1)^n\Gamma(n + 1/2)/\Gamma(1/2)$$

It is not necessary for an asymptotic series to be converging. The expansion of the function $f(x)$ into the converging series of eqn. 0.10 means that the error $\delta_N(x)$ tends to zero at fixed x and $N \to \infty$. The expansion into the asymptotic series means that $\delta_N(x)$ tends to zero for fixed N and $x \to \alpha$ and more rapidly than $\varphi_N(x)$. Note that the converging series can be a nonasymptotic one. For example, the expansion of $f(x)$ into the Fourier series, i.e. into a series in the functions $\varphi_n(x) = \exp(inx)$ is not asymptotic because the sequence $\varphi_n(x) = \exp(inx)$ is not asymptotic. Of course, there are series which are converging and asymptotic at the same time. For example, the Taylor series of an analytical function is both converging for $|x| < R$ (where R is the convergence radius of the series) and asymptotic as $x \to 0$.

Since the error $\delta_N(x)$ of a finite sum of the asymptotic series should not necessarily tend to zero at any fixed x, as $N \to \infty$, the asymptotic series makes it possible to evaluate $f(x)$ at a given x with a limited accuracy only. The condition of eqn. 0.10 guarantees only that the accuracy reached is the greater, the nearer to α the variable x is.

One can say that the definition of the converging series is of a static nature. It is possible to describe the function under consideration as accurately as necessary at any fixed x from the convergence domain of the series. The definition of the asymptotic series is of a dynamic nature. To reach a given accuracy for the function $f(x)$ one should take x close to α and the more accuracy that is required, the closer to α the variable x should be. Therefore, when one speaks of asymptotic expansions one should indicate with respect to which parameters and limiting transitions this asymptotic is obtained. For example, one should not speak of a scattered-field asymptotic, but of its short-wave asymptotic, or of the asymptotic with respect to a small thickness of the skin layer, etc.

The asymptotic expansion of eqn. 0.10 does not determine uniquely the function $f(x)$. Two noncoinciding functions $f_1(x)$ and $f_2(x)$ with the same asymptotic expansions can exist. In fact, if a function $\Psi(x)$ tends to zero more rapidly than any function $\varphi_n(x)$

$$\Psi(x) = o[\varphi_n(x)] \qquad (x \to \alpha)$$

then the right-hand part of eqn. 0.10 is not changed if $\Psi(x)$ is added to the left-hand part, and for $f_1(x)=f(x)+\Psi(x)$, as before, the error is

$$\delta_N(x)=f_1(x)-\sum_{n=0}^{N} C_n\varphi_n(x)$$
$$=f(x)-\sum_{n=0}^{N} C_n(x)\varphi_n(x)+\Psi(x)$$
$$=o[\varphi_n(x)]$$

For example, adding $\exp(-x)$ to the integral $\Phi(x)$: $\Phi_1(x)=\Phi(x)+\exp(-x)$, then $\Phi_1(x)$ is also expanded into the series of eqn. 0.6 since as $x\to\infty$ the additional term $\exp(-x)$ falls off more rapidly than any power of $1/x$ and therefore it does not affect the asymptotic of $\Phi(x)$.

The opposite statement is not valid. If the function $f(x)$ is expanded into the asymptotic series of eqn. 0.10, the coefficients C_n are determined uniquely according to the following recurrence equations

$$C_o=\lim_{x\to a}\left[\frac{f(x)}{\varphi_o(x)}\right]$$

$$C_{n+1}=\lim_{x\to a}\left[f(x)-\sum_{m=0}^{n} C_m\varphi_m(x)\right]\Big/\varphi_{n+1}(x) \qquad \text{for} \qquad n\geqslant 1$$

and it follows from the definition of the asymptotic series that the limits on the righthand sides of these relations exist.

Although the asymptotic series is an approximate description of a function, the uniqueness of its coefficients makes it possible to introduce the concepts of an exact and approximate asymptotic. Namely, let the function $f(x)$ be expressed approximately as $\tilde{f}(x)$, the function $f(x)$ being expanded into the asymptotic series eqn. 0.10, while $\tilde{f}(x)$ is expanded into the series

$$\tilde{f}(x)\simeq\sum_{n=0}^{\infty}\tilde{C}_n\varphi_n(x)$$

with the same functions $\varphi_n(x)$ as in eqn. 0.10. Then the approximation $\tilde{f}(x)$ has the exact first N terms of asymptotics, if for $n=0, 1, \ldots, N-1$ the equality $\tilde{C}_m=C_m$ holds. For example, in the problem where the field of a point source is reflected from a smooth, fully illuminated surface the reflected wave obtained according to the Kirchhoff approximation (an approximation of physical optics) has a leading term which coincides with that of the short-wave asymptotic (the exact asymptotic), while the subsequent terms do not coincide.

In conclusion, note that one can add and subtract asymptotic expansions in the same system of functions according to the standard rules, while the expansions in powers of x^{-n} can be multiplied and divided.

Chapter 1

Asymptotic contribution of critical point in one-dimensional integral

Consider an integral

$$I(\lambda)=\int_{-\infty}^{\infty} \exp\,[i\lambda\varphi(x)]f(x)\,dx \tag{1.1}$$

and treat the problem of how to find its asymptotic as $\lambda\to\infty$. Let $f(x)$ and $\varphi(x)$ be infinitely differentiable functions with $\varphi(x)$ being strictly monotonic, i.e. $\varphi'(x)$ does not vanish. Also assume that $f(x)$ and its derivatives tend to zero as $|x|\to\infty$ so rapidly that $I(\lambda)$ and the integrals $I_n(\lambda)$ converge absolutely as $x\to\infty$ where $I_n(\lambda)$ are the integrals obtained from $I(\lambda)$ when integrating by parts, see eqn. 1.2. I prove that under these conditions $I_n(\lambda)$ tends to zero more rapidly than any power function of λ^{-1}. Indeed,

$$I(\lambda)=(1/i\lambda)\int_{-\infty}^{\infty}\frac{f(x)}{\varphi'(x)}\,d\exp\,[i\lambda\varphi(x)]$$

$$=(i/\lambda)\int_{-\infty}^{\infty}\exp\,[i\lambda\varphi(x)]\,\frac{d}{dx}\left[\frac{f(x)}{\varphi'(x)}\right]dx$$

Repeating the procedure obtains that, for any $n>1$,

$$I=I_n(\lambda)=(i/\lambda)^n\int_{-\infty}^{\infty}\exp\,[i\lambda\varphi(x)]f_n(x)\,dx \tag{1.2}$$

where $f_n(x)$ is defined by the recurrence formula

$$f_{n+1}(x)=\frac{d}{dx}\left[\frac{f_n(x)}{\varphi'(x)}\right],\qquad f_0(x)=f(x) \tag{1.3}$$

It follows from eqn. 1.2 and the absolute convergence of the integral $I_n(\lambda)$ that for any $n>1$ $I(\lambda)=O(\lambda^{-n})$ as $\lambda\to\infty$, i.e. $I(\lambda)$ falls off more rapidly than any power function of λ^{-1}.

The question arises of when is this estimation not valid and the integral falls off as a power of λ^{-1}. Obviously, this is a case where the previously formulated conditions are not satisfied, i.e. $\varphi'(x)$ has stationary points where $\varphi'(x)=0$ or $f(x)$ and $\varphi(x)$ or their derivatives have singular points (discontinuity and branch points or poles: in these cases one cannot integrate many times by parts) or the integral is taken over a finite interval (in this case when integrating by parts the nonintegral summands appear).

Call the singular points of $f(x)$, $\varphi(x)$, stationary points of $\varphi(x)$ and the integration domain boundaries the *critical points*. As will be shown in Section 1.3 the asymptotic of $I(\lambda)$ is the sum of contributions corresponding to each critical point x_k. Each summand is of the form of $A \exp[i\lambda\varphi(x_k)]$ where A is expanded in an asymptotic series in powers of λ^{-1} (powers can be fractional), the series coefficients being dependent on the values of $f(x)$, $\varphi(x)$ and their derivatives at $x = x_k$. This dependence is determined by the type of a critical point and this Chapter presents expressions of the amplitude A for critical points that frequently occur in applications.

1.1 Asymptotic contribution of integration domain boundary

The simplest case, where $I(\lambda)$ can be expanded into an asymptotic series in powers of λ^{-1} with nonzero coefficients, is when $\varphi(x)$ is a strictly monotonic function and $f(x)$ and $\varphi(x)$ are infinitely differentiable but the integral is taken over a finite interval

$$I(\lambda) = \int_a^b \exp[i\lambda\varphi(x)] f(x)\, dx \tag{1.4}$$

To find the asymptotic of the integral, integrate many times by parts. The only difference is that nonintegral terms appear to give the sought-for asymptotic.

$$\begin{aligned} I = I(\lambda) &= (-i/\lambda) \int_a^b \frac{f(x)}{\varphi'(x)}\, d \exp[i\lambda\varphi(x)] \\ &= (i/\lambda)\left[\exp[i\lambda\varphi(a)]\frac{f(a)}{\varphi'(a)} - \exp[i\lambda\varphi(b)]\frac{f(b)}{\varphi'(b)}\right] \\ &\quad + (i/\lambda)\int_a^b f_1(x) \exp[i\lambda\varphi(x)]\, dx \end{aligned}$$

where $f_1(x)$ is defined by recurrence formula of eqn. 1.3. Repeating the same procedure n times obtains

$$\begin{aligned} I(\lambda) = {} & \frac{\exp[i\lambda\varphi(a)]}{\varphi'(a)} \sum_{m=1}^{n} (i/\lambda)^m f_{m-1}(\alpha) \\ & - \frac{\exp[i\lambda\varphi(b)]}{\varphi'(b)} \sum_{m=1}^{n} (i/\lambda)^m f_{m-1}(\beta) \\ & + (i/\lambda)^n \int_a^b \exp[i\lambda\varphi(x)] f_n(x)\, dx \end{aligned}$$

Since the integral on the right-hand side of this expression is of the order of $O(\lambda^{-n-1})$ (to prove this integrate by parts once more) one obtains the following asymptotic

$$I(\lambda)=\int_a^b \exp\,[i\lambda\varphi(x)f(x)]\;dx$$

$$\simeq\frac{\exp\,[i\lambda\varphi(a)]}{\varphi'(a)}\sum_{m=1}^{\infty}(i/\lambda)^m f_{m-1}(a)$$

$$-\frac{\exp\,[i\lambda\varphi(b)]}{\varphi'(b)}\sum_{m=1}^{\infty}(i/\lambda)^m f_{m-1}(b) \qquad (1.5)$$

where $f_0(x)=f(x)$ and the rest of $f_n(x)$ are defined by eqn. 1.3.

In what follows the asymptotics of $I(\lambda)$ are sought for the case when $f(x)$ vanishes identically outside the vicinity of the integration interval boundary that does not include the region near the second boundary. First, one of the sums on the right-hand side of eqn. 1.5 vanishes. Secondly, one can weaken the condition for $\varphi(x)$ requiring $\varphi'(x)$ not to vanish only in the region supp (f) where $f(x)\neq 0$, i.e. in the integration region.

The theorem formulated below is an obvious consequence of eqn. 1.5:

Theorem 1.1: Let an integral $I(\lambda)$ is defined by eqn. 1.4 where f and φ are infinitely differentiable functions, the function f being vanished outside a vicinity of the boundary $x=c$ of the integration interval and identically vanished near the second boundary. Let the derivative $\varphi'(x)$ be bounded in magnitude from below in the region supp (f). *Then $I(\lambda)$ is expanded into an asymptotic series*

$$I(\lambda)=\pm\frac{\exp\,[i\lambda\varphi(c)]}{\varphi'(c)}\sum_{m=1}^{\infty}(i/\lambda)^m f_{m-1}(c) \qquad (1.6)$$

where the minus sign is taken when c is an upper boundary of the integration interval, otherwise take the plus sign. The functions f_n are defined by the recurrence eqn. 1.3.

1.2 Asymptotic contribution of phase function stationary point

The integral $I(\lambda)$ is also expanded into an asymptotic series when the phase function $\varphi(x)$ has a stationary point where $\varphi'(x)$ vanishes.

Assume that integration is carried over the entire line $-\infty<x<\infty$ and let the functions $f(x)$ and $\varphi(x)$ be infinitely differentiable and $\varphi(x)$ have a single nondegenerate stationary point $x=x_0$ (i.e. the point at which $\varphi'(x_0)=0$, $\varphi''(x_0)\neq 0$). In this case it is proven that the asymptotic is specified by general formulas of eqns. 1.10 and 1.11c. The first two terms of this asymptotic are of the form of eqns. 1.11a and 1.11b, respectively.

First, assume that x_0 is a minimum point, i.e. $\varphi''(x_0)>0$. The simplest example is $\varphi(x)=x^2$, $f(x)=1$. In this case the integral $I(\lambda)$ is evaluated analytically

$$I(\lambda)=\int_{-\infty}^{\infty}\exp\,(i\lambda x^2)\,dx=\sqrt{(\pi/\lambda)}\,\exp\,(i\pi/4)=\sqrt{(i\pi/\lambda)} \qquad (1.7)$$

To prove this formula shift the integration path to the line $\arg(x) = \pi/4,\ 5\pi/4$ and then after a variable change, $x = (1/\sqrt{\lambda}) \exp(i\pi/4)t$, the problem is reduced to the evaluation of the probability integral

$$I(\lambda) = (1/\sqrt{\lambda}) \exp(i\pi/4) \int_{-\infty}^{\infty} \exp(-t^2)\, dt = (1/\sqrt{\lambda}) \exp(i\pi/4)\Phi$$

where

$$\Phi = \int_{-\infty}^{\infty} \exp(-s^2)\, ds$$

It remains to verify that $\Phi = \sqrt{\pi}$. To do this write Φ^2 in terms of the two-fold integral

$$\Phi^2 = \int_{-\infty}^{\infty} \exp(-s^2)\, ds \int_{-\infty}^{\infty} \exp(-t^2)\, dt = \iint_{-\infty}^{\infty} \exp[-(s^2 + t^2)]\, ds\, dt$$

and evaluate it in polar co-ordinates $s = \rho \cos \varphi$, $t = \rho \sin \varphi$ to obtain

$$\Phi^2 = \int_0^{2\pi} d\psi \int_0^{\infty} \exp(-\rho^2)\rho\, d\rho = 2\pi \int_0^{\infty} \exp(-\rho^2)\rho\, d\rho = \pi$$

Next, let $\varphi(x) = x^2$ and $f(x)$ be arbitrary infinitely differentiable functions. Prove that in this case

$$I(\lambda) = \int_{-\infty}^{\infty} \exp(i\lambda x^2) f(x)\, dx$$

$$\simeq \sum_{n=0}^{\infty} \frac{\Gamma(n + 1/2)}{\Gamma(2n + 1)} i^{n+1/2} f^{(2n)}(0) \lambda^{-n-1/2} \tag{1.8}$$

First, this equation is discussed. As $\lambda \to \infty$ the principal term of the asymptotic of $I(\lambda)$ is determined by the value $f(0)$ of the nonexponential function at the point $x = 0$ only, while the subsequent terms of the asymptotic are determined by the even derivatives $f^{(2n)}(0)$. The question arises of how to explain that the asymptotic of $I(\lambda)$ taken over the entire line $-\infty < x < \infty$ is determined by the local behaviour of $f(x)$ near zero, i.e. by the values of the function f and its even derivatives at $x = 0$.

The point is that for large λ and $|x| \neq 0$ the function $\exp(i\lambda x^2)$ oscillates rapidly, while $f(x)$ varies slowly over the oscillation period. Therefore, when integrating, e.g. an expression for $\mathrm{Re}\,[I(\lambda)]$ (i.e. $\cos \lambda x^2$), one obtains that integrals over neighbouring halfwaves where $\cos \lambda x^2 > 0$ and $\cos \lambda x^2 < 0$ almost compensate each other and the initial integral almost vanishes. Near zero the situation is different; there is no total compensation.

To simplify the problem, set $f(x) = 1$. The function $\cos \lambda x^2$ is positive for $|x| < \sqrt{(\pi/2\lambda)}$ and the integral over this interval is approximately equal to $1.95/\sqrt{\lambda}$. In each of the intervals $\sqrt{(\pi/2\lambda)} < x < \sqrt{(3\pi/2\lambda)}$ and $-\sqrt{(3\pi/2\lambda)} < x < -\sqrt{(\pi/2\lambda)}$ the function $\cos \lambda x^2$ is negative (Figure 1.1), the integral over each of these intervals is equal to $-0.57/\sqrt{\lambda}$. The subsequent oscillations of $\cos \lambda x^2$

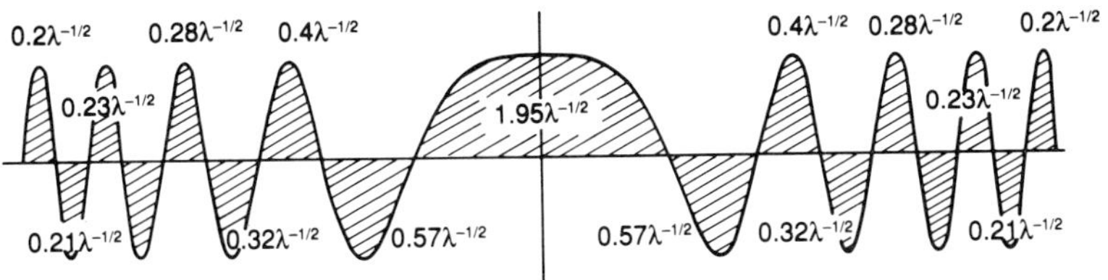

Figure 1.1 Function cos (λx^2)

are shown in Figure 1.1. It is seen that with x increasing the halfwaves compensate each other more and more exactly, but there is no compensation for the first halfwaves.

The situation does not change qualitatively if exp $(i\lambda x^2)$ is multiplied by $f(x)$ slowly varying over the λ^{-1}-scale; this is why the value $f(0)$ is singled out in the integral of eqn. 1.8 up to the factor of the order of $1/\sqrt{\lambda}$, i.e. of the order of the first halfwave area.

Eqn. 1.5 can be interpreted in a similar way. Here the halfwaves near the boundary of the integration interval are not compensated for $x=a$ and $x=b$ because the integration domain is bounded. Obviously, the area of these halfwaves is of the order of $1/\lambda$.

To prove eqn. 1.8 let us begin with some formal mathematical manipulations are made without justifying them (in particular, without proving the convergence of the integrals). Replacing $f(x)$ by its Taylor series and integrating term by term,

$$I(\lambda)=\int_{-\infty}^{\infty} \exp\,(i\lambda x^2) \sum_{m=0}^{\infty} \frac{x^m f^{(m)}(0)}{\Gamma(m+1)}\, dx$$

$$= \sum_{m=0}^{\infty} \frac{f^{(m)}(0)}{\Gamma(m+1)} \int_{-\infty}^{\infty} x^m \exp\,(i\lambda x^2)\, dx$$

Now to obtain eqn. 1.8 it is sufficient to note that the integral on the right-hand side vanishes for odd values of m; for even m use the relation

$$\int_{-\infty}^{\infty} x^{2n} \exp\,(i\lambda x^2)\, dx = i^{n+1/2}\Gamma(n+1/2)\lambda^{-n-1/2}$$

which can be derived from eqn. 1.7 by differentiating n times with respect to λ.

Now eqn. 1.8 is rigorously substantiated. It is sufficient to prove it for an even function $f(x)$ only, since any function $f(x)$ can be represented as a sum of even and odd functions

$$f(x)=f_+(x)+f_-(x), \qquad f_\pm(x)=[f(x)\pm f(-x)]/2$$

and for an odd function $f(x)=f_-(x)$ the integral of eqn. 1.4 vanishes identically.

Assume that as $|x|\to\infty$ the function $f(x)$ and all its derivatives tend to zero so rapidly that $I(\lambda)$ and integrals obtained from it when integrating by parts converge absolutely as $|x|\to\infty$. Let us show that if $f(0)=f'(0)=\ldots f^{(2N-1)}(0)=0$, then $I(\lambda)=0(\lambda^{-N-1/2})$. Indeed, setting $f(x)=x^{2N}\psi_0(x)$ one can integrate $I(\lambda)$ by parts N times, each time integrating x exp $(i\lambda x^2)$ and differentiating the remaining nonexponential factor

$$
\begin{aligned}
I(\lambda) &= \int_{-\infty}^{\infty} \exp(i\lambda x^2) x^{2N} \psi_0(x)\, dx \\
&= (-i/2\lambda) \int_{-\infty}^{\infty} x^{2N-1} \psi_0(x) d \exp(i\lambda x^2) \\
&= (1/\lambda) \int_{-\infty}^{\infty} \exp(i\lambda x^2) x^{2N-2} \psi_1(x)\, dx \\
&= (1/\lambda)^N \int_{-\infty}^{\infty} \exp(i\lambda x^2) \psi_N(x)\, dx = 0(\lambda^{-N-1/2})
\end{aligned}
$$

where functions ψ_n are defined by the recurrence equation

$$
\psi_{n+1}(x) = (i/2)\left[(2N - 2n - 1)\psi_n(x) + x\frac{d\psi_n(x)}{dx}\right]
$$

It follows from the estimation obtained that if the function $f(x)$ is approximated by a finite linear combination of $f_n(x)$

$$
f(x) = \sum_{n=0}^{N-1} C_n f_n(x) + \tilde{f}(x)
$$

in such a way that, first, eqn. 1.8 is valid for each function $f_n(x)$ and, secondly, the function $\tilde{f}(x)$ and its $(2N-1)$ first derivatives vanish at $x=0$, it is proved that for the function $f(x)$ the first N terms of the asymptotics of $I(\lambda)$ will be specified by eqn. 1.8.

It is convenient to choose $f_n(x)$ in the form

$$
f_n(x) = x^{2n} \exp(-ax^2)
$$

for certain $a>0$. For $f_0(x)$,

$$
\begin{aligned}
I(\lambda) &= \int_{-\infty}^{\infty} \exp(i\lambda x^2 - ax^2)\, dx \\
&= \sqrt{(\pi/\lambda)} \exp(i\pi/4)/\sqrt{(1 + ia/\lambda)}
\end{aligned}
$$

Expanding the right-hand side in a power series in a/λ obtains, first, that eqn. 1.8 for $f(x)=f_0(x)$ is valid and, secondly, the series on the right-hand side of the equation is not only asymptotic, but absolutely converging as well for $\lambda > a$.

Differentiating the relation n times with respect to a gives

$$
\begin{aligned}
&\int_{-\infty}^{\infty} \exp(i\lambda x^2) x^{2n} \exp(-ax^2)\, dx \\
&= -\frac{d^n}{da^n} \int_{-\infty}^{\infty} \exp(i\lambda x^2 - ax^2)\, dx = \frac{\Gamma(n+1/2)\exp(i\pi/4) i^n}{(\lambda + ia)^{n+1/2}} \\
&= \frac{\Gamma(n+1/2) i^n}{\lambda^{n+1/2}} (1 + ia/\lambda)^{-n-1/2} \exp(i\pi/4)
\end{aligned}
$$

and expanding again $(1+ia/\lambda)^{-n-1/2}$ into a power series in λ^{-1}, it is obtained that eqn. 1.8 is valid for $f(x)=f_n(x)=x^{2n}\exp(-ax^2)$ as well.

To approximate an arbitrary even function $f(x)$ by a linear combination of $f_n(x)$ it is sufficient to take the coefficients of the Taylor series expansion for $f(x)\exp(ax^2)$ as C_n.

The case of an arbitrary phase function $\varphi(x)$ with a single stationary point $x=x_0$ where $\varphi'(x_0)=0$ and $\varphi''(x_0)>0$ is reduced to the considered case of $\varphi(x)=x^2$ by the variable change

$$\xi=\xi(x)=\text{sign}\,(x-x_0)\sqrt{|\varphi(x)-\varphi(x_0)|} \tag{1.9}$$

For $x\neq x_0$,

$$\frac{d\xi}{dx}=\frac{\varphi'(x)\,\text{sign}\,(x-x_0)}{2\sqrt{|\varphi(x)-\varphi(x_0)|}}>0$$

As $x\to x_0$ the function $d\xi/dx$ remains bounded from below. Indeed, set $\varphi(x)=\varphi(x_0)+(x-x_0)^2\psi(x)$, where $\psi(x_0)>0$. Then $\xi(x)=(x-x_0)\sqrt{[\psi(x)]}$ and

$$\frac{d\xi}{dx}=\sqrt{[\psi(x)]}+\frac{\psi'(x)(x-x_0)}{2\sqrt{[\psi(x)]}}$$

Obviously, $\xi'(x)$ remains positive at $x\simeq x_0$. Denoting the function inverse of $\xi(x)$ by $x(\xi)$,

$$\begin{aligned}I(\lambda)&=\int_{-\infty}^{\infty}\exp\,[i\lambda\varphi(x)]f(x)\,dx\\&=\exp\,[i\lambda\varphi(x_0)]\int_{-\infty}\exp\,(i\lambda\xi^2)f[x(\xi)]\,\frac{dx}{d\xi}\,d\xi\end{aligned}$$

and to find the asymptotic of the integral on the right-hand side of the above expression use eqn. 1.8. It is sufficient to restrict study to the case of the finite function $f(x)$ identically vanishing for large x. The following theorem is valid.

Theorem 1.2: Let an integral $I(\lambda)$ be defined by eqn. 1.1 where $\varphi(x)$ and $f(x)$ are infinitely differentiable functions, $f(x)$ is finite and $\varphi(x)$ has a single stationary point x_0 at which $\varphi''(x)\neq 0$ in the region supp (f)*. Then as $\lambda\to\infty$ the integral $I(\lambda)$ is expanded into an asymptotic series*

$$I(\lambda)\simeq\exp\,[i\lambda\varphi(x_0)]\sum_{n=0}^{\infty}C_n\lambda^{-n-1/2} \tag{1.10}$$

where

$$C_0=\sqrt{(\pi)}\exp\,(i\delta\pi/4)f[x(\xi)]\left.\frac{dx}{d\xi}\right|_{\xi=0}$$

$$=\sqrt{[2\pi/|\varphi''(x_0)|]}\exp\,(i\delta\pi/4)f(x_0) \tag{1.11a}$$

$$C_1=\sqrt{(\pi/2)}\frac{\exp\,(3i\delta\pi/4)}{|\varphi''|^{3/2}}\left[f''-\frac{\varphi'''f'}{\varphi''}-\frac{\varphi^{(4)}f}{4\varphi''}+\frac{5(\varphi''')^2f}{12(\varphi'')^2}\right] \tag{1.11b}$$

for $n > 1$

$$C_n = \frac{\Gamma(n+1/2)\exp\left[i\pi\delta(n+1/2)/2\right]}{\Gamma(2n+1)} \frac{d^{2n}}{d\xi^{2n}}\left[f[x(\xi)]\frac{dx}{d\xi}\right]_{\xi=0} \tag{1.11c}$$

$\delta = \text{sign}\ [\varphi''(x_0)]$, $x(\xi)$ is a function inverse to $\xi(x)$ specified by eqn. 1.9 and all the derivatives are taken at the point $x = x_0$.

Eqns. 1.11*a* and 1.11*c* for $\varphi''(x_0) > 0$ are proved above. The case of $\varphi''(x_0) < 0$ is reduced to that of $\varphi''(x_0) > 0$, if we factor the constant $\exp[i\lambda\varphi(x_0)]$ out of the integral sign and then subject the integral to complex conjugation.

The coefficients C_n are expressed in terms of the derivatives of $f(x)$ and $\varphi(x)$ at $x = x_0$. To write the corresponding formulas and to obtain, in particular, the explicit eqn. 1.11*b* for C_1 one needs to expand the function $\xi(x) = \text{sign}\,(x - x_0)\sqrt{[\varphi(x) - \varphi(x_0)]}$ into a power series in $(x - x_0)$ and then find the expansion of the inverse function $\xi(x)$ into the Taylor series. To avoid these cumbersome manipulations use the following formal procedure.

Write $I(\lambda)$ as

$$I(\lambda) = \exp[i\lambda\varphi(x_0)] \int_{-\infty}^{\infty} \exp\left[i\lambda(x - x_0)^2\varphi''(x)/2\right] h(\lambda, x)\, dx \tag{1.12}$$

where

$$h(\lambda, x) = f(x)\exp\{i\lambda[\varphi(x) - \varphi(x_0) - (x - x_0)^2\varphi''(x_0)/2]\}$$

Here, the argument of the exponential function is expanded into a power series starting from $(x - x_0)^3$

$$i\lambda[\varphi(x) - \varphi(x_0) - (x - x_0)^2\varphi''(x_0)/2]$$
$$= i\lambda(x - x_0)^3 \sum_{n=3}^{\infty} \frac{(x - x_0)^{n-3}\varphi^{(n)}(x)}{\Gamma(n+1)}$$

Therefore expanding $f(x)$ and $\exp\{i\lambda[\varphi(x) - \varphi(x_0) - (x - x_0)\varphi''(0)/2]\}$ into a power series obtains

$$h(\lambda, x) = \sum_{p=0}^{\infty}\sum_{q=0}^{\infty} \lambda^p (x - x_0)^{3p+q} C_{p,q}$$

where $C_{p,q}$ are expressed in terms of the derivatives of $f(x)$ and $\varphi(x)$ at $x = x_0$. Substituting this double series into eqn. 1.12 and integrating term by term according to eqn. 1.8:

$$\int_{-\infty}^{\infty} \exp[i\lambda\varphi(x)] f(x)\, dx$$
$$\simeq \exp[i\lambda\varphi(x_0)] \sum_{p,q}{}' \Gamma[(3p + q + 1)/2]$$
$$\times (2i\delta/|\varphi''|)^{(3p+q+1)/2} \lambda^{-(p+q+1)/2} C_{p,q}$$

where the primed sum means that the summation is taken with respect to even $p + q \geq 0$ only. The leading term of the asymptotic corresponds to the indices $q = p = 0$, the next one is a sum of three terms with $p = 0$, $q = 2$; $p = q = 1$ and

$p=2$, $q=0$. Determining $C_{2,0}$, $C_{1,1}$ and $C_{0,2}$ obtains eqn. 1.11*b* for C_1. Thus Theorem 1.2 is proved.

Since in most cases deriving asymptotic expansions for the integrals of eqn. 1.4 form follows the same logical scheme, even for more complicated situations, here is a reminder of the main steps of the derivation. Start with writing the simplest (model) integral (eqn. 1.7) with the same properties of the phase function φ, as in the general case considered (i.e. the function with a single stationary point). In the example the integral was evaluated analytically; in more complicated cases $I(\lambda)$ depends on the parameters of a problem and after scaling (reducing the number of parameters by a simple manipulation such as variable change) cannot be further simplified. The integrals obtained are named, for example, the Fresnel integral or the Airy function. The problem arises as to how a general case of an integral with arbitrary f and φ can be reduced to a model integral (in the example, to the integral with $\varphi=x^2$ and $f=1$). When going from an arbitrary phase function to a model one, the nondegenerate and reversible transformation of an integration variable is used. To find the asymptotic of $I(\lambda)$ with a model phase function φ and an arbitrary nonexponential function f a rather wide set of functions f_n is constructed for which the asymptotic is known and by which any function f can be approximated. To construct such a set one has to determine the first function f_0 and then find the subsequent functions f_n differentiating with respect to the parameters of the problem.

1.3 Isolated critical points and partition of unity

The approaches used in the preceding Sections appear to be inapplicable when there are a few critical points. Let, for example, the phase function $\varphi(x)$ in the integral of eqn. 1.4 have a stationary point. Then the integral cannot be integrated by parts because even after one integration the nonexponential function $f_1(x)=d[f(x)/\varphi'(x)]/dx$ of the integral obtained tends to infinity at the point where $\varphi'(x)=0$. However, if $f(x)$ is not equal to zero only in the vicinity of the point a so small that $\varphi'(x)\neq 0$ in it, then $f_1(x)$ and all subsequent functions $f_n(x)$ are regular functions of x.

Similarly, when the phase function $\varphi(x)$ has stationary points $x=x_0, x_1, \ldots, x_N$, the variable change $\xi=\xi(x)$ transforming the function into a quadratic one in a vicinity of $x=x_0$ (see eqn. 1.9) appears to be degenerate at other stationary points. However, if $f(x)$ is not equal to zero only in vicinity of $x=x_0$ so small that $\xi(x)$ is regular and an inverse variable change is possible, then the nonexponential function $f[x(\xi)]\ dx/d\xi$ in the integral over ξ is regular since $f[x(\xi)]$ vanishes identically in vicinities of those points where $dx/d\xi$ is infinite.

Therefore it is reasonable in the integral $I(\lambda)$ with critical points $x_0, x_1, \ldots, x_N$ to represent $f(x)$ as a sum

$$f(x)=f_0(x)+f_1(x)+\cdots+f_N(x)+\tilde{f}(x)$$

where each of functions $f_k(x)$ vanishes outside a rather small vicinity of the critical point x_k and there are no critical points at all in the region where $\tilde{f}(x)\neq 0$

(i.e. in the region supp $(\tilde{f})$). Then the integral $I(\lambda)$ is written as a sum

$$I(\lambda)=\sum_k I_k+\tilde{I}=\sum_k \int_a^b \exp\,[i\lambda\varphi(x)]f_k(x)\,dx + \int_a^b \exp\,[i\lambda\varphi(x)]\tilde{f}(x)\,dx$$

Dealing now with a single critical point $x=x_k$ in each integral I_k one can evaluate its asymptotics depending on the specific nature of this critical point. Since there are no critical points in the region supp $(\tilde{f})$ (i.e. $\varphi'(x)\neq 0$ and the functions $\tilde{f}(x)$ and $\varphi(x)$ are infinitely differentiable) one can evaluate $\tilde{I}(\lambda)$ by repeated integration by parts and prove that it falls off more rapidly than any power function of λ^{-1}.

The conventional method of representing $f(x)$ as such a sum is as follows. Let δ be so small that a distance between any neighbouring critical points x_k and x_{k+1} is more than 2δ. Choose $\varepsilon>0$ with $\varepsilon<\delta$ and introduce an infinitely differentiable even function $g(x)$, equal to unity for $|x|<\varepsilon$ and to zero for $|x|>\delta$ (such a function is constructed in what follows). Then the function $g_k(x)=g(x-x_k)$ is equal to unity in an ε-vicinity of the point x_k (i.e. for $|x-x_k|<\varepsilon$) and to zero outside a larger δ-vicinity of this point. Set

$$\tilde{g}(x)=1-g(x-x_0)-g(x-x_1)-\cdots-g(x-x_N)$$

Obviously, the function $\tilde{g}(x)$ vanishes identically in an ε-vicinity of each of the critical points $x_0, x_1, \ldots, x_N$. The schematic plot of the function $g(x)$ is given in Figure 1.2*a*. In Figure 1.2*b* the solid curves correspond to $g_0(x)$ and $g_1(x)$ while $\tilde{g}(x)$ is given by the dashed line. The expansion

$$1=g_0(x)+g_1(x)+\cdots+g_N(x)+\tilde{g}(x)$$

is termed the *partition of unity* while each of the functions $g_k(x)$ is termed the *cutting function*. Now,

$$\begin{aligned}f(x)&=f(x)[g_0(x)+g_1(x)+\cdots+g_N(x)+\tilde{g}(x)]\\&=f_0(x)+f_1(x)+\cdots+f_N(x)+\tilde{f}(x)\end{aligned}$$

where, obviously, the functions $f_k(x)=f(x)g_k(x)$ for $k=0, 1, 2, \ldots, N$ and $\tilde{f}(x)=f(x)\tilde{g}(x)$ have the properties required. It remains now to construct the function $g(x)$ with the previously mentioned properties. In so doing consider a

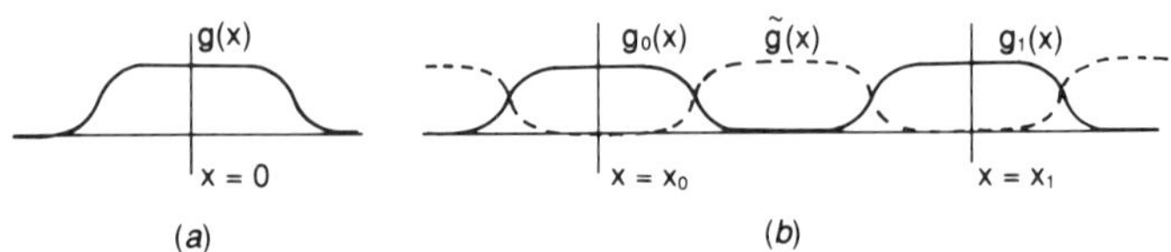

Figure 1.2 Partition of unity

function

$$e(s)=\begin{cases}0 & \text{for } s\leq 0\\ \exp(-1/s) & \text{for } s>0\end{cases}$$

This function and all its derivatives tend to zero as $s\to+0$. Therefore, $e(s)$ is an infinitely differentiable function equal to zero for $s<0$. Set

$$h(x)=\int_x^\infty e(\delta-s)e(s-\varepsilon)\,ds$$

As $e(\delta-s)\equiv 0$ for $s>\delta$, so the function $h\equiv 0$ for $x>\delta$. Since $e(s-\varepsilon)\equiv 0$ for $s<\varepsilon$, so $h(x)$ is constant for $x<\varepsilon$

$$h(x)=\int_x^\infty e(\delta-s)e(s-\varepsilon)\,ds=\int_\varepsilon^\delta e(\delta-s)e(s-\varepsilon)\,ds=h_0$$

Therefore the function $h(x)/h_0$ is an infinitely differentiable function equal to zero for $x>\delta$ and to unity for $x<\varepsilon$, i.e. it can be taken as a sought-for function $g(x)$. The particular form of $g(x)$ is of no influence on the asymptotics of the integrals

$$\begin{aligned}I_k(\lambda)&=\int_{-\infty}^\infty \exp\,[i\lambda\varphi(x)]f_k(x)\,dx\\ &=\int_{-\infty}^\infty \exp\,[i\lambda\varphi(x)]f(x)g(x-x_k)\,dx\end{aligned}$$

since $g(x-x_k)$ is identically equal to unity in a vicinity of the critical point $x=x_k$ and therefore the values of f_k and its derivatives at these points (in terms of which the asymptotic of I_k is expressed) coincide with the corresponding values of $f(x)$.

Thus, expanding unity, represent $I(\lambda)$ as a sum of the integrals I_k and an integral $\tilde{I}$ falling off more rapidly than any power of λ^{-1}. Each of the integrals I_k corresponds to the critical point $x=x_k$; it is an integral over a small vicinity of such a point and has an asymptotic determined by the values of functions f, φ and their derivatives at $x=x_k$. For the integration domain boundary this asymptotic is specified by theorem 1.1 and for a nondegenerate stationary point of the phase function by theorem 1.2.

In what follows the contributions of other critical points in asymptotics of the corresponding integrals are found.

1.4 Branch point contribution

Consider a case when the critical point $x=x_0$ is at the same time a branch point both for the phase function $\varphi(x)$ and the nonexponential factor $f(x)$ of the integral of eqn. 1.4. It is sufficient to consider an integral of the form

$$I(\lambda)=\int_0^\infty \exp\,[i\lambda\varphi(x)]f(x)\,dx \tag{1.13}$$

where $x=0$ is a branch point of the functions $\varphi(x)$ and $f(x)$, no other critical points being in the region supp $[f(x)]$. The model integral is of the form

$$I_+ = I_+(\lambda, \alpha, \beta) = \int_0^\infty \exp(i\lambda x^\alpha) x^\beta \, dx$$

where $\alpha > 0$. For the integral to converge at $x=0$ it is required that $\beta > -1$, while at $x=\infty$ the condition $\beta < \alpha - 1$ should be satisfied. Changing the variables $\lambda x^\alpha = t$, $x = (t/\lambda)^{1/\alpha}$, reduce the integral I_+ to the form

$$I_+ = (1/\alpha)\lambda^{-(\beta+1)/\alpha} \int_0^\infty \exp(it) t^{(\beta-\alpha+1)/\alpha} \, dt$$

Shift the integration path to the positive imaginary halfaxis by setting $t = i\xi$, $0 < \xi < \infty$, obtain the expression for I_+ in terms of the gamma function

$$I_+ = (1/\alpha)(i/\lambda)^{(\beta+1)/\alpha} \int_0^\infty \exp(-\xi) \xi^{(\beta-\alpha+1)/\alpha} \, d\xi$$
$$= (1/\alpha)(i/\lambda)^{(\beta+1)/\alpha} \Gamma[(\beta+1)/\alpha]$$

In the same manner an integral with the phase function $-i\lambda x^\alpha$ is evaluated, but here shift the integration to the negative imaginary halfaxis. The result can be written as a unified equation

$$I_+ = \int_0^\infty \exp(\pm i\lambda x^\alpha) x^\beta \, dx = (1/\alpha)(\pm i/\lambda)^{(\beta+1)/\alpha} \Gamma[(\beta+1)/\alpha] \qquad (1.14)$$

where $(\pm i)^\mu$ means $\exp(\pm i\pi\mu/2)$.

Theorem 1.4: Let in the integral

$$I_\pm(\lambda) = \int_0^\infty \exp(\pm i\lambda x^\alpha) f(x) \, dx \qquad \alpha > 0 \qquad (1.15)$$

infinitely differentiable function $f(x)$ and all its derivatives fall off at infinity no slower than $x^{\alpha-1}$ and let the function $f(x)$ be expanded into the asymptotic series as $x \to 0$

$$f(x) \simeq \sum_n a_n x^{\beta_n} \qquad (1.16)$$

which can be differentiated term by term and where $\beta_1 > -1$, $\beta_{n+1} > \beta_n$ and $\beta_n \to \infty$ as $n \to \infty$. Then as $\lambda \to \infty$ the integral of eqn. 1.15 is expanded into an asymptotic series

$$I_\pm(\lambda) \simeq (1/\alpha) \sum_n a_n (\pm i/\lambda)^{(\beta_n+1)/\alpha} \Gamma[(\beta_n+1)/\alpha] \qquad (1.17)$$

When proving the theorem one can set $\alpha = 1$ in eqn. 1.15 since the integration variable change $\tau = x^\alpha$ reduces the case of arbitrary $\alpha > 0$ to this particular case. To derive eqn. 1.17 formally it is sufficient to replace the function $f(x)$ in eqn. 1.15 by the series of eqn. 1.16 and then using eqn. 1.14 to integrate term by term. To prove theorem 1.4 rigorously single out the terms decreasing at infinity by setting

$$f(x)=\sum_{n=1}^{m} a_n x^{\beta_n}+f_1(x)$$

where $\beta_m<0$ and $\beta_{m+1}\geq 0$. Integrate these summands term by term and the remainder $f_1(x)$ by parts. This results in

$$I_\pm(\lambda)=\sum_{n=1}^{m} a_n(\pm i/\lambda)^{\beta_n}\Gamma(\beta_n+1)\pm(i/\lambda)f_1(0)$$

$$\pm(i/\lambda)\int_0^\infty \exp(\pm i\lambda x)f'_1(x)\,dx \tag{1.18}$$

Here, $f_1(0)\neq 0$ only if $\beta_{m+1}=0$. Then $f_1(0)=a_{m+1}$ and eqn. 1.18 can be written as

$$I_\pm(\lambda)=\sum_{n=1}^{m+1} a_n(\pm i/\lambda)^{\beta_n}\Gamma(\beta_n+1)\pm(i/\lambda)\int_0^\infty \exp(\pm i\lambda x)f'_1(x)\,dx$$

For small x the function $f'_1(x)$ is expanded into a series

$$f'_1(x)=\sum_{n=m+1}^{\infty} a_n\beta_n x^{\beta_n-1}$$

and falls off, as well as $f(x)$, as a power function as $x\to\infty$.

In the series written one can again single out the terms decreasing at infinity for which $0<\beta_n<1$ by setting

$$f_1(x)=\sum_{n=m+1}^{p} a_n\beta_n x^{\beta_n-1}+f_2(x)$$

where $\beta_p<1$ and $\beta_{p+1}\geq 1$. Substitute this expression into eqn. 1.18. Integrating the sum term by term and the integral (with $f_2(x)$) by parts,

$$I_\pm(\lambda)=\sum_{n=1}^{p} a_n(\pm i/\lambda)^{\beta_n+1}\Gamma(\beta_n+1)+(i/\lambda)^2 f_2(0)$$

$$+(i/\lambda)^2\int_0^\infty \exp(\pm i\lambda x)f'_2(x)\,dx$$

where

$$f_2(x)=\sum_{n=p+1}^{\infty} a_n\beta_n(\beta_n-1)x^{\beta_n-2}$$

and the summation in $f_2(x)$ is taken over all those indices n for which $\beta_n>1$. Again, one can single out a summand decreasing at infinity, i.e. the summand with $\beta_n<2$. Then one can integrate these summands term by term and integrate the remainder by parts. This procedure leads to theorem 1.4.

The case of the phase function $\varphi(x)=x^\alpha\varphi_0(x)$ (where $\varphi_0(x)$ is the analytical function for small x and $\varphi_0(0)\neq 0$) is reduced to theorem 1.4 by the variable change $t=x|\varphi_0(x)|^{1/\alpha}$. It is required, of course, that the variable change be nondegenerate in the region supp (f) (i.e. $\varphi(x)$ is required to have no stationary points in this region for $x>0$) and the integral to converge as $x\to\infty$. Both these conditions are satisfied if when using partition of unity one introduces the cutting function $g(x)$ into the integral, the function being equal to zero outside a small vicinity of $x=0$.

One can write the explicit expressions for the coefficients of the asymptotic expansion of $I_{\pm}(\lambda)$ with the phase function $\varphi(x)=x^{\alpha}\varphi_0(x)$ without changing the variable to t (which requires cumbersome manipulations when evaluating the coefficients of the expansion $f[x(t)]\ dx/dt$ into a power series in t). In so doing one needs to write $\varphi(x)$ as

$$\varphi(x)=\varphi_0(0)x^{\alpha}+\varphi_1(x)$$

and then

$$I(\lambda)=\int_0^{\infty}\exp\,[i\lambda\varphi(x)]f(x)\,dx$$
$$=\int_0^{\infty}\exp\,[i\lambda x^{\alpha}\varphi_0(0)\{\exp\,[i\lambda\varphi_1(x)]f(x)\}\,dx \qquad (1.19)$$

after which the expression in parentheses is expanded in a series in powers of x using eqn. 1.17, with $\lambda|\varphi_0(0)|$ instead of λ, the expression I_+ being taken for $\varphi_0(0)>0$ and I_- for $\varphi_0(0)<0$.

Eqns. 1.17 and 1.19 describe not only the branch points of phase functions, but also the case when a nondegenerate stationary point of the phase function (for $\alpha=2$) coincides with a branch point of the nonexponential factor as well as the case of a degenerate stationary point (α is an integer more than two).

1.5 One-sided and two-sided integrals

Section 1.4 considered one-sided integrals taken over the region $x>0$ on one side of the critical point $x=0$. Two-sided integrals (eqn. 1.1) where there is a single critical point $x=x_0$ in the region supp $[f(x)]$ can be, obviously, reduced to a sum of the one-sided integrals:

$$I(\lambda)=\int_{-\infty}^{\infty}\exp\,[i\lambda\varphi(x)]f(x)\,dx$$
$$=\left(\int_{-\infty}^{x_0}+\int_{x_0}^{\infty}\right)\exp\,[i\lambda\varphi(x)]f(x)\,dx$$
$$=\int_0^{\infty}\exp\,[i\lambda\varphi(x_0+\xi)]f(x_0+\xi)\,d\xi$$
$$+\int_0^{\infty}\exp\,[i\lambda\varphi(x_0-\xi)]f(x_0-\xi)\,d\xi$$

Therefore an asymptotic expansion of the two-sided integral is expressed as a sum of asymptotic expansions of two one-sided integrals. When summing some terms of the expansion can cancel. For example, if $\varphi(x) = x^2$ and $f(x)$ is expanded into the Taylor series about zero

$$f(x) = \sum_{n=0}^{\infty} [x^n/\Gamma(n+1)] f^{(n)}(0)$$

then according to eqn. 1.17,

$$\int_0^{\infty} \exp(i\lambda x^2) f(x)\, dx \simeq \frac{1}{2} \sum_{n=0}^{\infty} (i/\lambda)^{(n+1)/2} \frac{\Gamma[(n+1)/2]}{\Gamma(n+1)} f^{(n)}(0)$$

$$\int_{-\infty}^{0} \exp(i\lambda x^2) f(x)\, dx \simeq \frac{1}{2} \sum_{n=0}^{\infty} (i/\lambda)^{(n+1)/2} (-1)^n \frac{\Gamma[(n+1)/2]}{\Gamma(n+1)} f^{(n)}(0)$$

and the terms with odd ns cancel when summing.

If $\varphi'(x) \neq 0$, the derivatives of $f(x)$ or $\varphi(x)$ are continuous up to the order of n, while $f^{(n)}(x)$ or $\varphi^{(n)}(x)$ have jump discontinuities at $x = x_0$, then the asymptotic contribution due to the point x_0 to the two-sided integral starts from a term of the order of λ^{-n-1} or λ^{-n}, i.e. when summing the asymptotics of one-sided integrals the first n terms cancel. Here the appropriate equations are given.

If $f(x)$ is infinitely differentiable at $x = x_0$ and $\varphi(x)$ is continuous with its derivatives up to the order of $(n-1)$ (where $n > 1$),

$$I(\lambda) = (i/\lambda)^n \frac{\exp[i\lambda\varphi(x_0)]}{[\varphi'(x_0)]^{n+1}} \Delta\varphi^{(n)}(x_0) f(x_0) + 0(\lambda^{(-n-1)}) \qquad (1.20)$$

where

$$\Delta\varphi^{(n)}(x_0) = \varphi^{(n)}(x_0+0) - \varphi^{(n)}(x_0-0)$$

is the discontinuity magnitude for $d^n\varphi/dx^n$ at $x = x_0$. If the function $\varphi(x)$ has a corner point $x = x_0$, i.e. its first derivative is discontinuous, then

$$I(\lambda) \simeq (i/\lambda) \exp[i\lambda\varphi(x_0)] f(x) \left[\frac{1}{\varphi'(x_0+0)} - \frac{1}{\varphi'(x_0-0)}\right] \qquad (1.21)$$

If $\varphi(x_0)$ is infinitely differentiable at $x = x_0$ and $f(x)$ is continuous with all its derivatives up to the order of $(n-1)$ at $x = x_0$, while the derivative $f^{(n)}(x_0)$ has a discontinuity, then

$$I(\lambda) = (i/\lambda)^{n+1} \exp[i\lambda\varphi(x_0)] \frac{\Delta f^{(n)}(x_0)}{[\varphi'(x_0)]^{n+1}} + 0(\lambda^{-n-2}) \qquad (1.22)$$

where

$$\Delta f^{(n)}(x_0) = f^{(n)}(x_0+0) - f^{(n)}(x_0-0)$$

is the discontinuity magnitude for $f^{(n)}(x)$ at $x = x_0$.

To prove these equations one should repeatedly integrate by parts according to eqns. 1.2 and 1.3 until the integrand becomes discontinuous, and then integrate by parts over the intervals $x < x_0$ and $x > x_0$ once again. The nonintegral term discontinuity gives eqns. 1.20–1.22.

Finally, the situation is possible when asymptotic expansions of one-sided integrals completely cancel and the critical point under consideration (the branch point of $f(x)$) does not contribute to $I(\lambda)$ at all. For such a situation to be a case it is required, first that the values of $f(x)$ in the vicinity of $x = x_0$ for $x < x_0$ to be obtained by analytical continuation from the region $x > x_0$ through the upper or lower halfplanes, i.e. the function $f(x)$ in a vicinity of $x = x_0$ to be a boundary value of a complex variable function $f(z)$ analytical in the upper and lower halfplanes. This means that the expansion of the function $f(x)$ in the region $x > x_0$

$$f(x) = \sum a_n (x - x_0)^{\beta_n}$$

should determine its expansion in the region $x < x_0$.

$$f(x) = \sum a_n \exp(\pm i\pi\beta_n)|x - x_0|^{\beta_n}$$

where the sign $\pm$ corresponds to analytical continuations in the upper or lower halfplanes. Further, it is required the function $\varphi(x)$ to be analytical in a complex vicinity of $x = x_0$ and $\varphi'(x) \neq 0$. Then the point $x = x_0$ does not contribute asymptotically into $I(\lambda)$, if $f(z)$ is analytical in the upper halfplane and $\varphi'(x_0) > 0$, or if $f(z)$ is analytical in the lower halfplane and $\varphi'(x_0) < 0$.

To prove this statement, let $\varphi'(x)$ be positive in the vicinity of $x = x_0$. Set $z = x + iy$, $\varphi(z) = u + iv$. For small $y > 0$,

$$\varphi(z) = \varphi(x + iy) = \varphi(x) + iy\varphi'(x) + 0(y^2)$$

therefore for small $y > 0$ $v = y\,\varphi'(x) + 0(y^2) > 0$. If $f(x)$ is a finite infinitely differentiable function at $x = x_0$, and in the vicinity of this point $|x - x_0| < \varepsilon$ it coincides at $y = 0$ with the values of the function $f(z) = f(x + iy)$ analytical in the upper halfplane $y > 0$, then the integration in $I(\lambda)$ in this vicinity can be shifted to the upper halfplane (see dashed line in Figure 1.3) in such a way that the integration contour C_+ is specified by an infinitely differentiable function $y = y(x)$. The exponential on this contour is of the form

$$\begin{aligned} \exp[i\lambda\varphi(z)] &= \exp\{i\lambda[u(z) + iv(z)]\} \\ &= \exp[i\lambda u(z)]\exp[-\lambda v(z)] \end{aligned}$$

and as $\lambda \to \infty$ it remains to be bounded in magnitude since $v(z) \geq 0$.

Since the contour C_+ is separated from $x = x_0$ and $f(z)$ is analytical on the contour in the vicinity of x_0, then all the derivatives $f^{(n)}(z)$ are bounded on C_+

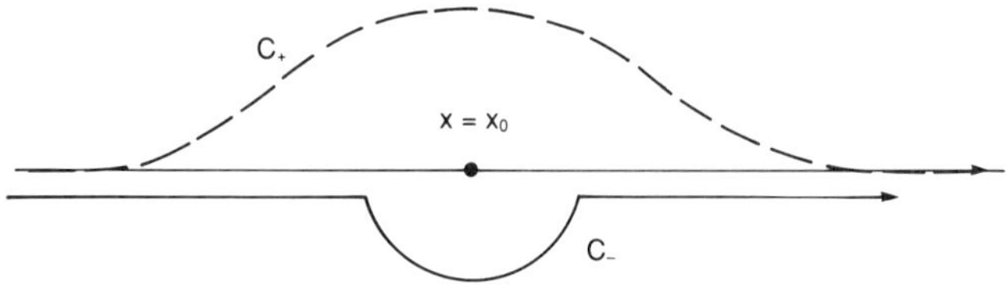

Figure 1.3 Integration contours in vicinity of critical point x_0

and one can evaluate the integral $I(\lambda)$ integrating by parts

$$I(\lambda) = (i/\lambda)^n \int_{C_+} \exp\,[i\lambda u(z)]\,\exp\,[-\lambda v(z)] f_n(z)\,dz$$

where $f_n(z)$ is defined by recurrence eqn. 1.3 and is bounded on C_+ as well. Therefore $I(\lambda) = 0(\lambda^{-n})$. Since n is arbitrary here the integral falls off more rapidly than any power of λ^{-1}.

In a similar way one can prove that the branch point $x = x_0$ does not contribute into $I(\lambda)$, if $\varphi'(x) < 0$ and $f(x)$ can be analytically continuated into the lower halfplane. In this case the integration contour is displaced to the lower halfplane.

1.6 Asymptotic contribution of pole of nonexponential factor

Assume that in integral of eqn. 1.1 the nonexponential function $f(x)$ has a pole at $x = x_0$, i.e. it can be represented as

$$f(x) = [a/(x - x_0)] + f_1(x) \tag{1.23}$$

where $f_1(x)$ is regular at x_0.

Before considering the asymptotic of eqn. 1.1 integral due to the pole contribution at $x = x_0$, i.e. the asymptotic of the integral

$$I(\lambda) = \int_{-\infty}^{\infty} \exp\,[i\lambda\varphi(x)] f(x) g(x)\,dx$$

where $f(x)$ is of the form of eqn. 1.23 and $g(x)$ is the cutting function one needs to define how to understand the diverging integral

$$I(\lambda) = \int_{-\infty}^{\infty} \exp\,[i\lambda\varphi(x)]\,[g(x)/(x - x_0)]\,dx$$

Note that such integrals arise when sought-for functions are written as contour integrals over a complex variable, the functions $f_1(x)$ and $\varphi(x)$ being usually analytical at x close to x_0.

Three different definitions of the integral $I(\lambda)$ can be indicated. First, one can understand the integral in a sense of the Cauchy principal value

$$I_0(\lambda) = \lim_{\varepsilon \to 0} \left[\int_{-\infty}^{x_0 - \varepsilon} + \int_{x_0 + \varepsilon}^{\infty} \right] \exp\,[i\lambda\varphi(x)]\,[a/(x - x_0)]\,g(x)\,dx$$

Secondly, the pole can be passed around in the upper halfplane

$$I_+(\lambda) = \int_{C_+} \exp\,[i\lambda\varphi(x)]\,[a/(x - x_0)]\,g(x)\,dx$$

where the contour C_+ is shown in Figure 1.3. Thirdly, the pole $x=x_0$ can be passed around in the lower halfplane along the contour C_-.

$$I_-(\lambda)=\int_{C_-} \exp\,[i\lambda\varphi(x)]\,[a/(x-x_0)]g(x)\,dx$$

Naturally, all these definitions are interrelated

$$I_\pm(\lambda)=I_0(\lambda)\mp ia\pi \exp\,[i\lambda\varphi(x_0)] \tag{1.24}$$

Now find the conditions under which the point $x=x_0$ does not contribute into the integral asymptotics. To avoid the pole influence one should move off the integration contour from the pole, but it has to be done in such a way that the integrand remains bounded. As seen in Section 1.5, in doing so for $\varphi'(x)>0$ it is necessary to shift the integration contour to the upper halfplane. Therefore for $\varphi'(x)>0$ the integral $I_+(\lambda)$ will tend to zero more rapidly than any power of λ^{-1}, and it follows from eqn. 1.24 that $I_-(\lambda)\simeq 2\pi ia \exp\,[i\lambda\varphi(x_0)]$. Thus, for $\varphi'(x)>0$

$$I_+(\lambda)\simeq 0$$

$$I_-(\lambda)\simeq -2\pi ia \exp\,[i\lambda\varphi(x_0)] \tag{1.25a}$$

When $\varphi'(x)<0$ the function $\exp\,[i\lambda\varphi(x)]$ is bounded in the lower halfplane. Shifting the integration contour to this halfplane one obtains for $\varphi'(x)<0$

$$I_-(\lambda)\simeq 0$$

$$I_+(\lambda)\simeq -2\pi a \exp\,[i\lambda\varphi(x_0)] \tag{1.25b}$$

1.7 Integrals of functions with large parameter in argument

Consider the asymptotics of integrals of the form

$$I(\lambda)=\int_{-\infty}^{\infty} f(x)\Phi[\lambda h(x)]\,dx \tag{1.26}$$

as $\lambda\to\infty$. Such integrals arise, for example, when solving axially symmetrical problems of diffraction and wave propagation, in these cases $\Phi(x)=I_\nu(x)$ where I_ν is the Bessel function. This particular case was considered in many papers (see References 20 and 21 and the references therein).

Consider, first, the case $h(x)=x$. Here the integral of eqn. 1.26 is reduced to the sum of two one-sided integrals of the form

$$I_1(\lambda)=\int_0^{\infty} \psi(x)\Phi(\lambda x)\,dx \tag{1.27}$$

To find the asymptotic of $I_1(x)$ as $\lambda\to\infty$ let us begin with the formal mathematics. Assume the function $\Phi(x)$ to fall off as $x\to\infty$. Then when increasing $\Phi(\lambda x)$ is concentrated in the vicinity of $x=0$ and the less $\psi(x)$ is in this vicinity (i.e. the more rapidly it tends to zero as $x\to 0$), the more rapidly $I_1(\lambda)$ falls off as $\lambda\to\infty$.

Let $\psi(x) = x^p$. On setting $\lambda x = \xi$,

$$I_1(\lambda) = \int_0^\infty x^p \Phi(\lambda x)\, dx = \lambda^{-p-1} \int_0^\infty \xi^p \Phi(\xi)\, d\xi = C(p)\lambda^{-p-1}$$

where

$$C(p) = \int_0^\infty \xi^p \Phi(\xi)\, d\xi$$

If $\psi(x)$ is expanded in an asymptotic series in growing powers of x (as $x \to 0$)

$$\psi(x) \simeq \sum_{n=0}^{\infty} a_n x^{p_n} \tag{1.28}$$

then integrating eqn. 1.27 term by term obtains

$$I_1(\lambda) \simeq \sum_{n=0}^{\infty} a_n C(p_n) \lambda^{-p_n - 1} \tag{1.29}$$

Now formulate the conditions for $\Phi(z)$ which will justify asymptotic eqn. 1.29. The function $\Phi(x)$ is required to be analytical for real $z \neq 0$, to have a branch point $\Phi(z) \simeq z^\nu$ as $z \to 0$, and to have an asymptotic expressed as the sum of two exponentials (as $z \to \infty$)

$$\Phi(z) = \Phi_+(z) + \Phi_-(z) = A_+ \exp(iz) + A_- \exp(-iz) \tag{1.30}$$

where $A_\pm$ are asymptotic series

$$A_\pm \simeq \sum_n \beta_n^\pm z^{-n-\mu}, \qquad \mu > 0 \tag{1.31}$$

Let it be required that the functions $\Phi_\pm(z)$ with their asymptotic expansions be analytically continuous (for sufficiently large $|z|$) in the upper or lower halfplane, respectively, where the functions fall off exponentially, the function Φ_+ being continuous into the sector $0 < \text{Im}\, z \leq tg\alpha |\text{Re}\, z|$, while the function Φ_- into the sector $-tg\alpha |\text{Re}\, z| \leq \text{Im}\, z < 0$ where $\alpha > 0$. Then for any exponent p and sufficiently large M the integral

$$\int_M^\infty x^p \Phi(x)\, dx$$

is determined by the expressions

$$\begin{aligned}\int_M^\infty x^p \Phi(x)\, dx &= \int_M^\infty x^p \Phi_+(x)\, dx + \int_M^\infty x^p \Phi_-(x)\, dx \\ &= \int_M^{M + \infty \exp(i\alpha)} z^p \Phi_+(z)\, dz \\ &+ \int_M^{M + \infty \exp(-i\alpha)} z^p \Phi_-(z)\, dz \end{aligned} \tag{1.32}$$

integrating along the rays $z = M + \rho \exp(\pm i\alpha)$ $(0 < \rho < \infty)$ and where the integrals converge absolutely since Φ_+ and Φ_- fall off exponentially. The following theorem is valid:

Theorem 1.7: Let

$$I(\lambda) = \int_0^\infty \psi(x) g(x) \Phi(\lambda x)\, dx \tag{1.33}$$

Here, $\psi(x)$ is an infinitely differentiable function for $x > 0$ which is expanded into the asymptotic series of eqn. 1.28 as $x \to 0$ where $p_{n+1} > p_n$ and $p_n \to \infty$ as $n \to \infty$, $g(x)$ is a cutting function which singles out a vicinity of zero, $\Phi(x)$ has asymptotics of eqns. 1.30 and 1.31 and does not increase more rapidly (as $x \to 0$) than x^ν where $\nu + p_0 > -1$. Then $I(\lambda)$ is expanded (as $\lambda \to \infty$) into the asymptotic series

$$I(\lambda) = \sum_{n=0}^{\infty} a_n C(p_p) \lambda^{-p_n - 1} \tag{1.34}$$

where

$$C(p) = \int_0^\infty y^p \Phi(y)\, dy = \int_0^M y^p \Phi(y)\, dy + \int_M^\infty y^p \Phi(y)\, dy$$

and as $y \to \infty$ the second integral is understood in a sense of eqn. 1.32.

Before proving the theorem the initial integral (eqn. 1.26) is reduced to the particular case considered in theorem 1.7.

Let $h(x)$ have zeros $x_1, x_2, \ldots, x_k$ and denote a cutting function which singles out a point x_k by $g_k(x) = g(x - x_k)$. Set $g_0(x) = 1 - \sum_k g(x - x_k)$. In the region supp (g_0) the function h is bounded in magnitude from below and for the function Φ in the integral

$$I_0(\lambda) = \int_0^\infty g_0(x) f(x) \Phi[\lambda h(x)]\, dx$$

one can use the asymptotic of eqn. 1.30

$$I_0(\lambda) = \int_0^\infty f(x) g_0(x) \{A_+[\lambda h(x)] \exp[i\lambda h(x)] + A_-[\lambda h(x)] \exp[-i\lambda h(x)]\}\, dx$$

The asymptotic of the integral obtained is evaluated by the standard method of stationary phase or its modifications considered below which are applicable when critical points are close to each other. Therefore it is sufficient to find an asymptotic contribution into the integral $I(\lambda)$ due to zeros of function $h(x)$.

Let $x = 0$ be a root of the function $h(x)$ and let $g(x)$ be a cutting function which singles out a vicinity of this point. Representing the integral over this vicinity in the form of two one-sided integrals,

$$\int_0^{\infty} f(x)g(x)\Phi[\lambda h(x)]\ dx$$

where $h(x)\to 0$ as $x\to 0$. The region supp (g) can be chosen sufficiently small in order $h(x)$ to be strictly monotonic. Assume $h(x)$ to have a power singularity as $x\to 0$, $h(x)=x^p h_0(x)$, where $h_0(x)$ is analytical. Considering $y=h(x)$ as a new independent variable,

$$I(\lambda)=\int_0^{\infty} \psi_1(y)g_1(y)\Phi(\lambda y)\ dy \tag{1.35}$$

where $\psi_1(y)=f[x(y)]\ dx/dy$, $g_1(y)=g[x(y)]$ and $x(y)$ is the function inverse to $y(x)$ and expanding into a series in powers of $y^{1/p}$. If $f(x)$ is expanded into a series in fractional powers of x starting from some x^{ξ} then $\psi(y)$ is expanded into a series in fractional powers of y starting with $y^{(\xi+1-p)/p}$ and the conditions of theorem 1.7 are satisfied.

To prove the theorem, it follows from the asymptotic of eqn. 1.28 that for any integer N

$$\psi(x)=\sum_{n=0}^{N} a_n x^{p_n}+x^{p_{n+1}}\psi_1(x) \tag{1.36}$$

where $\psi_1(x)$ is bounded for bounded $|x|$. Therefore the integral of eqn. 1.33 can be written as

$$\int_0^{\infty} x^{p_{n+1}}\psi_1(x)g_1(x)\Phi(\lambda x)\ dx+\sum_{n=0}^{N} a_n \int_0^{\infty} x^{p_n}g_1(x)\Phi(\lambda x)\ dx$$

Theorem 1.7 is an obvious consequence of the written relation and the following lemmas.

Lemma 1: For sufficiently large $q=q(N)$ the integral

$$I=\int_0^{\infty} g(y)\psi(y)y^q\Phi(\lambda y)\ dy$$

(where $\psi(y)$ is bounded in the region supp (g)) falls off more rapidly than any given power function of λ^{-N}.

Lemma 2: The following asymptotic equality is valid

$$I=\int_0^{\infty} g(y)y^q\Phi(\lambda y)\ dy\simeq\lambda^{-q-1}C(q)$$

in other words, the difference between the left-hand and right-hand sides of this expression falls off more rapidly than any given power function of λ^{-1}.

To prove the lemmas, we start with lemma 1. Choose a positive $\alpha < 1$ and set

$$I = I_1 + I_2 = \int_0^{\lambda^{-\alpha}} g(y)\psi(y)y^q\Phi(\lambda y)\, dy + \int_{\lambda^{-\alpha}}^{\infty} g(y)\psi(y)y^q\Phi(\lambda y)\, dy$$

The integral I_1 is estimated in magnitude. Since $\Phi(y) = 0(y^{-\mu})$ as $y \to \infty$ and $\Phi(y) = 0(y^{\nu})$ as $y \to 0$ the uniform estimation is valid

$$\Phi(y) < \text{const.}\, y^{\nu}(1+y)^{-\mu-\nu}$$

Taking into account that $g(y)\psi(y)$ is bounded one obtains the estimation for I_1

$$I_1 \leq \text{const.} \int_0^{\lambda^{-\alpha}} y^q(\lambda y)^{\nu}(1+\lambda y)^{-\nu-\mu}\, dy \leq \text{const.}\, \lambda^{-\alpha q - \alpha - \mu + \alpha\mu}$$

It is clear that for sufficiently large q this expression falls off more rapidly than any given power function of λ^{-1}.

Consider now the integral I_2. In this integral the function Φ has an argument $\lambda y > \lambda^{1-\alpha} \gg 1$, therefore for Φ asymptotic eqn. 1.27 is valid and

$$I_2 = I_{21} + I_{22} = \int_{\lambda^{-\alpha}}^{\infty} g(y)\psi(y)y^q A_+(\lambda y) \exp(i\lambda y)\, dy$$

$$+ \int_{\lambda^{-\alpha}}^{\infty} g(y)\psi(y)y^q A_-(\lambda y) \exp(i\lambda y)\, dy$$

As $\lambda \to \infty$ the asymptotics of the integrals I_{21} and I_{22} are evaluated by repeated integrations by parts, the leading terms of the asymptotics are estimated as

$$\lambda^{-1} g(y)\psi(y) \cdot (\lambda y)^{-\mu}|_{y=\lambda^{-\alpha}} \leq \text{const.}\, \lambda^{-\alpha q - 1 + \mu(\alpha - 1)}$$

which for sufficiently large q also falls off more rapidly than any given power of λ^{-1}.

To prove lemma 2 write

$$I = \int_0^{\infty} g(y)y^q\Phi(\lambda y)\, dy = \int_0^{\infty} y^q\Phi(\lambda y)\, dy - \int_0^{\infty} [1 - g(y)]y^q\Phi(\lambda y)\, dy$$

where integrals on the right-hand side are understood in the sense of eqn. 1.32. The first of these integrals is equal to $\lambda^{-q-1}C(q)$. Since $1 - q(y) \equiv 0$ for small $|y|$, when evaluating the second integral the function Φ can be replaced by its asymptotic eqn. 1.27 and integrating by parts shows that the integral tends to zero more rapidly than any power of λ^{-1}.

Thus lemmas 1 and 2 are proved and, hence, theorem 1.7.

Chapter 2

Two critical points merging in a one-dimensional integral

In the previous Chapter the algorithm to find asymptotics of one-dimensional integrals is formulated. According to the algorithm the critical points are determined and asymptotic contributions of these points are summed. However, the algorithm is inapplicable when the critical points approach each other. Consider, for instance, an integral

$$I(\lambda, \alpha) = \int_{-\infty}^{\infty} \exp\left[i\lambda\left(x^3/3 - \alpha x\right)\right] dx$$

For $\alpha < 0$ the phase function grows monotonically, its derivative is positive and $I(\lambda, \alpha)$ falls off more rapidly than any power of λ^{-1}. For $\alpha > 0$ the phase function has two stationary points, $x = \pm\sqrt{\alpha}$. Summing asymptotic contributions of these points yields

$$I(\lambda, \alpha) \simeq \left[\begin{array}{ll} O(\lambda^{-n}) & \text{for } \alpha < 0 \text{ and arbitrary } n \\ 2\sqrt{(2\pi/\lambda)}\alpha^{-1/4} \cos\left[(2\lambda\alpha^{3/2}/3) - \pi/4\right] + O(\lambda^{-3/2}) & \text{for } \alpha > 0 \end{array}\right.$$

It is clear that this asymptotic fails as $\alpha \to 0$ since even its leading term tends to infinity while the integral $I(\lambda, \alpha)$ remains bounded. The situation is quite general. When critical points in the integral $I(\lambda, \alpha)$ approach each other the asymptotic contribution of some of these points tends to infinity. Two problems naturally arise.

The first is to find asymptotics of the integral $I(\lambda, \alpha)$ applicable under these conditions, i.e. uniform over a distance between critical points. The second is to estimate the applicability range for simpler nonuniform asymptotics found in the previous Chapter. As will be shown, the analytical form (Ansatz) of the uniform asymptotic is determined by the types of critical points tending to each other and is expressed in terms of the appropriate special function Φ. As a rule, the argument of Φ is a power function of the dimensionless combination $\lambda(\varphi_1 - \varphi_2)$ where φ_1 and φ_2 are the values of the phase function at critical points approaching each other (the only exception is the case when a stationary point tends to infinity, see Section 2.7). Going from the uniform asymptotic to the nonuniform appears to be equivalent to replacing the function Φ by its asymptotic; this is possible only for sufficiently high values of the argument, i.e. for high $\lambda(\varphi_1 - \varphi_2)$. The results for the applicability range of the nonuniform asymptotic obtained by this approach are given in Section 2.8.

This Chapter considers the most frequent cases of two critical points merging, while the next one treats the case of three critical points.

2.1 Stationary point near integration domain boundary

Consider an integral

$$I(\lambda, \alpha)=\int_{-\infty}^{\alpha} \exp [i\lambda\varphi(x)]f(x)g(x)\, dx \tag{2.1}$$

where $\varphi(x)$ and $f(x)$ have the required number of continuous derivatives, the phase function $\varphi(x)$ has the one stationary point $x=x_0$ with positive second derivative, $g(x)$ is the cutting function which singles out the points x_0 and α, i.e. it is a finite infinitely differentiable function identically equal to unity in an interval including these points.

It is easy to write the asymptotic of $I(\lambda, \alpha)$ with fixed α and $\lambda\to\infty$. The boundary of the integration domain $x=\alpha$ and the stationary point $x=x_0$ are critical points. Taking into account that the stationary point $x=x_0$ is inside the integration domain only for $\alpha>x$ and retaining only the leading term when evaluating each critical point contribution, one gets for $\lambda\gg 1$

$$\begin{aligned} I(\lambda, \alpha) &\simeq \exp [i\lambda\varphi(x_0)]\sqrt{[2\pi i/\lambda\varphi''(x_0)]}f(x_0)\chi(\alpha-x_0) \\ &\quad -(i/\lambda)\frac{f(\alpha)\exp [i\lambda\varphi(\alpha)]}{\varphi'(\alpha)}+O(\lambda^{-3/2}) \end{aligned} \tag{2.2}$$

Here, and in what follows, the unit step function (the Heaviside function) is denoted by

$$\chi(\xi)=\begin{bmatrix} 1, & \xi>0 \\ 0, & \xi<0 \end{bmatrix} \tag{2.3}$$

The asymptotic of eqn. 2.2 is inapplicable as $\alpha\to x_0$ since the first summand is discontinuous at $x=x_0$, while the second tends to infinity as $\alpha\to 0$ because its denominator includes $\varphi'(\alpha)$ and $\varphi'(x_0)=0$.

This Section constructs the asymptotic of the integral $I(\lambda, \alpha)$ uniform over α as $\lambda\to\infty$.

For the case in question the model integral is

$$\begin{aligned} I(\lambda, \alpha) &= \int_{-\infty}^{\alpha} \exp (i\lambda x^2)\, dx = [1/\sqrt{(\lambda)}]\int_{-\infty}^{\sqrt{(\lambda)}\alpha} \exp (i\xi^2)\, dx \\ &= \sqrt{(\pi i/\lambda)}F[\sqrt{(\lambda)}\alpha] \end{aligned} \tag{2.4}$$

where

$$F(s)=\frac{1}{\sqrt{(\pi i)}}\int_{-\infty}^{s} \exp (it^2)\, dt \tag{2.5}$$

is the Fresnel integral. The factor $\sqrt{(i\pi)}$ is introduced to normalise the integral. (It follows from eqn. 1.7 that $F(\infty)=1$.)

In what follows one needs to find the asymptotic of $F(\xi)$ for $|\xi|\gg 1$. In the case of $\xi\ll 0$ one can easily evaluate the asymptotic by integrating by parts (see

Introduction) while the case of $\xi \gg 0$ is reduced to that of $\xi \ll 0$ by the relation

$$F(\xi)+F(-\xi)=\frac{1}{\sqrt{(\pi i)}}\left[\int_{-\infty}^{\xi} \exp(it^2)\,dt+\int_{\xi}^{\infty} \exp(it^2)\,dt\right]=1$$

As a result for $|\xi| \gg 1$ the Fresnel integral $F(\xi)$ has the asymptotic expansion

$$F(\xi)\simeq\chi(\xi)+\frac{1}{2\pi}\exp(i\xi^2)\times\sum_{s=0}^{\infty}\frac{\exp[(is\pi/2)-3i\pi/4]}{\xi^{2s+1}}\Gamma(s+1/2) \tag{2.6}$$

Consider an integral which differs from eqn. 2.4 by the cutting function $g(x)$ which singles out the points $x=0$ and $x=\alpha$

$$I(\lambda, \alpha)=\int_{-\infty}^{\alpha} \exp(i\lambda x^2)g(x)\,dx$$

The integral is asymptotically equivalent to eqn. 2.4, i.e. the difference

$$\Delta I=I_0-I=\int_{-\infty}^{\alpha} \exp(i\lambda x^2)[1-g(x)]\,dx$$

falls off more rapidly than any power of λ^{-1} as $\lambda\to\infty$. Indeed, since $g\equiv 1$ in an interval enclosing the point $x=0$ the function x^2 is strictly monotone in the region supp $(1-g)$, i.e. in the region where $1-g\neq 0$. Integrating many times by parts and taking into account that $g\equiv 1$ in the vicinity of the point $x=\alpha$ and therefore the nonintegral terms (arising when integrating by parts) vanish,

$$\Delta I=(i/\lambda)^n\int_{-\infty}^{\alpha} \exp(i\lambda x^2)f_n(x)\,dx=O(\lambda^{-n})$$

where n is arbitrary and

$$f_0(x)=1-g(x), \qquad f_{n+1}(x)=-\frac{1}{2}\frac{d}{dx}\left[\frac{f_n(x)}{x}\right]$$

Thus,

$$\int_{-\infty}^{\alpha} \exp(i\lambda x^2)g(x)\,dx\simeq\sqrt{(i\pi/\lambda)}F[\sqrt{(\lambda)}\alpha] \tag{2.7}$$

Introduce an arbitrary analytical function $f(x)$ under the integral sign, i.e. consider eqn. 2.1 for $\varphi(x)=x^2$. To prove that

$$I(\lambda, \alpha)=\int_{-\infty}^{\alpha} \exp(i\lambda x^2)f(x)g(x)\,dx$$

$$=\lambda^{-1}P(\lambda^{-1}, \alpha)\exp(i\lambda\alpha^2)+\frac{1}{\sqrt{(\lambda)}}Q(\lambda^{-1}, \alpha)F[\sqrt{(\lambda)}\alpha] \tag{2.8}$$

where P and Q are the asymptotic power series in λ^{-1} with regular coefficients as $\alpha \to 0$, it is sufficient to prove that for any $N > 0$

$$I(\lambda, \alpha) = \lambda^{-1} P_N(\lambda^{-1}, \alpha) \exp(i\lambda\alpha^2) + Q_N(\lambda^{-1}, \alpha) F[\sqrt{(\lambda)}\alpha]/\sqrt{\lambda} + O(\lambda^{-N}) \tag{2.9}$$

where P_N and Q_N are the polynomials in λ^{-1} regular over α coefficients and the estimation $O(\lambda^{-N})$ is uniform as $\alpha \to 0$. Set

$$f(x) = \sum_{n=0}^{2N} f_n x^n + x^{2N+1} f^*(x)$$

and verify that

$$I^*(\lambda, \alpha) = \int_{-\infty}^{\alpha} \exp(i\lambda x^2) x^{2N+1} f^*(x) g(x)\, dx = \lambda^{-1} P^*(\lambda^{-1}, \alpha) \exp(i\lambda\alpha^2) + O(\lambda^{-N})$$

where P is the polynomial in λ^{-1}.

To prove the equation it is sufficient to integrate N times by parts, each time integrating $x \exp(i\lambda x^2)$ and differentiating the remainder of the integrand. Nonintegral terms give $\lambda^{-1} P^*(\lambda^{-1}, \alpha) \exp(i\lambda\alpha^2)$ and the resulting integral is estimated by $O(\lambda^{-N})$ uniformly over α as $\alpha \to 0$. Therefore it remains to prove eqn. 2.9 for

$$f(x) = \sum_{n=0}^{2N} f_n x^n$$

i.e. to prove that

$$I_n(\alpha) = \int_{-\infty}^{\alpha} \exp(i\lambda x^2) x^n g(x)\, dx = \lambda^{-1} A_n(\lambda^{-1}, \alpha) \exp(i\lambda\alpha^2) + \frac{B_1(\lambda^{-1}, \alpha)\, F[\sqrt{(\lambda)}\alpha]}{\sqrt{(\lambda)}} \tag{2.10}$$

for certain polynomials $A_n(\lambda^{-1}, \alpha)$, $B_n(\lambda^{-1}, \alpha)$. For even $n = 2p$, one obtains eqn. 2.9 by differentiating eqn 2.7 p times with respect to λ and taking into account the following relation

$$\frac{\partial}{\partial \lambda} F[\sqrt{(\lambda)}\alpha] = \frac{\alpha}{2\sqrt{(\lambda i \pi)}} \exp(i\lambda\alpha^2)$$

For $n = 1$,

$$I_1(a) = \int_{-\infty}^{\alpha} x \exp(i\lambda x^2) g(x)\, dx = (1/2i\lambda) \int_{-\infty}^{\alpha} g(x) d[\exp(i\lambda x^2)]$$
$$= (1/2i\lambda) \exp(i\lambda\alpha^2) + (i/2\lambda) \int_{-\infty}^{\alpha} \exp(i\lambda x^2) g'(x)\, dx$$

where the integral tends to zero more rapidly than any power of λ^{-1} (as $\lambda\to\infty$) since $g'\equiv 0$ in the vicinities of the points $x=0$ and $x=\alpha$. Therefore

$$I_1(\alpha)\simeq(1/2i\lambda)\exp(i\lambda\alpha^2)$$

Differentiating this expression p times with respect to λ obtains eqn. 2.10 for $n=2p+1$ and $B_{2p+1}\equiv 0$. Thus, eqn. 2.8 is proved.

It remains to show how to evaluate P and Q in the asymptotic expansion eqn. 2.8. To find them fix $\alpha\neq 0$ and set $\lambda\to\infty$. Replacing the Fresnel integral $F[\surd(\lambda)\alpha]$ by its asymptotic (eqn. 2.6) as $\surd(\lambda)\alpha\to\infty$ obtains

$$I(\lambda,\alpha)\simeq\{\lambda^{-1}P(\lambda^{-1},\alpha)+\lambda^{-1/2}Q^{-1}(\lambda^{-1},\alpha)T[\surd(\lambda)\alpha]\}\exp(i\lambda\alpha^2)$$
$$+[1/\surd(\lambda)]Q(\lambda,\alpha)\chi(\alpha) \qquad (2.11)$$

where $T[\surd(\lambda)\alpha]$ is the asymptotic series on the right-hand side of eqn. 2.6.

$$T[\surd(\lambda)\alpha]=\sum_{n=0}^{\infty}t_n\lambda^{-n-1/2}\alpha^{-2n-1}$$
$$t_n=(1/2\pi)\exp[(-in\pi/2)-3\pi i/4]\Gamma(n+1/2) \qquad (2.12)$$

On the other hand, the same nonuniform asymptotic can be easily found when directly applying the results of Section 1.5 to $I(\lambda,\alpha)$.

For $\alpha>0$ the integral of eqn. 2.8 has two critical points, the stationary point $x=0$ and the boundary of the integration domain $x=\alpha$. For $\alpha<0$ the only critical point is an end of the integration domain interval. Taking into account contributions of these points obtains

$$I(\lambda,\alpha)\simeq\sum_{n=0}^{\infty}c_n\lambda^{-n-1/2}\chi(\alpha)+\exp(i\lambda\alpha^2)\sum_{n=0}^{\infty}d_n(\alpha)\lambda^{-n-1} \qquad (2.13)$$

Here the first summand is a contribution of the stationary critical point $x=0$ and the second is that of the integration domain boundary, $x=\alpha$. The coefficients c_n and $d_n(\alpha)$ are evaluated according to theorems 1.1 and 1.2

$$c_n=\frac{i^n\surd(\pi i)}{2^{2n}\Gamma(n+1)}f^{(2n)}(0)$$
$$d_n=\frac{i^{n-1}}{2}\frac{d^n}{ds^n}\left[\frac{f(\surd s)}{\surd s}\right]\bigg|_{s=\alpha^2}$$
$$=\sum_{p=0}^{\infty}\gamma_{p,n}f^{(p)}(\alpha)\alpha^{-2n-1+p} \qquad (2.14)$$

where $\gamma_{p,n}$ are constant coefficients independent of the function $f(x)$.

Equations 2.11 and 2.13 are asymptotic expansions of the same function $I(\lambda,\alpha)$ and therefore should coincide. Comparing the coefficients of $\chi(\alpha)$ and $\exp(i\lambda\alpha^2)$,

$$\frac{Q(\lambda^{-1},\alpha)}{\surd(\lambda)}=\sum_{p=0}^{\infty}c_n\lambda^{-n-1/2}$$

$$\lambda^{-1}P(\lambda^{-1},\alpha)=\sum_{n=0}^{\infty}d_n\lambda^{-n-1}-\left[\sum_{n=0}^{\infty}c_n\lambda^{-n-1/2}\right]\left[\sum_{n=0}^{\infty}t_n\lambda^{-n-1/2}\alpha^{-2n-1}\right]$$

Let

$$Q = \sum_{n=0}^{\infty} q_n \lambda^{-n}, \qquad P = \sum_{n=0}^{\infty} p_n \lambda^{-n}$$

Then for q_n and p_p

$$q_n = c_n, \qquad p_n = d_n(\alpha) - \sum_{m=0}^{n} c_n t_{n-m} \alpha^{-2(n-m)-1} \tag{2.15}$$

As $\alpha \to 0$ the coefficients $d_n(\alpha)$ tend to infinity as α^{-2n-1}. The second summand on the right-hand side of eqn. 2.15 also contains negative powers of α. However, as shown, the function $P(\lambda^{-1}, \alpha)$ and hence the coefficients p_n of its expansion into a power series in λ^{-1} are regular as $\alpha \to 0$. Therefore the poles of the first and second summands in eqn. 2.15 should cancel. One can verify this by writing explicit expressions for p_n and using eqns. 2.15 and 2.14 for d_n, c_n and t_n, though the calculations become more cumbersome of course, when n increases. In particular, for the first terms,

$$\begin{aligned}
p_0 &= (i/2)[f(0) - f(\alpha)]/\alpha \\
p_1 &= [f(0) - f(\alpha) + \alpha f'(\alpha) - \alpha^2 f''(0)/2]/\alpha^3 \\
p_2 &= (i/8\alpha^5)[\alpha^2 f''(\alpha) - 3\alpha f'(\alpha) \\
&\qquad + 3f(\alpha) - 3f(0) + (\alpha^2/2)f''(0) - (\alpha^4/8)f^{(4)}(0)]
\end{aligned}$$

This method of finding amplitude coefficients p_n and q_n in the uniform asymptotic is based on the procedure: replacing the Fresnel integral $F[\sqrt{(\lambda)}\alpha]$ by its asymptotic as $\sqrt{(\lambda)}\alpha \to \infty$, the expression obtained from eqn. 2.8 is equated to the nonuniform asymptotic of the integral $I(\lambda, \alpha)$ obtained independently.

This approach is termed the asymptotic matching method. The method makes it possible to find not only the amplitudes, but also the phase factor ψ as well as the arguments ξ of special functions in terms of which the uniform asymptotic is expressed. The amplitudes p_n in the uniform asymptotic remain bounded when critical points approach each other and are expressed in terms of amplitude factors of the nonuniform asymptotic which tend to infinity. Therefore for p_n one obtains expressions of the form of $(\infty - \infty)/0$ or (∞/∞). Analysis of these uncertainties and regularities of p_n (as well as ψ, ξ) turns out to be a reliable method to control if the ansatz (the analytical expression) for uniform asymptotic is chosen correctly and mathematical manipulations performed to construct the asymptotic are correct.

The case of the phase function $\varphi(x) = x^2$ has been considered. In the general case the following theorem holds

Theorem 2.1: Let the phase function $\varphi(x)$ in eqn. 2.1 have a nondegenerate stationary point $x = x_0$ close to the endpoint of the integration interval $x = \alpha$. Let $g(x)$ be a cutting function which singles out vicinity of the points α, x_0 where there are no other critical points. Let functions $\varphi(x)$ and $f(x)$ be infinitely differentiable in the region supp *(g). Then for $\varphi''(x) > 0$ the asymptotic of $I(\lambda, \alpha)$ uniform over a distance between α and x_0 has the form (as $\lambda \to \infty$)*

$$\begin{aligned}
I(\lambda, \alpha) &\simeq P(\lambda^{-1}, \alpha) \exp[i\lambda\varphi(\alpha)] \\
&\quad + \exp[i\lambda\varphi(x_0)] Q(\lambda^{-1}, \alpha) F[\sqrt{(\lambda)}\xi_0]
\end{aligned} \tag{2.16}$$

where $P(\lambda^{-1}, \alpha)$ and $Q(\lambda^{-1})$ are the asymptotic series in powers of λ^{-1}, the coefficients of the series $P(\lambda^{-1}, \alpha)$ being the regular functions of α, $F[\sqrt{(\lambda)}\xi_0]$ is the Fresnel integral and $\xi_0 = \xi(\alpha)$ where

$$\xi(x) = \text{sign}\,(x - x_0)\sqrt{|\varphi(x) - \varphi(x_0)|} \tag{2.17}$$

The series $P(\lambda^{-1}, \alpha)$ and $Q(\lambda^{-1})$ are expressed in terms of the nonuniform asymptotic of $I(\lambda, \alpha)$

$$I(\lambda, \alpha) \simeq C(\lambda) \exp[i\lambda\varphi(x_0)]\chi(\alpha - x_0) + D(\lambda, \alpha) \exp[i\lambda\varphi(\alpha)] \tag{2.18}$$

as $\lambda \to \infty$ and for α fixed by means of

$$Q(\lambda^{-1}, \alpha) = C(\lambda)$$
$$P(\lambda^{-1}, \alpha) = D(\lambda, \alpha) - Q(\lambda^{-1}, \alpha)T[\sqrt{(\lambda)}\xi_0] \tag{2.19}$$

where $T[\sqrt{(\lambda)}\xi_0]$ is the series of eqn. 2.12. If $\varphi''(x_0) < 0$ the Fresnel integral and series T in eqns. 2.16 and 2.19 are replaced by their complex conjugate.

To prove eqn. 2.16 one needs to reduce $I(\lambda, \alpha)$ to the integral

$$I(\lambda, \alpha) = \exp[i\lambda\varphi(x_0)] \int_{-\infty}^{\xi_0} \exp(i\lambda\xi^2) f[x(\xi)] \frac{\partial x(\xi)}{\partial \xi} d\xi$$

by change of variable (eqn. 2.17) and then to apply eqn. 2.8.

Eqn. 2.19 is proved by asymptotic matching. Finally, the case when x_0 is a maximum of the function $\varphi(x)$ follows from eqns. 2.16 and 2.19 by complex conjugation. In particular, when evaluating leading terms of the asymptotic series $P(\lambda^{-1}, \alpha)$ and $Q(\lambda^{-1}, \alpha)$ one gets

$$\begin{aligned} I(\lambda, \alpha) &= \int_{-\infty}^{\alpha} \exp[i\lambda\varphi(x)] f(x)\, dx \\ &\approx \sqrt{\{2\pi i/[\lambda|\varphi''(x_0)|]\}} \exp\{i[\pi \,\text{sign}\, \varphi''(x_0)/4 \\ &\quad + \lambda\varphi(x_0)]\} F_1\{\text{sign}\,(\alpha - x_0)\sqrt{[\lambda|\varphi(\alpha) - \varphi(x_0)|]}\} \\ &\quad + (i/\lambda) \exp[i\lambda\varphi(\alpha)] P(\alpha) + O(\lambda^{-3/2}) \end{aligned} \tag{2.20a}$$

where

$$\begin{aligned} P(\alpha) = \{& f(x)\varphi'(\alpha)\,\text{sign}\,\varphi''(x_0) - f(\alpha)\sqrt{[2|\varphi''(x_0)||\varphi(\alpha)} \\ & - \varphi(x_0)]|\,\text{sign}\,(\alpha - x_0)\}/\{\sqrt{|2\varphi''(x_0)|}\varphi'(\alpha)\sqrt{|\varphi(\alpha)} \\ & - \varphi(x_0)|\,\text{sign}\,(\alpha - x_0)\} \end{aligned} \tag{2.20b}$$

F_1 is the Fresnel integral if $\varphi''(x_0) > 0$, and the complex conjugate Fresnel integral if $\varphi''(x_0) < 0$. The function $P(\alpha)$ remains regular as $x \to \alpha$ though both the numerator and denominator tend to zero as $(x - \alpha)^2$:

$$\begin{aligned} P(\alpha) = \frac{\varphi''' f}{3(\varphi'')^2} - \frac{f'}{\varphi''} + (\alpha - x_0)\Bigg[& \frac{\varphi^{(4)} f}{8(\varphi'')^2} - \frac{5(\varphi''')^2 f}{24(\varphi'')^3} \\ & + \frac{f'\varphi'''}{2(\varphi'')^2} - \frac{f''}{\varphi''}\Bigg] + O(\alpha - x_0)^2 \end{aligned} \tag{2.20c}$$

Eqn. 2.16 for the uniform asymptotic makes it possible to estimate the applicability range for simpler nonuniform asymptotics (eqn. 2.18). Since going from eqns. 2.16 to 2.18 is reduced to replacing the Fresnel integral $F\{\pm\sqrt{[\lambda|\varphi(\alpha)-\varphi(x_0)|]}\}$ by its asymptotic as $\lambda\to\infty$ the condition to apply eqn. 2.16 is the possibility of such a procedure, i.e. sufficiently high values of the argument $\sqrt{\{\lambda[\varphi(\alpha)-\varphi(x_0)]\}}$ of the Fresnel integral. The numerical estimations depend on the accuracy required.

The Fresnel integral values at $\xi=\pm\sqrt{\pi}$ are equal to $F(\pm\sqrt{\pi})=0.5\pm(0.621+0.091i)$ accurate within 10^{-3}. The sum of $\chi(\xi)$ and the first summand in eqn. 2.6 gives the value $F(\pm\sqrt{\pi})\cong 0.5\pm(0.612+0.112i)$. The absolute error of this approximation is of the order of 0.025, while the relative error is of order of 2.5% for $\xi=\sqrt{\pi}$ and of 15% for $\xi=-\sqrt{\pi}$. When increasing, the absolute error falls off as $|\xi|^{-3/2}$, while the relative error falls as $|\xi|^{-3/2}$ for $\xi>0$ and as $|\xi|^{-1}$ for $\xi<0$.

Probably, for $|\xi|>\sqrt{\pi}$ replacing the Fresnel integral by the leading terms of its asymptotic eqn. 2.6 is admissible for the most part of practical applications. Therefore the condition to apply the nonuniform asymptotic can be taken in the form of the inequality

$$|\lambda\varphi(\alpha)-\lambda\varphi(x_0)|>\pi$$

This is the requirement for the phase variation (exponential argument) in the integral $I(\lambda, \alpha)$ to be not less than π when going from the stationary point x_0 to the endpoint of the integration interval α. In other words, it is required that the distance between these two points contains not less than one halfwave of the exponential.

Getting ahead, note that the applicability range for the nonuniform asymptotic is similar for other cases of critical-point merging, to be treated in what follows.

2.2 Stationary point near pole

Consider an integral

$$I_{\pm}(\lambda, \alpha)=\int_{-\infty\pm i0}^{\infty\pm i0} \exp[i\lambda\varphi(x)]\,\frac{h(x)}{x-\alpha}\,dx \tag{2.21}$$

where the pole $x=\alpha$ of the integrand is close to the stationary point $x=x_0$ of the phase function. This Section will show that the asymptotic of the integral $I(\lambda, \alpha)$ uniform over a distance between critical points $x=x_0$ and $x=\alpha$ is of the form of eqn. 2.28. The designation $\pm\infty\pm i0$ in the integration limits means that the pole $x=\alpha$ is passed around in the upper halfplane for the integral I_+ and in the lower halfplane for the integral I_-. These integrals are related to the integral I_0 (understood in the sense of the Cauchy principal value) by eqn. 1.24.

Consider, for definity, the integral $I_-(\lambda, \alpha)$ and write its nonuniform asymptotic applicable as $\lambda\to\infty$ and $\alpha\neq x_0$ fixed. Eqn. 2.21 has two critical points, the stationary-phase point $x=x_0$ and the pole $x=\alpha$; here the pole contributes into the asymptotic of $I(\lambda, \alpha)$ only for $\varphi'(\alpha)>0$. Let the point $x=x_0$ be a minimum point of the function $\varphi(x)$, then $\varphi''(x) > 0$ in a vicinity of x_0. Here $\varphi'(x) < 0$ for $x < x_0$ and $\varphi'(x) > 0$ for $x > x_0$. Taking into account these inequalities and using eqn. 1.25 and 1.10,

$$I_{-}(\lambda, \alpha) = 2\pi i \exp[i\lambda\varphi(\alpha)]h(\alpha)\chi(\alpha - x) + C \exp[i\lambda\varphi(x_0)] \tag{2.22a}$$

where C is an asymptotic series

$$C = \sum_{n=0}^{\infty} (i/\lambda)^{n+1/2} c_n, \; c_o = \sqrt{[2\pi/\varphi''(x_0)]}\, h(x_0)/(x_0 - \alpha) \tag{2.22b}$$

Recall that χ is the Heaviside function, i.e. $\chi(\alpha - x_0) = 0$ for $\alpha < x_0$ and $\chi(\alpha - x_0) = 1$ for $\alpha > x_0$. Asymptotic eqn. 2.22*a* is inapplicable as $\alpha \to x_0$, since the first summand in eqn. 2.22*a* is discontinuous at $\alpha = x_0$ and since even the first term c_0 of the asymptotic series C tends to infinity as $(x_0 - \alpha)^{-1}$ as $\alpha \to x_0$.

To find that special function in terms of which the uniform asymptotic of the integral $I_{-}(\lambda, \alpha)$ is expressed consider the model integral

$$G(\lambda, \alpha) = \int_{-\infty - i0}^{\infty - i0} \exp(i\lambda x^2) \frac{dx}{x - \alpha}$$

After the variable change $\xi = \sqrt{(\lambda)}x$,

$$G(\lambda, \alpha) = \Phi[\sqrt{(\lambda)}\alpha]$$

where

$$\Phi(s) = \int_{-\infty - i0}^{\infty - i0} \exp(i\xi^2) \frac{d\xi}{\xi - s}$$

It is shown that $\Phi(s)$ is expressed in terms of the conjugate Fresnel integral. For $\Phi(s)$,

$$\Phi'(s) = \int_{-\infty - i0}^{\infty - i0} \exp(i\xi^2) \frac{d\xi}{(\xi - s)^2} = -\int_{-\infty - i0}^{\infty - i0} \exp(i\xi^2) d\left[\frac{1}{\xi - s}\right]$$

$$= \int_{-\infty - i0}^{\infty - i0} \left[2i + \frac{2is}{\xi - s}\right] \exp(i\xi^2) d\xi = 2i\sqrt{(\pi i)} + 2is\Phi(s)$$

Therefore the function $\Phi(s)$ satisfies the equation

$$\Phi'(s) - 2is\Phi(s) = 2\sqrt{\pi} i^{3/2}$$

Integrating the equation obtains

$$\Phi(s) = 2\sqrt{\pi} i^{3/2} \exp(is^2) \int_{-\infty}^{s} \exp(-it^2)\, dt + A \exp(is^2)$$

$$= 2\pi i F^*(s) \exp(is^2) + A \exp(is^2)$$

where A is the integration constant to be determined and $F^*(s)$ is the complex conjugate Fresnel integral. Expressing A in terms of Φ and F^* and returning to $G(\lambda, \alpha)$ obtains

$$A = \exp(-i\lambda\alpha^2)\int_{-\infty - i0}^{\infty - i0} \exp(i\lambda x^2)\frac{dx}{x-\alpha} - 2\pi i F^*[\sqrt{(\lambda)}\alpha]$$

The left-hand side of the relation is constant while the right-hand side depends on λ and α and tends to zero as $1/\sqrt{(\lambda)}$ as $\lambda \to \infty$ and $\alpha < 0$. Therefore $A = 0$ and

$$\begin{aligned} G(\lambda, \alpha) &= \int_{-\infty - i0}^{\infty - i0} \exp(i\lambda x^2)\frac{dx}{x-\alpha} \\ &= 2\pi i \exp(i\lambda\alpha^2) F^*[\sqrt{(\lambda)}\alpha] \end{aligned} \tag{2.23}$$

The integral of eqn. 2.21 with an arbitrary function $h(x)$ and the phase function $\varphi(x) = x^2$ is reduced to the integral $G(\lambda, \alpha)$. Setting $f(x, \alpha) = [h(x) - h(\alpha)]/(x-\alpha)$ obtains

$$\begin{aligned} &\int_{-\infty - iO}^{\infty - iO} \exp(i\lambda x^2)\frac{h(x)}{x-\alpha}dx \\ &\quad = h(\alpha)\int_{-\infty - iO}^{\infty - iO} \exp(i\lambda x^2)\frac{dx}{x-\alpha} \\ &\qquad + \int_{-\infty - iO}^{\infty - iO} \exp(i\lambda x^2) f(x, \alpha)\, dx \\ &\quad = h(\alpha) 2\pi i \exp(i\lambda\alpha^2) F^*[\sqrt{(\lambda)}\alpha] + \sum_{n=0}^{\infty} (i/\lambda)^{n+1/2} a_n \end{aligned} \tag{2.24}$$

where

$$a_n = \frac{\Gamma(n+1/2)}{\Gamma(2n+1)} f^{(2n)}(0, \alpha)$$

are the regular functions of α. The case of arbitrary phase function $\varphi(x)$ with the minimum point $x = x_0$ where $\varphi''(x_0) > 0$ is reduced to the one considered by the variable change

$$\xi = \xi(x) = \text{sign}(x - x_0)\sqrt{|\varphi(x) - \varphi(x_0)|} \tag{2.25}$$

Denoting the function inverse to $\xi(x)$ by $x(\xi)$,

$$\begin{aligned} I(\lambda, \alpha) &= \int_{-\infty - i0}^{\infty - i0} \exp[i\lambda\varphi(x)]\frac{h(x)}{x-\alpha}dx \\ &= \exp[i\lambda\varphi(x_0)]\int_{-\infty - i0}^{\infty - i0} \exp(i\lambda\xi^2)\frac{h[x(\xi)]}{x(\xi)-\alpha}\frac{\partial x}{\partial \xi}d\xi \end{aligned}$$

The denominator of the integral vanishes at $\xi = \xi(\alpha)$. Writing a nonexponential factor in the obtained integral as $g(\xi)/[\xi - \xi(\alpha)]$ and using eqn. 2.24 gives

$$I_-(\lambda, \alpha) = \exp [i\lambda\varphi(x_0)] \int_{-\infty - i0}^{\infty - i0} \exp (i\lambda\xi^2) \frac{g(\xi)}{\xi - \xi(\alpha)} d\xi$$
$$= g[\xi(\alpha)] 2\pi i \exp [i\lambda\varphi(\alpha)] F^*[\sqrt{\lambda}\xi(\alpha)]$$
$$+ \exp [i\lambda\varphi(x_0)] \sum_{n=0}^{\infty} (i/\lambda^{n+1/2}) d_n \qquad (2.26)$$

Here, $d_n = d_n(\alpha)$ are regular as $\xi(\alpha) \rightarrow 0$, i.e. as $\alpha \rightarrow x_0$. To find the expressions for d_n and $g[\xi(\alpha)]$ use asymptotic matching. Fix α and set $\lambda \rightarrow \infty$ in eqn. 2.26. Replacing the conjugate Fresnel integral $F^*[\sqrt{(\lambda)}\xi]$ by its asymptotic as $\lambda \rightarrow \infty$

$$F^*[\sqrt{(\lambda)}\xi] \simeq \chi(\xi) + \exp (-i\lambda\xi^2) \sum_{n=0}^{\infty} \frac{\Gamma(n + 1/2) \exp [(i\pi n/2) + 3i\pi/4]}{2\pi\lambda^{n+1/2}\xi^{2n+1}}$$

obtains

$$I_-(\lambda, \alpha) = 2\pi i g[\xi(a)] \exp [i\lambda\varphi(\alpha)] \chi(\alpha - x_0)$$
$$+ \exp [i\lambda\varphi(x_0)] \sum_{n=0}^{\infty} (i/\lambda)^{n+1/2} \left[d_n - \frac{\Gamma(n + 1/2) g[\xi(\alpha)]}{\xi^{2n+1}} \right]$$

Equating the expression to nonuniform asymptotic eqn. 2.22 obtains $g[\xi(\alpha)] = h(\alpha)$, and taking into account that for $\varphi''(x_0) > 0$ the relation sign $[\xi(\alpha)] =$ sign $[\varphi'(\alpha)]$ holds, gives the following expression for d_n

$$d_n = c_n + \frac{\Gamma(n + 1/2) h(\alpha)}{|\varphi(a) - \varphi(x_0)|^{n+1/2} \operatorname{sign} [\varphi'(\alpha)]} \qquad (2.27)$$

Eqns. 2.26 and 2.27 give the uniform asymptotic of the integral $I_-(\lambda, \alpha)$ for $\varphi''(x_0) > 0$. In general the following theorem is valid.

Theorem 2.2: Let function $\varphi(x)$ in the integral

$$I_\pm(\lambda, \alpha) = \int_{-\infty \pm i0}^{\infty \pm i0} \exp [i\lambda\varphi(x)] \frac{h(x)}{x - \alpha} dx$$

have a nondegenerate stationary point $x = x_0$ close to the pole $x = \alpha$. Let $h(x)$ be a finite function, analytical in the vicinity of the points α and x_0 and in the region supp *(h) there are no critical points different from α and x_0. Then as $\lambda \rightarrow \infty$ the asymptotic of $I_\pm(\lambda, \alpha)$ uniform over a distance between α and x_0 has the form*

$$I_\pm(\lambda, \alpha) = \mp 2\pi i h(\alpha) \exp [i\lambda\varphi(\alpha)]$$
$$\times F_1\{\mp\sqrt{(\lambda)} \operatorname{sign} [\varphi'(\alpha)] \sqrt{|\varphi(x_0) - \varphi(\alpha)|}\}$$
$$+ \exp [i\lambda\varphi(x_0)] \sum_{n=0}^{\infty} \lambda^{-n-1/2} d_n(\alpha)$$
$$\times \exp \{i\pi[n/2 + 1/4] \operatorname{sign} [\varphi''(x_0)]\} \qquad (2.28)$$

Here, F_1 is the Fresnel integral for $\varphi''(x_0)<0$ and complex conjugate Fresnel integral for $\varphi''(x_0)>0$. The coefficients $d_n(\alpha)$ are regular as $\alpha\to x_0$ and are expressed according to eqn. 2.27 in terms of $c_n=c_n(\alpha)$ where c_n are the coefficients of the asymptotic contribution of the stationary point $x=x_0$ into the nonuniform asymptotic of $I_\pm$

$$I_\pm(\lambda,\alpha)\simeq\mp 2\pi i h(\alpha)\exp[i\lambda\varphi(\alpha)]\chi\{\mp\operatorname{sign}[\varphi'(\alpha)]\}$$
$$+\exp[i\lambda\varphi(x_0)]\sum_{n=0}^{\infty}\lambda^{-n-1/2}c_n(\alpha)$$
$$\times\exp\{i\pi(n/2+1/4)\operatorname{sign}[\varphi''(x_0)]\}\tag{2.29a}$$

In particular,

$$d_o=\frac{\sqrt{(2\pi)}h(x_0)}{\sqrt{|\varphi''(x_0)|}(x_0-\alpha)}+\frac{\sqrt{\pi}\operatorname{sign}\varphi''(x_0)h(\alpha)}{\operatorname{sign}[\varphi'(\alpha)]\sqrt{|\varphi(x_0)-\varphi(\alpha)|}}\tag{2.29b}$$

For small $\alpha-x_0$

$$d_0=\frac{\sqrt{(2\pi)}}{\sqrt{|\varphi''(x_0)|}}\left\{h'-\frac{h\varphi'''}{6\varphi''}+(\alpha-x_0)\left[\frac{h''}{2}-\frac{h\varphi'''}{6\varphi''}+\frac{h(\varphi''')^2}{24(\varphi'')^2}-\frac{h\varphi^{(4)}}{24\varphi''}\right]\right\}$$
$$+O(\alpha-x_0)^2\tag{2.29c}$$

In the last expression the functions h, φ and their derivatives are taken at the point x_0. This theorem is proved for the integral I_- when $\varphi''(x_0)>0$. To prove it for the integral I_+ one should use the equality $1-F^*(s)=F^*(-s)$ and eqn. 1.24 relating I_+ and I_-. The case of $\varphi''(x_0)<0$ is reduced to that of $\varphi''(x)>0$ by complex conjugation and one should take into account that the direction of a pole bypassing is changed when performing this procedure.

Let the pole $x=\alpha$ in the integral eqn. 2.21 be a complex one. Then the nonuniform asymptotics of $I(\lambda,\alpha)$ includes only contribution of the stationary point $x=x_0$, i.e. it is of the form

$$I(\lambda,\alpha)\simeq\exp[i\lambda\varphi(x_0)]\sum_{n=0}^{\infty}c_n\lambda^{-n-1/2}$$
$$\times\exp\{i\pi[n/2+1/4]\operatorname{sign}[\varphi''(x_0)]\}\tag{2.30}$$

Suppose the functions $\varphi(x)$ and $h(x)$ be analytically extended into a vicinity of the point x_0 on the complex plane. One can show that under these conditions the uniform asymptotic is described, as previously, by eqn. 2.28 and if the pole $x=\alpha$ is passed around from below, i.e. $\operatorname{Im}\alpha>0$, the formula for $I_-(\lambda,\alpha)$ is to be used. If $\operatorname{Im}\alpha<0$ use the formula for $I_+(\lambda,\alpha)$. The expression

$$\operatorname{sign}[\varphi'(\alpha)]\sqrt{|\varphi(\alpha)-\varphi(x_0)|}$$

means the analytical extension of this function from the real halfline $\alpha>x_0$. For example, for $\varphi''(x_0)>0$,

$$\operatorname{sign}[\varphi'(\alpha)]\sqrt{|\varphi(\alpha)-\varphi(x_0)|}=\sqrt{\{[(\alpha-x_0)^2\varphi''(x_0)/2]+\cdots\}}$$
$$=(\alpha-x_0)\sqrt{\{[\varphi''(x_0)/2]+(\alpha-x_0)\varphi'''(x_0)+\cdots\}}$$

where the series under the radical sign converges and is not equal to zero for all sufficiently small complex values of $|\alpha-x_0|$.

In application, the case often occurs when $\varphi(x)$ is an even function with respect to the stationary point x_0, $\varphi(x_0 - c) = \varphi(x_0 + c)$ and the pole is set off by a purely imaginary quantity $\alpha = x_0 + i\xi$. Then the function $\varphi(x_0 + i\xi)$ is a real function, the relation $\varphi''_{\xi\xi}(x_0 + i\xi) = -\varphi''_{xx}(x_0)$ being held. For $\varphi''(x_0) < 0$ the Fresnel integral in eqn. 2.28 is written as $F\{\sqrt{[\varphi(x_0) - \varphi(x + i\xi)]}\} = F\{i\sqrt{[\varphi(x_0 + i\xi) - \varphi(x_0)]}\}$. Using the relation $F(is) = F^*(s)$ one can write eqn. 2.28 as

$$\begin{aligned} I(\lambda, x_0 + i\xi) &= \int_{-\infty}^{\infty} \exp[i\lambda\varphi(x)] \frac{h(x)}{(x - x_0 - i\xi)} dx \\ &= 2\pi i \operatorname{sign}(\xi) \exp[i\lambda\varphi(x_0 + i\xi)] \\ &\quad \times F^*\{-\sqrt{\lambda}|\varphi(x_0 + i\xi) - \varphi(x_0)|\} h(x_0 + i\xi) \\ &\quad + \exp[i\lambda\varphi(\xi_0)] \sum_{n=0}^{\infty} (i\lambda)^{-n-1/2} d_n \end{aligned}$$

Similarly, for $\varphi''(x_0) > 0$,

$$\begin{aligned} I(\lambda, x_0 + i\xi) &= \int_{-\infty}^{\infty} \exp[i\lambda\varphi(x)] \frac{h(x)}{(x - x_0 - i\xi)} dx \\ &= 2\pi i \operatorname{sign}(\xi) \exp[i\lambda\varphi(x_0 + i\xi)] \\ &\quad \times F\{-\sqrt{[\lambda|\varphi(x_0 + i\xi) - \varphi(x_0)|]}\} h(x_0 + i\xi) \\ &\quad + \exp[i\lambda\varphi(\xi_0)] \sum_{n=0}^{\infty} (i/\lambda)^{n+1/2} d_n \end{aligned}$$

In particular, for $\varphi(x) = x^2$, $h(x) = 1$,

$$\begin{aligned} -i &\int_{-\infty}^{\infty} \exp(i\lambda x^2) \frac{dx}{x - i\xi} dx \\ &= \xi \int_{-\infty}^{\infty} \exp(i\lambda x^2) \frac{dx}{x^2 + \xi^2} \\ &= 2\pi \operatorname{sign}(\xi) \exp(-i\lambda\xi^2) F[-\sqrt{\lambda}|\xi|] \end{aligned} \tag{2.31}$$

2.3 Two close stationary points

Consider an integral

$$I(\lambda, \alpha) = \int_{-\infty}^{\infty} \exp[i\lambda\varphi(x, \alpha)] f(x) g(x)\, dx \tag{2.32}$$

for the case when for $\alpha > 0$ the phase function has two close critical points x_1 and x_2 approaching the same limit x_0 as $\alpha \to 0$. Assume that $\varphi''_{xx}(x_1) < 0$ and $\varphi''_{xx}(x_2) > 0$, then $\varphi(x_1) > \varphi(x_2)$. Here, $g(x)$ is a cutting function for which there are no critical points, except x_1 and x_2, in the region supp (g), $g \equiv 1$ in the vicinity of x_0 including x_1 and x_2. The nonuniform asymptotic of such an integral is determined by the contributions due to the points x_1 and x_2 and is of the form

$$I(\lambda, \alpha)=\sqrt{(2\pi/\lambda)}\left[\frac{f(x_2)\exp[i\lambda\varphi(x_2, \alpha)+i\pi/4]}{\sqrt{\varphi''(x_2, \alpha)}}\right.$$

$$\left.+\frac{f(x_1)\exp[i\lambda\varphi(x_1, \alpha)-i\pi/4]}{\sqrt{|\varphi''(x_1, \alpha)|}}\right]+O(\lambda^{-1}) \tag{2.33}$$

This equation is applicable as $\lambda\to\infty$ and for fixed $\alpha>0$ fails as $\alpha\to 0$ when $|\varphi''(x_{1,2}, \alpha)|\to 0$. The model integral in this case is expressed in terms of the Airy function (see Section 6.2)

$$\int_{-\infty}^{\infty}\exp[i\lambda(x^3/3-\alpha x)]\,dx=\lambda^{-1/3}\int_{-\infty}^{\infty}\exp[i(t^3/3-\lambda^{2/3}t)]\,dt$$

$$=2\pi\lambda^{-1/3}Ai(-\lambda^{2/3}\alpha)$$

To reduce the general integral (eqn. 2.32) to the model one assume the function $\varphi(x, \alpha)$ to be analytical for small α and $x-x_0$, the relations

$$\varphi'''_{xxx}\neq 0, \qquad \varphi'_x=\varphi''_{xx}=0, \qquad \varphi''_{x\alpha}\neq 0 \tag{2.34}$$

held at $x=x_0$, $\alpha=0$. Under these conditions there exists a variable change $x=x(\tau)$ analytical and reversible for small α and $x-x_0$, depending parametrically on α and transforming $\varphi(x, \alpha)$ into a cube polynomial

$$\varphi(x, \alpha)=\varphi_0(\alpha)+\tau^3/3-\xi(\alpha)\tau \tag{2.35}$$

where $\varphi_0(\alpha)$ and $\xi(\alpha)$ are analytical functions of α.

The proof of this statement can be found in References 22 and 3. In what follows it will be shown how to find $\varphi_0(\alpha)$ and $\xi(\alpha)$.

After the variable change to τ the integral $I(\lambda, \alpha)$ is written as

$$I(\lambda, \alpha)=\exp(i\lambda\varphi_0)\int_{-\infty}^{\infty}\exp[i\lambda(\tau^3/3-\tau\xi)]f[x(\tau)]g[x(\tau)]\frac{dx(\tau)}{d\tau}\,d\tau$$

Set

$$f[x(\tau)]g[x(\tau)]\frac{dx(\tau)}{d\tau}=A_0+B_0\tau+(\tau^2-\xi)h(\tau)$$

Then the integral can be represented as

$$I(\lambda, \alpha)=\exp(i\lambda\varphi_0)\left[A_0\int_{-\infty}^{\infty}\exp[i\lambda(\tau^3/3-\tau\xi)]\,d\tau\right.$$

$$+B_0\int_{-\infty}^{\infty}\exp[i\lambda(\tau^3/3-\tau\xi)]\,\tau\,d\tau$$

$$\left.+(1/i\lambda)\int_{-\infty}^{\infty}h(\tau)d\exp[i\lambda(\tau^3/3-\tau\xi)]\right]$$

$$=\exp(i\lambda\varphi_0)[2A_0\pi\lambda^{-1/3}Ai(-\lambda^{2/3}\xi)$$

$$-2B_0\pi i\lambda^{-2/3}Ai'(-\lambda^{2/3}\xi)+C(\lambda, \xi)]$$

were $Ai'(x)$ is the derivative of the Airy function and

$$C(\lambda, \xi) = (i/\lambda) \int_{-\infty}^{\infty} h'(\tau) \exp [i\lambda (\tau^3/3 - \tau\xi)] \, d\tau = O(\lambda^{-4/3})$$

Transforming the integral $C(\lambda, \xi)$ in the same way as $I(\lambda, \alpha)$ and repeating this procedure, one obtains the asymptotic expansion

$$I(\lambda, \alpha) = \exp (i\lambda\varphi_0)[2A\pi\lambda^{-1/3}Ai(-\lambda^{2/3}\xi) - 2iB\pi\lambda^{-2/3}Ai'(-\lambda^{2/3}\xi)] \tag{2.36}$$

Here A and B are asymptotic series

$$A = \sum_{n=0}^{\infty} (i/\lambda)^n A_n(\alpha), \qquad B = \sum_{n=0}^{\infty} (i/\lambda)^n B_n(\alpha) \tag{2.36a}$$

the coefficients of which are analytical functions of α.

To find the functions $\varphi_0(\alpha)$, $\xi(\alpha)$ and the leading terms $A_0(\alpha)$ and $B_0(\alpha)$ of the asymptotic series A and B use asymptotic matching. Fix $\alpha > 0$ and set $\lambda \to \infty$. Then the functions $Ai(-\lambda^{2/3}\xi)$ and $Ai'(-\lambda^{2/3}\xi)$ in eqn. 2.36 can be replaced by the leading terms of their asymptotics

$$Ai(-\lambda^{2/3}\xi) \simeq \frac{\exp (2i\lambda\xi^{3/2}/3 - i\pi/4) + \exp [-2i\lambda\xi^{3/2}/3 + i\pi/4]}{2\sqrt{\pi}\lambda^{1/6}\xi^{1/4}}$$

$$Ai'(-\lambda^{2/3}\xi) \simeq -\frac{\exp (2i\lambda\xi^{3/2}/3 + i\pi/4) + \exp (-2i\lambda\xi^{3/2}/3 - i\pi/4)}{2\sqrt{\pi}\lambda^{-1/6}\xi^{-1/4}}$$

and eqn. 2.35 is transformed into the expression

$$\begin{aligned} I(\lambda, \alpha) \simeq & \exp [i\lambda (\varphi_0 + 2\xi^{3/2}/3)]\sqrt{(\pi/i\lambda)}(A_0\xi^{-1/4} - B_0\xi^{1/4}) \\ & + \exp [i\lambda (\varphi_0 - 2\xi^{3/2}/3)]\sqrt{(i\pi/\lambda)}(A_0\xi^{-1/4} + B_0\xi^{1/4}) + O(\lambda^{-3/2}) \end{aligned}$$

This expression should coincide with nonuniform asymptotic eqn. 2.33. Comparing the phase factors and taking into account that $\varphi(x_1, \alpha) > \varphi(x_2, \alpha)$ one obtains two equations for two unknown functions, φ_0 and ξ,

$$\varphi_0 + 2\xi^{3/2}/3 = \varphi(x_1, \alpha), \qquad \varphi_0 - 2\xi^{3/2}/3 = \varphi(x_2, \alpha)$$

from which it follows that

$$\begin{aligned} \varphi_0 &= [\varphi(x_1, \alpha) + \varphi(x_2, \alpha)]/2 \\ \xi &= \{(3/4)[\varphi(x_1, \alpha) - \varphi(x_2, \alpha)]\}^{2/3} \end{aligned} \tag{2.36b}$$

In particular, as $\alpha \to 0$ and $x_{1,2} \to x_0$,

$$\begin{aligned} \varphi_0 &= \varphi(x_0, 0) + O(\alpha^2) \\ \xi &= \alpha\varphi''_{x\alpha}(x_0, 0)[\varphi'''_{xxx}(x_0, 0)/2]^{-1/3} + O(\alpha^2) \end{aligned}$$

Comparing the amplitude factors at corresponding exponentials one obtains two equations for the two unknowns A_0 and B_0

$$\begin{aligned} A_0\xi^{-1/4} + B_0\xi^{1/4} &= \sqrt{2}f(x_2)/\sqrt{\varphi''(x_2, \alpha)} \\ A_0\xi^{-1/4} - B_0\xi^{1/4} &= \sqrt{2}f(x_1)/\sqrt{|\varphi''(x_1, \alpha)|} \end{aligned}$$

from which it follows that

$$A_0 = 2^{-1/2}\xi^{1/4}\left[\frac{f(x_2)}{\sqrt{\varphi''_{xx}(x_2,\alpha)}} + \frac{f(x_1)}{\sqrt{|\varphi''_{xx}(x_1,\alpha)|}}\right]$$

$$B_0 = -2^{-1/2}\xi^{-1/4}\left[\frac{f(x_1)}{\sqrt{|\varphi''_{xx}(x_1,\alpha)|}} - \frac{f(x_2)}{\sqrt{\varphi''_{xx}(x_2,\alpha)}}\right] \qquad (2.36c)$$

The subsequent terms of the asymptotic series A and B in eqn. 2.36 are evaluated in the same manner.

Thus, the following theorem has been proved:

Theorem 2.3: Let the phase function in eqn. 2.1 for $\alpha > 0$ *have two critical points* x_1 *and* x_2 $(\varphi(x_1) > \varphi(x_2))$ *which tend to a common limit* x_0 *as* $\alpha \to 0$ *and let the phase function satisfy the relations of eqn. 2.34 at* $x = x_0$ *and* $\alpha = 0$. *Let g be a cutting function which singles out a vicinity of the points* x_1 *and* x_2. *Then as* $\lambda \to \infty$ *the uniform over* α *asymptotics of* $I(\lambda, \alpha)$ *is expressed in terms of the Airy function and its derivative according to eqn. 2.36, where* $\varphi_0 = \varphi_0(\alpha)$ *and* $\xi = \xi(\alpha)$ *are of form of eqn. 2.36b and are analytical functions of* α, *while A and B are asymptotic series (eqn. 2.36a) with analytical (for small* α*) coefficients* $A_n(\alpha)$ *and* $B_n(\alpha)$, $A_0(\alpha)$ *and* $B_0(\alpha)$ *being of the form of eqn. 2.36c.*

The right-hand sides of eqn. 2.36 are defined for $\alpha > 0$ only and the point $\alpha = 0$ is a singular point for functions $\varphi(x_{1,2}, \alpha)$ and $\varphi''(x_{1,2}, \alpha)$. However, since the functions ξ, φ_0 and the coefficients of the asymptotic series in eqn. 2.36 are analytical functions of α these singularities should cancel. For example, the functions ξ and $\varphi''(x_{1,2}, \alpha)$ tend to zero as $\alpha \to 0$, while for A_0 there is an uncertainty of the type of 0/0. An uncertainty for B_0 is of the type $(\infty - \infty)/0$. Therefore it is difficult to compute numerically these coefficients according to eqn. 2.36*c* for small α.

In the region $\alpha < 0$ the functions $\varphi_0(\alpha)$, $\xi(\alpha)$, $A_n(\alpha)$ and $B_n(\alpha)$ are defined by analytical continuation of eqns. 2.36*b* and 2.36*c* from the region $\alpha > 0$ where the stationary points $x_{1,2}(\alpha)$ are specified, i.e. by a procedure that is also difficult for numerical computations. However, it is sufficient to consider only α close to zero since as $\lambda \to \infty$ for $\xi(\alpha) < 0$ the Airy function $Ai(-\lambda^{2/3}\xi)$ tends to zero as $\exp(-\lambda\xi^{3/2})$ and the asymptotic contribution eqn. 2.36 in the vicinity of the point $x = 0$ into the integral $I(\lambda, \alpha)$ becomes negligible.

But if $|\alpha|$ are sufficiently small the functions φ_0, A_0 and B_0 can be replaced by their values at $\alpha = 0$ and only the linear term in α can be retained in $\xi(\alpha)$. The resulting expressions coincide with the formulas obtained when replacing the function $f(x)$ in $I(\lambda, \alpha)$ by the linear function $f(0) + xf'(0)$ and the function $\varphi(x, \alpha)$ by the cube polynomial in x, respectively. It is convenient to take the value of the phase function derivative at $x = 0$ (with opposite sign) as a small parameter depending on the distance between two stationary points, i.e. to set

$$\varphi(x, \alpha) = \varphi(0, \alpha) - \alpha x + x^2\varphi''_{xx}(0, \alpha)/2 + x^3\varphi'''(0, \alpha)/6$$

$$f(x) = f_0 + xf_1$$

the summand $\varphi'' x^2/2$ for small α and x being neglected since $\varphi''(0, 0)=0$. After this the initial integral $I(\lambda, \alpha)$ is reduced to the Airy function and its derivative

$$I(\lambda, \alpha) \simeq 2\pi \exp [i\lambda\varphi(0, \alpha)] \{(2/\lambda\varphi''')^{1/3} f_0 Ai[-\lambda^{2/3}(2/\varphi''')^{1/3}\alpha] \\ - i(2/\lambda\varphi''')^{2/3} f_1 Ai'[-\lambda^{2/3}(2/\varphi''')^{1/3}\alpha]\} \tag{2.37}$$

This equation can be applied both for $\alpha > 0$ and for $\alpha < 0$, but for small α only. It is natural to call it the *local asymptotic* of the integral $I(\lambda, \alpha)$.

Thus, three different expressions are obtained for the asymptotics of the integral $I(\lambda, \alpha)$. The first is a uniform asymptotic (eqn. 2.36) which is applicable for all α, but, in fact, cannot be applied for $\alpha < 0$ when the functions determining it are defined by the analytical continuation from the region $\alpha > 0$. The second one is a local asymptotic (eqn. 2.37) applicable for sufficiently small α. The third one is a nonuniform asymptotic (eqn. 2.33) simpler than eqn. 2.36 and equivalent to it for those α for which the argument of the Airy function and its derivative are so large that one can replace these functions by their asymptotics. Practically, it is sufficient to require $|-\lambda^{2/3}\xi(\alpha)| > 1.8$.

It is convenient for numerical computations to use the approximation of $\tilde{I}(\lambda, \alpha)$ coinciding for $\alpha < 0$ with a local asymptotic (eqn. 2.37) and for $\alpha > 0$ with uniform asymptotic (eqn. 2.36) in which only the leading terms A_0, B_0 of the series A, B are retained. At the transit point from eqn. 2.37 to eqn. 2.36, $x=0$, the function $\tilde{I}$ is continuous, but not its derivative. The relative magnitude of the derivative discontinuity

$$\delta = \lim_{\varepsilon \to 0} \{[\tilde{I}'(\varepsilon) - \tilde{I}'(-\varepsilon)]/[\tilde{I}'(\varepsilon) + \tilde{I}'(-\varepsilon)]\}$$

tends to zero when λ increases.

Computing $I(\lambda, \alpha)$ according to eqns. 2.33, 2.36 and 2.37 for particular problems and comparing the results with direct computations of the integral one can formulate a simple empirical criterion to apply the asymptotic equations obtained. Namely, it is sufficient to require an inflection of the graph of $\tilde{I}(\lambda, \alpha)$ at $x=0$ to be small. When choosing the scales of α and $\tilde{I}$ so that the maximum slope of the graph is of the order of 45° the function $\tilde{I}$ is required to be graphically close to a function analytical for small α.

To illustrate the criterion, consider an integral arising when computing a field of the internal gravity-wave source moving in a layer of stratified fluid the density of which increases with depth

$$I(y, \omega) = \int_{-\infty}^{\infty} \exp \{iy[\omega\mu(\nu) - \nu]\} f(\nu)\, d\nu \tag{2.38}$$

Here, the functions $f(\nu)$ and $\mu(\nu)$ are determined when solving the corresponding spectral problem numerically [23]. The function $f(\nu)$ is an even function of ν while $\mu(\nu)$ is an odd one: for small ν it is expanded into a series

$$\mu(\nu) = \beta\nu - \gamma\nu^3 + \cdots$$

where $\beta = 0.3256$, $\gamma = 125$. For $\nu > 0$ the function $\mu(\nu)$ is convex and strictly monotonic, $\mu'(\nu) > 0$, $\mu''(\nu) < 0$, the function $\mu(\nu)$ tends to a finite

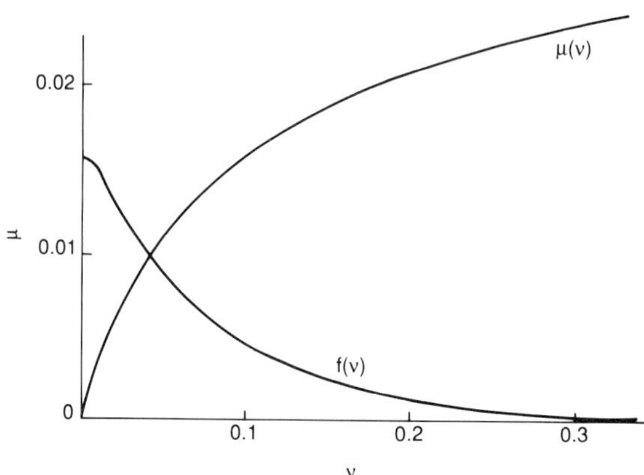

Figure 2.1 Function $f(\nu)$ and $\mu(\nu)$ in eqn. (2.38)

limit as $\nu \to \infty$. The graphs of the functions $f(\nu)$ and $\mu(\nu)$ are given in Figure 2.1.

The quantity y is a large parameter of the problem. For $\omega < \beta^{-1}$ the phase function $y[\omega\mu(\nu) - \nu]$ in the integral $I(y, \omega)$ falls off monotonically and $I(y, \omega)$ decreases when y grows exponentially. For $\omega > \beta^{-1}$ the phase function has stationary points $\nu = \pm\nu(\omega)$ which tend to zero as $\omega \to \beta^{-1}$. To reduce the problem to the case considered of the integral $I(\lambda, \alpha)$ set $\lambda = y$, $\alpha = \beta\omega - 1$.

In Figures 2.2–2.4 the graphs of the uniform (dot–dash line) and nonuniform (double dot–dash line) asymptotics are given for $y = 100$, 50 and 20 and $\omega > \beta^{-1} = 3.07$ as well as the graph of local asymptotics (broken line) defined for all ω and the graph of the integral $I(y, \omega)$ (solid line) obtained by numerical

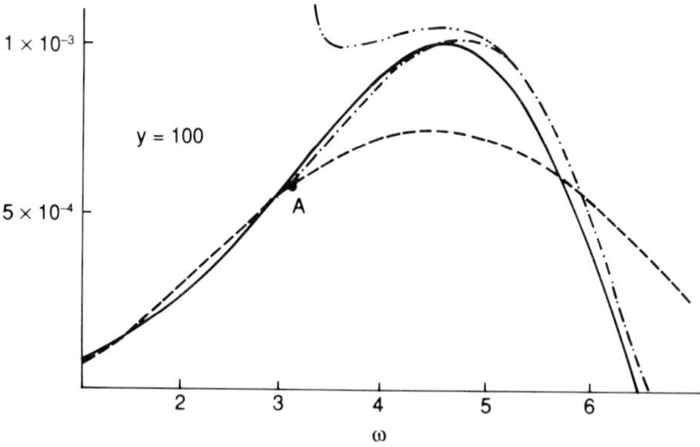

Figure 2.2 Integral $\mathcal{J}(y, \omega)$ for $y = 100$

———	exact value
–·–·–·–	uniform asymptotic
–··–··–··	nonuniform asymptotic
– – – – –	local asymptotic

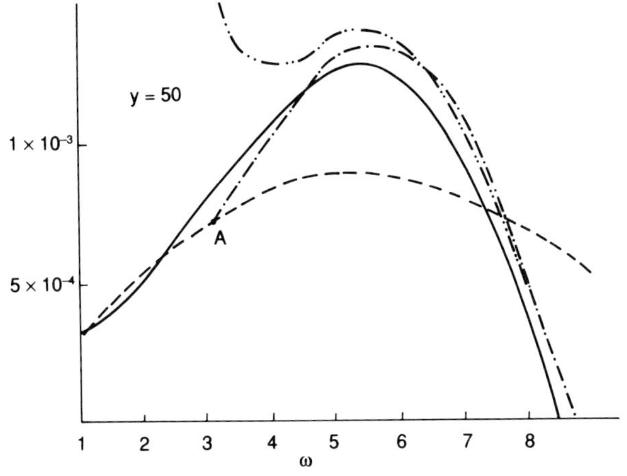

Figure 2.3 Integral $\mathcal{J}(y, \omega)$ for $y = 50$

——— exact value
–·–·–·– uniform asymptotic
–··–··–·· nonuniform asymptotic
– – – – – local asymptotic

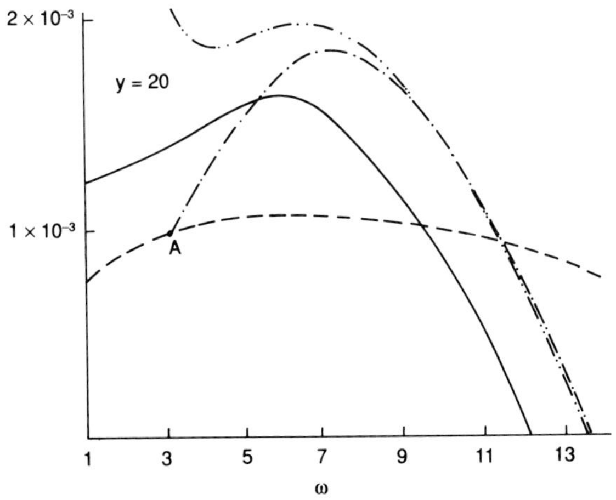

Figure 2.4 Integral $\mathcal{J}(y, \omega)$ for $y = 20$

——— exact value
–·–·–·– uniform asymptotic
–··–··–·· nonuniform asymptotic
– – – – – local asymptotic

integration. The approximation of $\tilde{I}$ is a curve composed of the broken line to the left of the point A corresponding to $\omega = 3.07$ and of the dot–dash line to the right of the point A. Note that since $f(\nu)$ and $\mu(\nu)$ are even and odd, respectively, in local and uniform asymptotics of $I(y, \omega)$, the summands with the Airy function derivatives vanish identically.

The numerical results given in Figures 2.1–2.4 were obtained by V.V. Bulatov and Yu.V. Vladimirov for this book.

For $y = 100$ the corner point of the function $\tilde{I}(y, \omega)$ is practically unnoticeable and the function is close to the exact value of the integral $I(y, \omega)$. The error of $\tilde{I}$ does not exceed 5% of the maximum of the integral I. Also, the local asymptotics close to $I(y, \omega)$ to the left of the corner point A is far from I to the right of A. At $y = 50$ the corner point A is more noticeable and the error of $\tilde{I}$ reaches 10%. At $y = 20$ one can, at best, speak only of rough qualitative agreement between exact (I) and approximate ($\tilde{I}$) values.

In all cases, the nonuniform asymptotic (eqn. 2.33) and the uniform one (eqn. 2.36) are merged just after the point where eqn. 2.36 reaches its maximum, i.e. the value of the Airy function argument in eqn. 2.36 is about -1.3.

Consider another aspect of the numerical comparison of $\tilde{I}$ and I. At $y = 50$ the asymptotic approximation $\tilde{I}$ based (as all our considerations) on the hypothesis of a rapidly oscillating integrand gives a quite satisfactory description of the integral I. But the question arises of which are the real oscillations of the integrand.

As seen from Figure 2.1 the nonexponential function $f(\nu)$ falls off rapidly when $|\nu|$ growing. For $\nu \simeq 0.17$ the function $f(\nu)$ is approximately ten times less than its maximum $f(0)$, therefore the integral over the interval $|\nu| < 0.17$ is the main part of $I(y, \omega)$. The integrals over the domains $\nu > 0.17$ and $\nu < -0.17$ contribute little to $I(y, \omega)$.

At $y = 50$ and for $\omega < 9$ a variation of the phase function $\varphi(\nu) = y[\omega\mu(\nu) - \nu]$ in the interval $|\nu| < 0.17$ does not exceed 4π. In other words, $\exp[i\varphi(\nu)]$ in the integrand has not more than two oscillations.

Thus, in the case considered when analysing conditions under which the asymptotic formulas are applicable (i.e. the integration domain incorporates many oscillations of the integrand) the word ‘many’ means ‘two’.

Similar estimations appear to be valid in all cases (known to me) where asymptotic expressions are compared with the exact values of the integrals under consideration.

This situation is quite general. For example, the geometrical diffraction theory (GDT) is constructed as a shortwave asymptotic which can be applied for the wave length λ essentially less than a typical length B of the problem in question. But the comparison with exact computations and experimental results shows that the GDT gives an adequate description up to small parameter (λ/B) values of the order of 1–2.

2.4 Close stationary and branch points

This Section considers the uniform asymptotic of the integral $I(\lambda, \alpha)$ when a branch point of the nonexponential factor is close to a stationary point of the phase function [24]. Set

$$I(\lambda, \alpha) = \int_{-\infty}^{\infty} \exp[i\lambda\varphi(x)] f(x)(x-\alpha)^{\mu} g(x)\, dx \tag{2.39}$$

Here, $\varphi(x)$ and $f(x)$ are analytical functions of x, $g(x)$ is a cutting function (neutraliser), $\varphi(x)$ has the single nondegenerate stationary point $x = x_0$ in the region supp (g). The function $(x-\alpha)^{\mu}$ is a function equal to $|x-\alpha|^{\mu}$ for $x > \alpha$ and to $|x-\alpha|^{\mu} \exp(i\pi\mu)$ or $|x-\alpha|^{\mu} \exp(-i\pi\mu)$ depending on the choice of the path around the branch point (in the upper or lower halfplane). Write, first, the nonuniform asymptotic of eqn. 2.39. Taking into account that the branch point $x = \alpha$ contributes to the asymptotic when passing around the point from above only for $\varphi'(x) < 0$ (Section 1.5) and when passing from below only for $\varphi'(x) > 0$, one obtains

$$I(\lambda, \alpha) \simeq \sqrt{[2\pi/\lambda|\varphi''(x_0)|]} \exp[i\lambda\varphi(x_0) + i\pi\delta/4] \sum_{n=0}^{\infty} (i/\lambda)^n c_n$$
$$+ [\lambda|\varphi'(\alpha)|]^{-\mu-1} \exp[i\lambda\varphi(\alpha)] \sum_{n=0}^{\infty} (i/\lambda)^n d_n \cdot \chi[\varepsilon\delta(x_0-\alpha)] \tag{2.40}$$

where $\delta = \text{sign}[\varphi''(x_0)]$, $\varepsilon = 1$ when passing the branch point from above and $\varepsilon = -1$ when passing it from below, and

$$c_0 = f(x_0)(x_0-\alpha)^{\mu}$$
$$d_0 = f(\alpha) \exp[-i\pi\varepsilon(\mu+1)/2] \Gamma(\mu+1)[1-\exp(2i\pi\mu\varepsilon)]$$
$$= 2\pi f(\alpha) \exp(i\pi\mu\varepsilon/2)/\Gamma(-\mu)$$

This asymptotic is inapplicable as $\alpha \to x_0$ when, first, $\varphi'(\alpha) \to 0$ and, secondly, when for Re $(\mu) < 0$ the coefficient c_0 tends to infinity, while for Re $(\mu) > 0$ and μ being not integer one of the subsequent coefficients c_n tends to infinity.

The uniform asymptotic of $I(\lambda, \alpha)$ is expressed in terms of the model integral I_0 coinciding (to a constant factor) with the parabolic cylinder function $D_{\mu}[(1+i)\sqrt{(\lambda)}\alpha]$.

$$I_0 = \int_{-\infty}^{\infty} \exp(i\lambda x^2)(x-\alpha)^{\mu}\, dx$$
$$= \lambda^{-(\mu-1)} \int_{-\infty}^{\infty} \exp(is^2)[s-\sqrt{(\lambda)}\alpha]^{\mu}\, ds$$

Shifting the integration contour to the straight line $s = \sqrt{2} \exp(4\pi i/4) g + \sqrt{(\lambda)}\alpha$ $(-\infty < g < \infty)$ and changing to the integration variable g gives

$$I_0 = p_{\mu}\lambda^{-(\mu+1)/2} D_{\mu}[(1+i)\sqrt{(\lambda)}\alpha] \exp(i\lambda\alpha^2/2) \tag{2.41}$$

where

$$p_{\mu} = \sqrt{\pi}\, 2^{-\mu/2} \exp[3i\pi\mu/4 + \pi i/4]$$

and $D_\mu(t)$ is the parabolic cylinder function (Section 6.3)

$$D_\mu(z)=\frac{2^{\mu+1/2}}{\sqrt{\pi}}\exp(-i\pi\mu/2)\exp(z^2/4)$$
$$\times\int_{-\infty+i0}^{\infty+i0}\exp(-2g^2+2igz)g^\mu\,dg$$

Here, one should assume that Re $(\mu)<1$, otherwise the initial integral diverges as $|x|\to\infty$. To avoid the divergence one should introduce (as in Section 2.1) a cutting function g under the integral sign. Then the exact equality in eqn. 2.41 is replaced by the asymptotic one, but the equation obtained

$$\int_{-\infty}^{\infty}\exp(i\lambda x^2)(x-\alpha)^\mu g(x)\,dx$$
$$\simeq p_\mu\lambda^{-\mu-1/2}\exp(i\lambda\alpha^2/2)D_\mu[(1+i)\sqrt{(\lambda)}\alpha] \tag{2.42}$$

is valid for all μ. Indeed, integrating a few times by parts the difference between the left- and right-hand sides of eqn. 2.42 (the last-mentioned coincides with eqn. 2.41), gives

$$I_1=\int_{-\infty}^{\infty}\exp(i\lambda x^2)(x-\alpha)^\mu[1-g(x)]\,dx$$
$$=(i/2\lambda)^n\int_{-\infty}^{\infty}\exp(i\lambda x^2)f_n(x)\,dx$$

where

$$f_n(x)=\frac{d}{dx}\left[\frac{f_{n-1}(x)}{x}\right],\qquad f_0(x)=[1-g(x)]\,(x-\alpha)^\mu$$

Since $f_n(x)\sim|x|^{\mu-2n}$ as $x\to\infty$ and $f_n\equiv 0$ for sufficiently small $|x|$ the integral I_1 falls off more rapidly than any power of λ for any fixed μ and $\lambda\to\infty$.

To proceed from eqn. 2.42 to the general case consider first an integral

$$I=\int_{-\infty}^{\infty}\exp(i\lambda x^2)(x-\alpha)^\mu f(x)g(x)\,dx$$

which differs from eqn. 2.42 by an analytical function $f(x)$ introduced under the integral sign. Next it is proved that the uniform asymptotic of the integral I is of the form

$$I\simeq\lambda^{-(\mu+1)/2}\exp(i\lambda\alpha^2/2)AD_\mu[-(1+i)\sqrt{(\lambda)}\alpha]$$
$$+\lambda^{-(\mu+2)/2}\exp(i\lambda\alpha^2/2)BD_{\mu+1}[-(1+i)\sqrt{(\lambda)}\alpha] \tag{2.43}$$

where A and B are asymptotic power series in λ^{-1}

$$A=\sum_{n=0}^{\infty}A_n(\alpha)\lambda^{-n},\qquad B=\sum_{n=0}^{\infty}B_n(\alpha)\lambda^{-n}$$

Prove, first, the equation for $f(x)=x^m$

$$\int_{-\infty}^{\infty} \exp(i\lambda x^2)x^m g(x)(x-\alpha)^\mu\, dx$$
$$\simeq \lambda^{-(\mu+1)/2} \exp(i\lambda\alpha^2/2)[A_m D_\mu[-(1+i)\sqrt{(\lambda)}\alpha]$$
$$+ (1/\sqrt{\lambda})B_m D_{\mu+1}[-(1+i)\sqrt{(\lambda)}\alpha]] \qquad (2.44)$$

For $m=0$, eqn. 2.44 coincides with eqn. 2.42, if one sets $A=A_0=p_\mu$, $B=B_0=0$. For $m=1$, when setting $x=x-\alpha+\alpha$ one obtains eqn. 2.44, if $A=A_1=\alpha p_\mu$ $B=B_1=p_{\mu+1}$. For the rest of m it is proven by induction. Let eqn. 2.44 be valid for some m. Then differentiating with respect to λ and using the relations

$$D'_\mu(z)=(zD_\mu/2)-D_{\mu+1},\; D'_{\mu+1}(z)=(zD_{\mu+1}/2)+(\mu+1)D_\mu$$

one obtains that it is valid for $m+2$ as well, and

$$A_{m+2}=[\alpha^2-i(\mu+1)/2\lambda]A_m-i\frac{\partial}{\partial\lambda}A_m+(1-i)(\mu+1)\alpha B_m/2\lambda$$

$$B_{m+2}=-i(\mu+2)(B_m/2\lambda)-i\frac{\partial}{\partial\lambda}B_m-(1-i)\alpha A_m/2$$

It follows from eqn. 2.44 that the equation is valid when $f(x)$ is a polynomial in x. To prove eqn. 2.43 in the general case it is sufficient to approximate $f(x)$ for any n by a polynomial $P_n(x)$ in such a way that the following estimate of the error δI

$$\delta I=\int_{-\infty}^{\infty} \exp(i\lambda x^2)[f(x)-P_n(x)]g(x)(x-\alpha)^\mu\, dx=0(\lambda^{-n})$$

should be valid uniformly over α as $\lambda\to\infty$. For such an approximation choose $P_n(x)$ in such a way that the difference $f(x)-P(x)$ vanishes with its derivatives up to the order of $2n$ at $x=0$, while at $x=\alpha$ it should vanish with the derivatives up to the order of $n+p$, where p is chosen to satisfy the condition $\mathrm{Re}\,(\mu+p)>0$. Then the difference can be written as

$$f(x)-P(x)=x^{2n}(x-\alpha)^{n+p}f_1(x)$$

with a regular function $f_1(x)$. Therefore one can integrate δI by parts n times, each time integrating $x\exp(i\lambda x^2)$ and differentiating the remaining part of the nonexponential factor. As a result, one obtains

$$(i/2\lambda)^n\int_{-\infty}^{\infty} \exp(i\lambda x^2)\left[\sum_{q=0}^{n}\psi_q(x)\frac{d^q g(x)}{dx^q}\right](x-\alpha)^{\mu+p}\, dx$$

where $\psi_q(x)$ are the analytical functions. Since $\mathrm{Re}\,(\mu+p)>-1$ the integral is estimated by a constant independent of α, and this results in the estimate of δI.

The case of the phase function $\varphi(x)=-x^2$ is reduced to the one considered by complex conjugation of eqn. 2.42, while the case of arbitrary phase function with a nondegenerate stationary point x_0 is reduced by the appropriate variable change

$$\xi = \xi(x) = \varepsilon \operatorname{sign}[\varphi'(x)]\sqrt{|\varphi(x) - \varphi(x_0)|} \tag{2.45}$$

where $\varepsilon = 1$ when a singular point α is passed around in the upper halfplane, otherwise $\varepsilon = -1$.

As a result, the following theorem obtains.

Theorem 2.4: Let the functions f and φ in the integral $I(\lambda, \alpha)$ specified by eqn. 2.39 be analytical and let φ have a nondegenerate stationary point x_0 close to α. Let g be a cutting function which singles out a vicinity of these points. Then the asymptotic of $I(\lambda, \alpha)$ uniform over a distance between x and α is of the form (as $\lambda \to \infty$)

$$\begin{aligned} I = {} & \exp\{i\lambda[\varphi(\alpha) + \varphi(x_0)]/2\}\lambda^{-(\mu+1)/2}\{AD_\mu[(1+i\delta)\sqrt{(\lambda)}\xi(\alpha)] \\ & + (1/\sqrt{\lambda})BD_{\mu+1}[(1+i\delta)\sqrt{(\lambda)}\xi(\alpha)]\} \end{aligned} \tag{2.46}$$

where $\xi(x)$ is defined by eqn. 2.45 and A and B are asymptotic power series in λ^{-1}.

$$A = \sum_{n=0}^{\infty} A_n(\alpha)\lambda^{-n}, \qquad B = \sum_{n=0}^{\infty} B_n(\alpha)\lambda^{-n}$$

with regular coefficients as $\alpha \to x_0$. In particular,

$$\begin{aligned} A_0(\alpha) = {} & \sqrt{(2\pi)} \exp[i\pi\mu(2\varepsilon + \delta)/4] \exp(i\pi\delta/2)2^{(\mu+1)/2} f(\alpha) \\ & \times \left|\frac{\sqrt{|\varphi(\alpha) - \varphi_0|}}{\varphi'(\alpha)}\right|^{\mu+1} \end{aligned} \tag{2.47a}$$

$$\begin{aligned} B_0(\alpha) = {} & \frac{\sqrt{(2\pi)} \exp[i\pi\mu(2\varepsilon + \delta)/4]2^{-(\mu+1)/2}}{\xi(\alpha)} \\ & \times \left[f_0 \frac{|x_0 - \alpha|^{\mu}[1/\sqrt{(\varphi_0'')}]}{|\varphi(\alpha) - \varphi_0|^{\mu/2}} \right. \\ & \left. - f(\alpha)2^{\mu+1/2} \frac{|\varphi(\alpha) - \varphi_0|^{(\mu+1)/2}}{|\varphi'(\alpha)|^{\mu+1}}\right] \end{aligned} \tag{2.47b}$$

Eqns. 2.47 are obtained by asymptotic matching, i.e. by equating the expressions obtained when replacing D_μ and D'_μ by their asymptotics as $\lambda \to \infty$ to non-uniform asymptotics (eqn. 2.40). As $\alpha \to x_0$,

$$\begin{aligned} A_0(\alpha) = {} & \sqrt{(2\pi)} \exp[i\pi\mu(2\varepsilon + \delta)/4] \exp(i\pi\delta/4)2^{\mu/2} f(\alpha)|\varphi_0''|^{-(\mu+1)/2} \\ & \times [1 - (\mu+1)(\alpha - x_0)\varphi_0'''/(3\varphi_0'') + O(\alpha - x_0)^2] \\ B_0(\alpha) = {} & \frac{\sqrt{(2\pi)}\varepsilon\delta \exp[i\pi\mu(2\varepsilon + \delta)/4]}{|\varphi_0''|^{1+\mu/2}} \left[f_0 \frac{(\mu+2)\varphi_0'''}{6\varphi_0''} - f_0'\right] \\ & + O(\alpha - x_0) \end{aligned} \tag{2.48}$$

In these equations $\varphi_0 = \varphi(x_0)$, $\varphi_0'' = \varphi''(x_0)$, $\varphi_0''' = \varphi'''(x_0)$, $f_0 = f(x_0)$, $f_0' = f'(x_0)$, $\delta = \operatorname{sign}[\varphi''(x_0)]$.

2.5 Close pole and branch points

Consider an integral

$$I(\lambda, \alpha)=\frac{i}{2\pi}\int_{-\infty+i0}^{\infty+i0} \exp[i\lambda\varphi(x)]\,\frac{(x+\alpha)^{\mu}}{x} f(x)g(x)\,dx/x \tag{2.49}$$

where the pole $x=0$ and the branch point $x=-\alpha$ are passed around in the upper halfplane (for definity), the functions $f(x)$ and $\varphi(x)$ are analytical in the vicinity of zero and g is the neutraliser, vanishing outside the vicinity.

Such integrals arise, for example, when solving a problem of exciting oscillation in stratified media using the variable separation method, in the case when appropriate eigenfunctions have branch points, in particular, when considering internal gravity waves in stratified fluids with average shear flows [25].

It follows from considerations (similar to those of Section 1.5) that for $\varphi'(x)>\alpha$ the critical points $x=0$ and $x=\alpha$ do not contribute to the asymptotic of $I(\lambda, \alpha)$. For $\varphi'(x)<0$ the nonuniform asymptotic of $I(\lambda, \alpha)$ (as $\lambda\to\infty$ and α fixed) includes the contributions of the pole $x=0$ and the branch point $x=\alpha$, the asymptotic being of the form

$$I(\lambda, \alpha)\simeq\exp[i\lambda\varphi(0)]\,\alpha^{\mu}f(0)+\exp[i\lambda\varphi(-\alpha)]\sum_{n=0}^{\infty} c_n\lambda^{-\mu-n-1} \tag{2.50}$$

Here, for $\alpha<0$ the term α^{μ} means an analytical continuation from a ray $\alpha>0$ through the upper halfplane

$$\alpha^{\mu}=\begin{cases}|\alpha|^{\mu} & \text{for } \alpha>0\\ |\alpha|\exp(i\pi\mu) & \text{for } \alpha<0\end{cases}$$

The coefficients $c_n=c_n(\alpha)$ of the asymptotic series are evaluated according to the algorithm of Section 1.4. In particular,

$$c_0=\frac{-i\exp(i\pi\mu/2)f(-\alpha)}{\Gamma(-\mu)\alpha|\varphi'(0)|^{\mu+1}} \tag{2.51}$$

The nonuniform asymptotic (eqn. 2.50) is inapplicable as $\alpha\to 0$.

In this case the integral

$$P_{\mu}(\lambda, \alpha)=\frac{i}{2\pi}\int_{-\infty+i0}^{\infty+i0} \exp(-i\lambda x)\,\frac{(x+\alpha)^{\mu}}{x}\,dx \tag{2.52}$$

absolutely convergent for $\mathrm{Re}(\mu)<0$, is a model integral.

Express the integral in terms of the incomplete gamma function (Section 6.4). For $\lambda=0$ the integration contour can be closed in the upper halfplane, therefore $P_{\mu}(0, \alpha)=0$. Besides,

$$\frac{\partial P_{\mu}}{\partial\lambda}=\frac{1}{2\pi}\int_{-\infty+i0}^{\infty+i0} \exp(-i\lambda x)(x+\alpha)^{\mu}\,dx$$

Shifting the integration path from the real axis to the cutting $x = -\alpha - i\xi (0 < \xi < \infty)$ gives

$$\frac{\partial P_\mu}{\partial \lambda} = (1/\pi) \exp(i\lambda\alpha + i\pi\mu/2) \sin(\pi\mu)\Gamma(\mu+1)\lambda^{-\mu-1}$$

$$= \frac{\exp(i\lambda\alpha + i\pi\mu/2)}{\Gamma(-\mu)\lambda^{\mu+1}}$$

Therefore

$$P_\mu(\lambda, \alpha) = \frac{\exp(i\pi\mu/2)}{\Gamma(-\mu)} \int_0^\lambda \exp(i\alpha\xi)\xi^{-\mu-1}\, d\xi$$

from which one obtains

$$P_\mu(\lambda, \alpha) = \exp(i\pi\mu/2)\lambda^{-\mu}\gamma^*(-i\alpha\lambda, -\mu)$$

$$= \alpha^\mu - \frac{\exp(i\pi\mu/2)|\alpha|^\mu}{\Gamma(-\mu)} [C(|\alpha|\lambda, -\mu) + i \operatorname{sign}(\alpha) S(|\alpha|\lambda, -\mu)] \qquad (2.53)$$

Here, $\gamma^*(\alpha, x)$ is the incomplete gamma function

$$\gamma^*(\alpha, x) = \frac{x^{-\alpha}}{\Gamma(\alpha)} \int_0^x \exp(-t)t^{\alpha-1}\, dt$$

and $C(x, \alpha)$, $S(x, \alpha)$ are the generalised Fresnel integrals

$$C(x, \alpha) = \int_x^\infty t^{\alpha-1} \cos t\, dt, \qquad S(x, \alpha) = \int_x^\infty t^{\alpha-1} \sin t\, dt$$

(see Reference 19, eqns. 6.5.4, 6.5.7, 6.5.8). In eqn. 2.53 for $\alpha < 0$ the quantity α^μ means an analytical continuation from the ray $\alpha > 0$ through the upper halfplane. The equation is derived when assuming $\operatorname{Re}(\mu) < 0$. However, it is easy to show that if $g(x)$ is a neutraliser identically equal to unity in the vicinity of the points $x = 0$ and $x = \alpha$, then

$$\frac{i}{2\pi} \int_{-\infty+i0}^{\infty+i0} \exp(-i\lambda x) \frac{(x+\alpha)^\mu}{x} g(x)\, dx \simeq P_\mu(\lambda, \alpha) \qquad (2.54)$$

where for $\operatorname{Re}(\mu) < 0$ the function $P_\mu(\lambda, \alpha)$ means eqn. 2.51 while for $\operatorname{Re}(\mu) > 0$ it means the analytical continuation specified by the right-hand side of eqn. 2.53. The proof of this statement is similar to that of eqn. 2.7 in Section 2.1.

In what follows the asymptotic of $P_\mu(\lambda, \alpha)$ as $\lambda \to \infty$ and α fixed is required. It follows from eqn. 2.53 that the asymptotic is of the form

$$P_\mu(\lambda, \alpha) \simeq \alpha^\mu - \frac{i \exp(i\lambda\alpha + i\pi\mu/2) \sin(\pi\mu)}{\pi\alpha\lambda^{\mu+1}} \times \sum_{n=0}^\infty (-i/\lambda\alpha)^n \Gamma(\mu+n+1) \qquad (2.55)$$

Treating the general case, perform two steps. First, consider an integral

$$I = \frac{1}{2\pi} \int_{-\infty + i0}^{\infty + i0} \exp(-i\lambda\xi) \frac{(\xi + \alpha)^\mu}{\xi} f(\xi) g(\xi)\, d\xi \tag{2.56}$$

which differs from expn. 2.54 by an analytical function $f(\xi)$ under the integral sign. Setting $f(\xi) = f(0) + \xi f_1(\xi)$ gives

$$I \simeq f(0) P_\mu(\lambda, \alpha) + (i/2\pi) \int_{-\infty + i0}^{\infty + i0} \exp(-i\lambda\xi)(\xi + \alpha)^\mu f_1(\xi) g(\xi)\, d\xi$$

The integral has the sole critical point $\xi = -\alpha$. Evaluating the asymptotic according to the algorithms of Sections 1.4 and 1.5 obtains

$$I \simeq f(0) P_\mu(\lambda, \alpha) + \exp(i\lambda\alpha) \sum_{n=0}^{\infty} c_n(\alpha) \lambda^{-n-\mu-1}$$

The initial integral (eqn. 2.49) is reduced to eqn. 2.56 by the variable change $\xi = \xi(x) = \varphi(0) - \varphi(x)$. This results in the following theorem.

Theorem 2.5: Let an integral $I(\lambda, \alpha)$ be specified by eqn. 2.49 where the pole $x = 0$ and branch point $x = \alpha$ are passed around in the upper halfplane. Let $f(x)$ and $\varphi(x)$ be analytical in the vicinity of zero and $g(x)$ be a neutraliser which singles out this vicinity. Then for $\varphi'(0) > 0$ the integral $I(\lambda, \alpha)$ falls off (as $\lambda \to \infty$) more rapidly than any power of λ^{-1}. For $\varphi'(0) < 0$ the asymptotics of $I(\lambda, \alpha)$ uniform for small α is of the form

$$I(\lambda, \alpha) = \exp[i\lambda\varphi(0)] f(0) \left[\frac{\alpha}{\varphi(-\alpha) - \varphi(0)} \right]^\mu P_\mu[\lambda, \varphi(-\alpha) - \varphi(0)] + \exp[i\lambda\varphi(-\alpha)] \sum_{n=0}^{\infty} d_n(\alpha) \lambda^{-n-1-\mu} \tag{2.57}$$

where P_μ is expressed in terms of the incomplete gamma function by eqn. 2.53 and has the asymptotic of eqn. 2.55. The coefficients $d_n(\alpha)$ are regular as $\alpha \to 0$ and are determined by asymptotic matching to the known nonuniform asymptotic (eqn. 2.50) of the integral $I(\lambda, \alpha)$. In particular,

$$d_0(\alpha) = \frac{-i \exp(i\pi\mu/2)}{\Gamma(-\mu)} \left[\frac{f(-\alpha)}{\alpha |\varphi'(-\alpha)|^{\mu+1}} - \frac{\alpha^\mu f(0)}{[\varphi(-\alpha) - \varphi(0)]^{\mu+1}} \right]$$

As $\alpha \to 0$ the function $d_0(\alpha)$ tends to the limit

$$\lim_{\alpha \to 0} d_0(\alpha) = \frac{i \exp(i\pi\mu/2)}{\Gamma(-\mu) |\varphi'(0)|^{\mu+1}} \left[f'(0) + \frac{(\mu + 1)\varphi''(0) f(0)}{2\varphi'(0)} \right] \tag{2.58}$$

2.6 Pole near integration domain boundary

Consider an integral

$$I(\lambda, \alpha, \beta) = \int_\alpha^\infty \frac{\exp[i\lambda\varphi(x)] f(x) g(x)}{(x - \beta + iO)}\, dx \tag{2.59}$$

where for $\beta > \alpha$ the pole $x = \beta$ is passed around in the upper halfplane and $g(x)$ is the cutting function, $\varphi(x)$ and $f(x)$ are analytical functions and $\varphi'(x)$ is bounded in magnitude from below (i.e. $|\varphi'(x)| > \text{const} > 0$) in the region supp (g). The integral diverges logarithmically as $\beta \to \alpha$. At the same time, as $\lambda \to \infty$ the coefficients of the nonuniform asymptotic expansion of $I(\lambda, \alpha, \beta)$ are of the form

$$I(\lambda, \alpha, \beta) \simeq \exp[i\lambda\varphi(\alpha)] \sum_{n=0}^{\infty} (i/\lambda)^n a_n \tag{2.60}$$

where $a_0 = f(\alpha)/[(\alpha - \beta)\varphi'(\alpha)]$, the coefficients being grown like power functions as $\beta \to \alpha$. Therefore, the asymptotics of eqn. 2.6 is inapplicable as $\beta \to \alpha$.

Here, the model integral in terms of which the uniform asymptotic is expressed has the form

$$\begin{aligned} I(\lambda, \alpha, \beta) &= \int_{\alpha}^{\infty} \frac{\exp(i\lambda x)}{(x - \beta + iO)}\, dx \\ &= \exp(i\lambda\alpha) \int_{0}^{\infty} \frac{\exp(i\xi)}{\xi + \lambda(\alpha - \beta) + iO}\, d\xi \\ &= \exp(i\lambda\alpha) F[\lambda(\alpha - \beta)] \end{aligned}$$

where $F(s)$ is expressed in terms of integral sine Si (z) and integral cosine Ci (z) or in terms of auxiliary functions $f(x)$ and $g(x)$ (see Reference 19, eqn. 5.2), or in terms of integral exponential function $E_1(z)$ (see Reference 19, eqn. 5.1):

$$\begin{aligned} F(s) &= \int_{0}^{\infty} \frac{\exp(i\xi)}{(\xi + s + iO)}\, d\xi = \int_{0}^{\infty} \frac{\exp(-\eta)}{(\eta - is)}\, d\eta \\ &= \int_{0}^{\infty} \frac{\exp(-\eta)(\eta + is)}{(\eta^2 + s^2)}\, d\eta \\ &= g(|s|) + i\,\text{sign}(s) f(|s|) \end{aligned} \tag{2.61}$$

where

$$f(t) = \int_{0}^{\infty} \frac{\exp(-\eta t)}{\eta^2 + 1}\, d\eta$$

$$g(t) = -\frac{df(t)}{dt} = \int_{0}^{\infty} \frac{\eta \exp(-t\eta)}{\eta^2 + 1}\, d\eta$$

or

$$F(s) = \exp(-is)\left\{-\text{Ci}(|s|) + i\,\text{sign}(s)\left[\frac{\pi}{2} - \text{Si}(|s|)\right]\right\} \tag{2.62a}$$

where

$$\mathrm{Ci}\,(\xi) = -\int_{\xi}^{\infty} \frac{\cos t}{t}\,dt = \gamma + \ln\,(\xi) + \int_{0}^{\xi} \frac{\cos t - 1}{t}\,dt$$

$$\mathrm{Si}\,(\xi) = \int_{0}^{\xi} \frac{\sin t}{t}\,dt$$

γ is the Euler constant, or

$$F(s) = \exp\,(-is)E_1(-is)$$

where

$$E_1(z) = \int_{z}^{\infty} \frac{\exp\,(-t)\,dt}{t} \tag{2.62b}$$

In the general case the following theorem is valid.

Theorem 2.6: The uniform asymptotic of eqn. 2.59 has the form

$$I(\lambda, \alpha, \beta) = \exp\,[i\lambda\varphi(\alpha)]\,[f(\beta)F\{\lambda[\varphi(\alpha) - \varphi(\beta)]\} + \sum_{n=0}^{\infty} (i/\lambda)^{n+1} b_n] \tag{2.63}$$

where $F(s)$ is defined by eqn. 2.61 and the coefficients b_n are regular as $a \to \beta$ and are determined by asymptotic matching to known nonuniform asymptotics (2.60).

In so doing the asymptotic of $F(s)$

$$F(s) = \sum_{n=0}^{\infty} (-1)^n (i/s)^{n+1} \Gamma(n+1)$$

is used. In particular,

$$b_0 = \frac{f(\alpha)}{\varphi'(\alpha)(a-\beta)} - \frac{f(\beta)}{\varphi(\alpha) - \varphi(\beta)}$$

Now the theorem is proven for the case of $\varphi'(x) > 0$, the case of $\varphi'(x) < 0$ being considered in the same way. After the variable change $\xi = \varphi(x)$ in eqn. 2.59 one obtains the integral

$$\int_{\varphi(\alpha)}^{\infty} \exp\,(i\lambda\xi) \frac{f[x(\xi)]g[x(\xi)]}{x(\xi) - \beta + iO} \frac{dx}{d\xi}\,d\xi$$

$$= \int_{\varphi(\alpha)}^{\infty} \exp\,(i\lambda\xi) \frac{f[x(\xi)]g[x(\xi)]h(\xi)}{\xi - \varphi(\beta) + iO}\,d\xi$$

Here,

$$h(\xi) = \frac{\xi - \varphi(\beta)}{x(\xi) - \beta} \frac{dx\,(\xi)}{d\xi}$$

and $h(\beta)=1$. Taking into account that $g(\alpha)=g(\beta)=1$, write the integral as

$$I_{+}=f(\beta)\cdot\int_{\varphi(\alpha)}^{\infty}\exp(i\lambda\xi)\frac{d\xi}{\xi-\varphi(\beta)+iO}$$

$$+\int_{\varphi(\alpha)}^{\infty}\exp(i\lambda\xi)\left[\frac{f[x(\xi)]g[x(\xi)]h(\xi)-f(\beta)}{\xi-\varphi(\beta)}\right]d\xi$$

where the nonexponential factor in the second integral is regular as $\xi\to\varphi(\beta)$. The first integral of this expression coincides with $\exp[i\lambda\varphi(\alpha)]\times F\{\lambda[\varphi(\alpha)-\varphi(\beta)]\}$. Evaluating the asymptotic of the second integral as $\lambda\to\infty$, one obtains eqn. 2.63.

2.7 Stationary point tending to infinity

Consider an integral

$$I(\lambda,\alpha)=\int_{0}^{\infty}\exp[i\lambda\varphi(\alpha,x)]f(x)g(x)\,dx \tag{2.64}$$

where the phase function $\varphi(\alpha, x)$ is expanded into the converging series (as $x\to\infty$)

$$\varphi(\alpha,x)=\alpha x+a_0+\frac{a_1}{x}+\frac{a_2}{x^2}+\cdots \tag{2.65}$$

where the coefficients a_n depend analytically on α for small α and $a_1(0)>0$. Let us assume that $f(x)$ in eqn. 2.64 is expanded into a series (as $x\to\infty$)

$$f(x)=\sum_{n=0}^{\infty}f_n x^{-\mu-n-1} \tag{2.66}$$

and $g(x)$ is a cutting function which singles out the vicinity of $x=\infty$. For the integral to converge as $x\to\infty$ the quantity Re μ is assumed to satisfy the non-equality Re $\mu>-1$.

The stationary point x_1 of the phase function $\varphi(\alpha, x)$ in eqn. 2.65 is determined from the equation

$$\varphi'(\alpha,x)=\alpha-\frac{a_1}{x^2}-\frac{2a_2}{x^3}-\cdots=0 \tag{2.67}$$

it tends to infinity as $\alpha\to 0$ and is expanded into a power series in $\sqrt{\alpha}$

$$x_1=x_1(\alpha)=\sqrt{(a_1/\alpha)}+\sum_{n=0}^{\infty}b_n\alpha^{n/2} \tag{2.68a}$$

which converges for sufficiently small α. The phase function at x_1 is also expanded into a power series in $\sqrt{\alpha}$

$$\varphi_1=\varphi(x_1)=\sqrt{(a_1\alpha)}+\sum_{n=0}^{\infty}c_n\alpha^{n/2}=\psi_0(\alpha)+\sqrt{\alpha}\,\psi_1(\alpha) \tag{2.69a}$$

where $\psi_0(\alpha)$ and $\psi_1(\alpha)$ are analytical for small α, $\psi_0(0)=a_0$, $\psi_1(0)=\sqrt{a_1}$.

Later it is required to have the value of the phase function at $x=x_2$ which is the second root of eqn. 2.67 tending to $-\infty$ as $\alpha\to 0$.

$$x_2=x_2(\alpha)=-\sqrt{(a_1/\alpha)}+\sum_{n=0}^{\infty}(-1)^n b_n\alpha^{n/2} \tag{2.68b}$$

$$\varphi_2=\varphi(x_2)=-\sqrt{(a_1\alpha)}+\sum_{n=0}^{\infty}(-1)^n c_n\alpha^{n/2}=\psi_0(\alpha)-\sqrt{\alpha}\,\psi_1(\alpha) \tag{2.69b}$$

$\psi_0(\alpha)$ and $\psi_1(\alpha)$ being the same as in eqn. 2.69*a*.

Write the nonuniform asymptotic of eqn. 2.64 fixing α and setting $\lambda\to\infty$. The second derivative of the phase function at the stationary point $x_1(\alpha)$ is of the form

$$\varphi''_{xx}[\alpha,\, x_1(\alpha)]=\alpha^{3/2}[p(\alpha)+\sqrt{\alpha}\,q(\alpha)]$$

where $p(\alpha)$ and $q(\alpha)$ are analytical for small α. Therefore for the nonuniform asymptotic of $I(\lambda,\alpha)$ one obtains

$$I(\lambda,\alpha)\simeq\frac{\alpha^{\mu/2-1/4}}{\sqrt{\lambda}}\exp\{i\lambda[\psi_0(\alpha)+\sqrt{\alpha}\,\psi_1(\alpha)]\}$$
$$\times\sum_{n=0}^{\infty}\lambda^{-n}\alpha^{-n/2}[P_n(\alpha)+\sqrt{\alpha}\,Q_n(\alpha)] \tag{2.70}$$

where $P_n(\alpha)$ and $Q_n(\alpha)$ are analytical for small $|\alpha|$ as well. This asymptotic is inapplicable for small α since the second derivative of the phase function at the stationary point $x_1(\alpha)$ tends to zero like $\alpha^{3/2}$ as $\alpha\to 0$. In the case under consideration the model integral is

$$I_0=\int_0^{\infty}\exp\,[i\lambda\,(\alpha x+1/x)]\,x^{-\mu-1}\,dx$$
$$=i\alpha^{\mu/2}\pi\exp\,(i\pi\mu/2)H_\mu^{(1)}(2\lambda\sqrt{\alpha}) \tag{2.71}$$

where $H_\mu^{(1)}$ is the Hankel function of the first kind (see, for example, Reference 26, eqn. 8.421 and Section 5.1). It is proven in what follows that in the general case the following theorem holds.

Theorem 2.7: The uniform (as $\alpha\to 0$) asymptotic of eqn. 2.64 where for sufficiently large x the functions f and φ are expanded into the series of eqns. 2.66 and 2.65 for $a_1>0$, Re $\mu>-1$ *and g(x) is a cutting function which singles out a vicinity of infinity has the form*

$$I(\lambda,\alpha)\simeq\alpha^{\mu/2}\exp\,(i\lambda\psi_0)[A(\lambda,\alpha)H_\mu^{(1)}(\lambda\sqrt{\alpha}\,\psi_1)$$
$$+\sqrt{\alpha}B(\lambda,\alpha)H_{\mu+1}^{(1)}(\lambda\sqrt{\alpha}\,\psi_1)] \tag{2.72}$$

where $H_\mu^{(1)}$ and $H_{\mu+1}^{(1)}$ are the Hankel functions. Here the phase functions $\psi_0=\psi_0(\alpha)$, and $\psi_1=\psi_1(\alpha)$ and the coefficients of the asymptotic series

$$A(\lambda,\alpha)=\sum_{n=0}^{\infty}\lambda^{-n}A_n(\alpha),\qquad B(\lambda,\alpha)=\sum_{n=0}^{\infty}\lambda^{-n}B_n(\alpha)$$

are analytical functions of α in a vicinity of $\alpha=0$ defined by asymptotic matching to the nonuniform asymptotic eqn. 2.70.

In particular,

$$\begin{aligned} A_0(\alpha) &= c_\mu P_0 \\ B_0(\alpha) &= ic_\mu Q_0 \\ A_1(\alpha) &= c_\mu Q_1 - B_0[(\mu+1)^2 - 1/4]/(2\psi_1(\alpha)) \\ B_1(\alpha) &= c_\mu\{2iP_1\psi_1(\alpha) + P_0[(\mu+1)^2 - 1/4]\}/(2\alpha\psi_1(\alpha)) \end{aligned} \tag{2.73}$$

where

$$c_\mu = \exp(i\pi\mu/2)\exp(i\pi/4)\sqrt{[\pi\psi_1(\alpha)/2]}$$

Here the functions $\psi_0(\alpha)$, $\psi_1(\alpha)$ and coefficients P_0, P_1, Q_0, Q_1 are the same as that in the nonuniform asymptotic (eqn. 2.70).

To prove eqn. 2.72, i.e. to reduce eqn. 2.64 to the model integral (eqn. 2.71), it is shown that there exists a nondegenerate (for sufficiently large x) variable change

$$\xi = \xi(x, \alpha) = x + d_0(\alpha) + d_1(\alpha)/x + d_2(\alpha)/x^2 + \cdots$$

transforming the phase function (eqn. 2.65) into the function

$$\psi(\alpha, x) = h_0(\alpha) + \alpha\xi + h_1(\alpha)/\xi \tag{2.74}$$

the series of eqn. 2.74 being convergent uniformly over α for sufficiently large x.

It is convenient to perform appropriate mathematical manipulations in the vicinity of $x=0$ rather than $x=\infty$. Therefore set $y=1/x$, $\eta=1/\xi$. In these variables,

$$\varphi(\alpha, y) = \alpha/y + a_0(\alpha) + a_1(\alpha)y + a_2(\alpha)y^2 + \cdots \tag{2.75}$$

One needs to find such a function

$$\eta = y + c_2y^2 + c_3y^3 + \cdots \tag{2.76}$$

for which

$$\varphi(\alpha, y) = \psi[\alpha, \eta(y)] = h_0(\alpha) + \alpha/\eta + h_1(\alpha)\eta \tag{2.77}$$

To solve the problem find, first, the functions $h_0(\alpha)$, $h_1(\alpha)$ in eqn. 2.77, then determine $\eta(y)$ from this equation and, finally, prove that $\eta(y)$ is represented by the series of eqn. 2.76 uniformly converging as $\alpha \to 0$.

Differentiating eqn. 2.77 one obtains $\psi_y(\alpha, y) = \psi'_\eta \eta'_y$ and since $\eta'_y > 0$ the stationary points

$$\eta_{1,2} = \eta(y_{1,2}) = \pm\sqrt{[\alpha/h_1(\alpha)]}$$

of the function ψ correspond to the stationary points $y_1 = [x_1(\alpha)]^{-1}$, $y_2(\alpha) = [x_2(\alpha)]^{-1}$ of the function φ (here x_1, x_2 are of form of eqn. 2.68), and the values of $\varphi_{1,2} = \varphi(\alpha, y_{1,2})$ at these points coincide with the values of $\psi(\eta)$ at the points $\eta_{1,2}$

$$\psi(\eta_{1,2}) = h_0(\alpha) \pm 2\sqrt{[\alpha h_1(\alpha)]} = \varphi_{1,2} = \psi_0(\alpha) \pm \sqrt{\alpha}\,\psi_1(\alpha)$$

From these relations one gets

$$\begin{aligned} h_0 &= h_0(\alpha) = \psi_0(\alpha) \\ h_1 &= h_1(\alpha) = \psi_1^2(\alpha)/4 \end{aligned} \tag{2.78}$$

Now find $\eta(y)$. From eqns. 2.77 and 2.78 one obtains

$$\eta = \{(\varphi - h_0) \pm \sqrt{[(\varphi - h_0)^2 - 4\alpha h_1]}\}/2h_1$$
$$= [\sqrt{(\varphi - \varphi_2)} \pm \sqrt{(\varphi - \varphi_1)}]^2/4h_1$$

In this expression the radical sign is chosen from the condition that η should tend to zero as $y \to 0$, hence, for $0 < y < y_1 = [x_1(\alpha)]^{-1}$ one has

$$\eta(y) = \{\sqrt{[\varphi(y) - \varphi_2]} - \sqrt{[\varphi(y) - \varphi_1]}\}^2/4h_1 \qquad (2.79)$$

To obtain eqn. 2.76 it is sufficient to expand the radicals using eqn. 2.75 and then to expand $\eta(y)$ into a power series in y.

Now it is shown that $\eta(y)$ defined by eqn. 2.79 for $0 < y < y_1$ can be analytically continued into a complex vicinity of $y = 0$ independent of α. To do this note that since $y = y_1$ is a nondegenerate stationary point of $\varphi(y)$ and $\varphi''(y) > 0$ eqn. 2.79 can be written as

$$\eta(y) = \frac{[\sqrt{[\varphi(y) - \varphi_2]} + (y - y_1)\sqrt{\{[\varphi(y) - \varphi_1]/(y - y_1)^2\}}]^2}{4h_1}$$

The function $\eta(y)$ is seen to be analytical in a vicinity of $y = y_1$ and it is analytically continuated into the region $y < 0$ according to the equation

$$\eta(y) = -[\sqrt{[\varphi_1 - \varphi(y)]} + (y_2 - y)\sqrt{\{[\varphi_2 - \varphi(y)]/(y - y_2)^2\}}]^2/4h_1$$

from which one can see that $\eta(y)$ is analytical in a vicinity of $y = y_2 = (x_2)^{-1}$ as well.

The singular points of $\eta(y)$ are either singular points of $\varphi(y)$ or branch points of the radicals (i.e. the roots of the equation $\varphi(y) = \varphi_{1,2}$). As we have seen, the pole $y = 0$ of the function $\varphi(y)$ is not a singular point of $\eta(y)$. There are no other singular points of $\eta(y)$ in a zero point vicinity independent of α. We have shown that the points $y = y_1$ and $y = y_2$ were not singular points of $\eta(y)$. Using eqn. 2.77 it is easy to see that in the vicinity of $y = 0$ there are no other roots of the equations $\varphi(\eta) = \varphi_{1,2}$.

Thus, it has been proved that $\eta(y)$ is analytical in the circle $|y| < C$ where C does not depend on α. On the boundary of the circle and therefore inside of it $\eta''(y)$ is bounded uniformly over α.

Since $\eta'(0) = 1$, then $\eta'(y) > 0$ in the interval $-C_2 < y < C_2$ where C_2 is independent of α. Returning to the variables $\xi = 1/\eta$, $x = 1/y$ one obtains that for sufficiently large x there exists the variable change $\xi = \xi(x)$ transforming the phase function $\varphi(\alpha, x)$ into eqn. 2.74 and the initial integral $I(\lambda, \alpha)$ (when eqn. 2.79 is taken into account) into the expression

$$I(\lambda, \alpha) = \exp[i\lambda\psi_0(\alpha)] \times \int_0^\infty \exp\{i\lambda[\alpha\xi + \psi_1^2(\alpha)/4\xi]\} f_1(\xi) g_1(\xi)\, d\xi \qquad (2.80)$$

where

$$f_1(\xi) = f[x(\xi)] \frac{dx(\xi)}{d\xi} = \sum_{n=0}^{\infty} f_n \xi^{-n-\mu-1} \qquad (2.81)$$

and $x(\xi)$ is the inverse function to $\xi(x)$, while $g_1(\xi)=g[x(\xi)]$ is the cutting function not equal to zero only for sufficiently large ξ.

To prove asymptotic eqn. 2.72 it is sufficient to prove it for the integral of eqn. 2.80. Formally, the cutting function $g_1(\xi)$ is omitted, $f_1(\xi)$ is replaced by the series of eqn. 2.81 and eqn. 2.80 is integrated term by term using eqn. 2.71. This results in

$$I(\lambda, \alpha)=\pi \exp [i\lambda\psi_0(\alpha)] \sum_{n=0}^{\infty} f_n \exp [i\pi(\mu+n+1)/2]$$
$$\times\left[\frac{2\sqrt{\alpha}}{\psi_1(\alpha)}\right]^{\mu+n} H_{\mu+n}^{(1)}[\lambda\sqrt{\alpha}\,\psi_1(\alpha)]$$

If in this equation the functions $H_{\mu+n}^{(1)}$ are expressed in terms of $H_{\mu}^{(1)}$ and $H_{\mu+1}^{(1)}$ subsequently using the identity valid for any cylindrical function $\mathcal{Z}_\nu$

$$\mathcal{Z}_{\nu+1}(z)=[2\nu\mathcal{Z}_\nu(z)/z]-\mathcal{Z}_{\nu-1}(z)$$

(see, for example, Reference 26, eqn. 9.1.27) to lower indices, one obtains eqn. 2.72.

To substantiate the asymptotic (eqn. 2.72) of eqn. 2.80 rigorously it is sufficient to do this for $-1<\mathrm{Re}\,(\mu)<0$, setting (in the case of more rapid decreasing of $f_1(\xi)$ as $\xi\to\infty$) that in the expansion eqn. 2.81 the few first coefficients of f_n are equal to zero.

To prove, first, eqn. 2.72 for $f_1(\xi)=\xi^{-\mu-1}$:

$$I_0=\int_0^\infty \exp [i\lambda(\alpha\xi+\psi_1^2/\xi)]\xi^{-\mu-1}g(\xi)\,d\xi$$
$$\simeq\exp [i\pi(\mu+1)/2](2\sqrt{\alpha}/\psi_1)^\mu H_\mu^{(1)}(2\sqrt{\alpha}\,\psi_1) \qquad (2.82)$$

One has

$$I_0=\int_0^\infty \exp [i\lambda(\alpha\xi+\psi_1^2/\xi)]\xi^{-\mu-1}\,d\xi$$
$$-\int_0^\infty \exp [i\lambda(\alpha\xi+\psi_1^2\xi)]\xi^{-\mu-1}[1-g(\xi)]\,d\xi \qquad (2.83)$$

The first integral coincides with the right-hand side of eqn. 2.82, therefore it is sufficient to verify that the second integral falls off more rapidly than any power of λ^{-1}. In this integral the cutting function $1-g(\xi)$ singles out a vicinity of $\xi=0$, therefore after the variable change $\eta=1/\xi$ one obtains the integral

$$\int_0^\infty \exp [i\lambda(\psi_1^2\eta+\alpha/\eta)]\,\eta^{\mu-1}h(\eta)\,d\eta$$

where the function $h(\eta)=1-g(1/\eta)$ singles out a vicinity of $\eta=\infty$. If α is sufficiently small, the phase function in the integral grows monotonically in the region supp (h) and its derivative is bounded in magnitude from below. Therefore the standard procedure to integrate many times by parts shows that the integral falls off more rapidly than any power of λ^{-1}.

Differentiating eqn. 2.82 many times with respect to a parameter ψ_1 and expressing the second derivative of $H_\mu^{(1)}$ in terms of the function and its first derivative proves eqn. 2.77 for $f_1(\xi)=P_n(\xi^{-1})\xi^{-\mu-1}$ where $P_n(\xi^{-1})$ is the polynomial in ξ^{-1}.

To prove eqn. 2.72 for an arbitrary function

$$f_1(\xi)=\xi^{-\mu-1}\sum_{n=0}^{\infty} f_n\xi^{-n}$$

it is sufficient to approximate the function

$$\sum_{n=0}^{\infty} f_n\xi^{-n}$$

by the polynomial $P_n\ (\xi^{-1})$ in such a way for the integral with the integrand

$$\Delta(\xi)=\sum_{n=0}^{\infty} f_n\xi^{-n}-P_n(\xi^{-1})$$

to be estimated by λ^{-N} uniformly over α where N is an arbitrary integer. Choose P_n in such a way that the difference should have zeros of the order of $2N$ at $\xi=\pm\psi_1/2\sqrt{\alpha}$

$$\sum_{n=0}^{\infty} f_n\xi^{-n}=P_n(\xi^{-1})+(\alpha-\psi_1^2/4\xi^2)^{2N}h(\xi^{-1},\alpha)$$

where h is regular as $\xi\to\infty$. Then the integral

$$\int_0^{\infty}\exp\left[i\lambda\left(\alpha\xi+\psi_1^2/\xi\right)\right]^{-\mu-1}\left[\sum^{\infty} f_n\xi^{-n}-P_n(\xi^{-1})\right]g(\xi)\,d\xi$$

$$=\int_0^{\infty}\exp\left[i\lambda\left(\alpha\xi+\psi_1^2/\xi\right)\right]\xi^{-\mu-1}(\alpha-\psi_1^2/4\xi^2)^{2N}h(\xi^{-1},\alpha)g(\xi)\,d\xi$$

can be integrated N times by parts, from which it follows that the integral falls off not slower than λ^{-N}.

For $\mu=0$ and $\alpha\to 0$ both the left- and right-hand sides of the asymptotic eqn. 2.72 grow as ln (α). However, the asymptotic remains valid. To prove this statement it is sufficient to consider the particular case of eqn. 2.80. When the first term $f_0\xi^{-1}$ is singled out in the expansion of eqn. 2.81 poorly converging at infinity the problem reduces to that of proving that at $\mu=0$ the asymptotic equality (eqn. 2.82) is uniform over α. But the difference between the right- and left-hand sides of the equality is the second integral in eqn. 2.83. It is easy to see that the given estimates of the integral behaviour as $\lambda\to\infty$ remain valid at $\mu=0$ uniformly over α and this results in eqn. 2.72 being applicable at $\mu=0$.

2.8 Applicability range for nonuniform asymptotics

The expressions for the uniform asymptotics of the integral $I(\lambda,\alpha)$ obtained in the preceding Sections make it possible to estimate the limits of application of the nonuniform asymptotic in the case when the critical points approach each other. Indeed, moving from the uniform asymptotic to the nonuniform one is equivalent to replacing the appropriate special function by its large argument asymptotic. Depending on the admissible error and on the type of critical points approaching (i.e. on the particular form of the function Φ) one can obtain the subsequent estimates for the argument Φ, i.e. for the difference between the values of the phase function at critical points, for which such procedure is possible.

These estimates for a stationary point close to the integration domain boundary were made in Section 2.1. In a similar way one can treat other types of critical point approaches. The results obtained in this way are given omitting appropriate mathematics. The relative error introduced when the contributions of two close critical points into $I(\lambda, \alpha)$ are evaluated, replacing the contributions by the leading terms of their nonuniform asymptotics, is shown. The error depends on the type of critical points approaching.

2.8.1 *Stationary point near integration domain boundary or pole*

The uniform asymptotic is expressed in terms of the Fresnel integral or the complex conjugate Fresnel integral. For $\lambda\Delta\varphi = \lambda\,(\varphi_1 - \varphi_2) \simeq 2\pi$ the error is of the order of 8%, then it falls off as $(\lambda\Delta\varphi)^{-1}$.

2.8.2 *Two close stationary points*

The uniform asymptotic is expressed in terms of the Airy function. For $\lambda\Delta\varphi = \pi$ the relative error is of the order of 3%, then it falls off as $(\lambda\Delta\varphi)^{-1}$.

2.8.3 *Close pole and integration domain boundary*

The uniform asymptotic is expressed in terms of the integral sine and cosine. For $\lambda\Delta\varphi = \pi$ the relative error is of the order of 5%, then it falls off as $(\lambda\Delta\varphi)^{-1}$.

2.8.4 *Close stationary and branch points of nonexponential factor, branching order is μ*

The uniform asymptotic is expressed in terms of the parabolic cylinder functions. For $\lambda\Delta\varphi = 2\pi$ the relative error is of the order of $2(|\mu|^2 + 1)$%, then it falls off as $(\lambda\Delta\varphi)^{-1}$.

2.8.5 *Close pole and branch point of nonexponential factor, branching order is μ*

The uniform asymptotic is expressed in terms of the incomplete gamma function. For $\lambda\Delta\varphi = 3\pi$ the relative error is of the order of $10(\mu + 1)$%, then it falls off as $(\lambda\Delta\varphi)^{-1}$.

2.8.6 *Stationary point tending to infinity in integral of eqn. 2.64 type*

The phase function is of the form $\varphi = \alpha x + a_0 + (a_1/x) + O(x^{-2})$ where $\alpha > 0$. There is a single critical point x_1 in the integration domain specified for small α by eqn. 2.68. The uniform asymptotic is expressed in terms of the Hankel function of the index μ, where $\mu + 1$ is of the order of the nonexponential term decreasing in eqn. 2.64 (see eqn. 2.66). The relative error is of the order of $(|\mu| + 1)/[8\lambda\sqrt{(\alpha a_1)}]$.

Chapter 3

Three close critical points in one-dimensional integral

In the case of isolated critical points the asymptotic of $I(\lambda)$ as $\lambda \to \infty$ (I is defined by eqn. 1.1) is a sum of contributions due to each of the critical points, i.e. it is expressed in terms of elementary functions. When there are two critical points, x_1 and x_2, the asymptotic of $I(\lambda)$ as $\lambda \to \infty$, uniform over a distance between these points, is expressed in terms of a special function, the form of which is determined by the type of critical point approach to each other and the argument of which is a power of the product $\lambda[\varphi(x_2) - \varphi(x_1)]$ and it tends to zero with a distance between these points.

When there are three critical points the distance between them is determined by two parameters, for example, by two differences, $\varphi(x_1) - \varphi(x_2)$ and $\varphi(x_1) - \varphi(x_3)$. Therefore, the uniform asymptotic of $I(\lambda)$ is expressed in terms of special functions of two independent variables. The form of the function is determined by the type of critical point approach. It is clear that the number of possible combinations of critical points in this case is essentially larger than for two critical points. In what follows only those simplest situations which frequently occur in oscillation theory and wave propagation are considered.

In all cases considered the uniform asymptotic of the integral $I(\lambda)$ follows the same scheme. First, the model integral is considered that is expressed in terms of the known special functions or is declared (after scaling its parameters) as a new special function.

When introducing a cutting function under the integral sign the exact equality is transformed into the asymptotic one. Since in the general case the initial integral $I(\lambda)$ differs from the model one, first by the phase function and, secondly, by the nonexponential term, hence the next step is to transform the phase function of the initial integral into that of the model integral by independent variable change. Then, if one of the critical points is a pole of the nonexponential factor, one can single out the pole as a common fraction and this immediately reduces $I(\lambda)$ to a sum of the model integral and the integral with two close critical points, considered in Chapter 2. If there is no pole, one should differentiate the model integral many times with respect to parameters of the problem. This procedure gives a set of nonexponential factors for which the suggested ansatz (analytic form of the uniform asymptotics) is valid and which is sufficiently wide to approximate any nonexponential function with given singularities (critical points). The last step when determining the uniform asymptotics is to use the asymptotic matching method to match the ansatz asymptotic to the nonuniform asymptotic of the initial integral $I(\lambda)$. This results in the equation system to determine the unknown arguments and coefficients in the uniform asymptotic. When solving the system both the leading and the subsequent terms of the uniform asymptotic are found.

3.1 Three close stationary points

Consider an integral

$$I(\lambda, a, b)=\int_{-\infty}^{\infty} \exp\,[i\lambda\varphi(x, a, b)]f(x)g(x)\,dx \tag{3.1}$$

where $g(x)$ is the cutting function, $f(x)$ and $\varphi(x, a, b)$ are the analytical functions in the region supp $[g(x)]$ and in this region $\varphi(x, a, b)$ has three close stationary points. The asymptotic of the integral $I(\lambda, a, b)$ uniform over a distance between these points is required.

In this case, the model integral has the form

$$\int_{-\infty}^{\infty} \exp\,[i\lambda\,(\alpha x+\beta x^2+x^4)]\,dx=\lambda^{-1/4}I(\lambda^{3/4}\alpha, \sqrt{\lambda}\beta) \tag{3.2}$$

where $I(p, q)$ is the Piercey integral (Section 6.7)

$$I(p, q)=\int_{-\infty}^{\infty} \exp\,[i(px+qx^2+x^4)]\,dx \tag{3.3}$$

Introducing the cutting function $g(x)$ under the integral sign this relation is transformed into the asymptotic equality

$$I(\lambda, \alpha, \beta)=\int_{-\infty}^{\infty} \exp\,[i\lambda\,(\alpha x+\beta x^2+x^4)]g(x)\,dx\simeq\lambda^{-1/4}I(\lambda^{3/4}\alpha, \sqrt{\lambda}\beta)$$

This formula is proved using the standard procedure. The difference between the left- and right-hand sides of the relation is

$$\int_{-\infty}^{\infty} \exp\,[i\lambda\,(\alpha x+\beta x^2+x^4)]\,[1-g(x)]\,dx$$

and if in the region supp $[1-g(x)]$ there are no critical points of the phase function, the integral falls off more rapidly than any power of λ^{-1}.

The next step is to introduce an arbitrary function $f(x)$ under the integral sign. The following relations are valid

$$\begin{aligned} I(\lambda, \alpha, \beta)&=\int_{-\infty}^{\infty} \exp\,[i\lambda\,(\alpha x+\beta x^2+x^4)]\,x^n g(x)\,dx \\ &\simeq\lambda^{-1/4}I(\lambda^{3/4}\alpha, \sqrt{\lambda}\beta)P_n(\alpha, \beta, \lambda^{-1}) \\ &\quad+(1/\sqrt{\lambda})I'_p(\lambda^{3/4}\alpha, \sqrt{\lambda}\beta)Q_n(\alpha, \beta, \lambda^{-1}) \\ &\quad+\lambda^{-3/4}I'_q(\lambda^{3/4}\alpha, \sqrt{\lambda}\beta)R_n(\alpha, \beta, \lambda^{-1}) \end{aligned} \tag{3.4}$$

where I'_p and I'_q are the derivatives of the Piercey integral and P_n, Q_n and R_n are polynomials of degree n in α, β, λ^{-1}. Eqn. 3.4 is easily proved by induction when applying the operator $(i/\lambda)d/d\alpha$ to both its sides and using the relation between the first and second derivatives of the Piercey integral

$$I''_{pp} = iI'_q$$

$$\begin{aligned} I''_{p,q} &+ ipI'_q/4 + iqI'_p/2 \\ &= -\int_{-\infty}^{\infty} [x^3 + px^2/4 + qx/2] \exp [i(px + qx^2 + x^4)]\, dx \\ &= (i/4) \int_{-\infty}^{\infty} d \exp [i(px + qx^2 + x^4)] = 0 \end{aligned}$$

It follows from eqn. 3.4 that for any function $f(x)$ analytical in the region supp $[g(x)]$ the integral

$$\int_{-\infty}^{\infty} \exp [i\lambda (\alpha x + \beta x^2 + x^4)] f(x) g(x)\, dx$$

can be asymptotically represented as

$$\begin{aligned} \int_{-\infty}^{\infty} &\exp [i\lambda (\alpha x + \beta x^2 + x^4)] f(x) g(x)\, dx \\ &\simeq \lambda^{-1/4} I(\lambda^{3/4}\alpha, \sqrt{\lambda}\beta) P(\alpha, \beta, \lambda^{-1}) \\ &\quad + (1/\sqrt{\lambda}) I'_p(\lambda^{3/4}\alpha, \sqrt{\lambda}\beta) Q(\alpha, \beta, \lambda^{-1}) \\ &\quad + \lambda^{-3/4} I'(\lambda^{3/4}\alpha, \sqrt{\lambda}\beta) R(\alpha, \beta, \lambda^{-1}) \end{aligned} \tag{3.5}$$

where P, Q, R are the asymptotic power series in λ^{-1} the coefficients of which are regular functions of α, β. Indeed, for any $n > 0$ the function $f(x)$ can be approximated by such a polynomial $F_n(x)$ that the integral of their difference approaches to zero more rapidly than λ^{-n}. Eqn. 3.5 for the polynomial $F_n(x)$ is an obvious consequence of eqn. 3.4.

To reduce the general integral of eqn. 3.1 to eqn. 3.5 one needs to perform the analytical and reversible variable change $x = x(\xi)$ which transforms the phase function $\varphi(x, \alpha, \beta)$ into the phase function of eqn. 3.5, i.e. into the polynomial of the fourth degree.

If $\varphi(x, a, b)$ has three stationary points x_1, x_2, x_3, i.e. the points where $\varphi'(x, a, b) = 0$, there exists a point in the interval (x_1, x_3) where

$$\frac{d^3\varphi(x, a, b)}{dx^3} = 0$$

Then one can shift the origin to this point and write the phase function as the Taylor series

$$\varphi(x, a, b) = \varphi_0 + ax + bx^2 + a_4x^4 + a_5x^5 + \cdots \tag{3.6}$$

where a, b are small and a_4 is not equal to zero.

Set $a_4 > 0$ (the case of $a_4 < 0$ is reduced to that of $a_4 > 0$ by complex conjugation). Under these conditions the sought-for variable change $x = x(\xi)$ exists that depends on a and b as parameters. One can write

$$\varphi[x(\xi)] = \psi_0(a, b) + \xi\alpha(a, b) + \xi^2\beta(a, b) + \xi^4$$

where $\psi_0(a, b)$, $\alpha(a, b)$, $\beta(a, b)$ are regular functions of a, b. One can find the proof of this statement in any monograph on the catastrophe theory, for example, References 28 and 29.

Changing in eqn. 3.1 to the independent variable ξ and using eqn. 3.5 one obtains the uniform asymptotic

$$\begin{aligned} I(\lambda, a, b) = \exp\,[i\lambda\varphi_0(a, b)]\,[&\lambda^{-1/4} I(\lambda^{3/4}\alpha, \sqrt{\lambda}\beta) P(\alpha, \beta, \lambda^{-1}) \\ &+ (1/\sqrt{\lambda}) I'_p(\lambda^{3/4}\alpha, \sqrt{\lambda}\beta) Q(\alpha, \beta, \lambda^{-1}) \\ &+ \lambda^{-3/4} I'_q(\lambda^{3/4}\alpha, \sqrt{\lambda}\beta) R(\alpha, \beta, \lambda^{-1}) \end{aligned} \tag{3.7}$$

where the functions $\psi_0(a, b)$, $\alpha(a, b)$, $\beta(a, b)$ and coefficients p, q, r of the power series $p = \sum \lambda^{-n} p_n$, $Q = \sum \lambda^{-n} q_n$ and $R = \sum \lambda^{-n} r_n$ are the regular functions of a, b. In principle, one can find $\psi_0(a, b)$, $\alpha(a, b)$, $\beta(a, b)$ and the leading terms of the series P, Q and R by the asymptotic matching method equating the nonuniform asymptotic obtained from eqn. 3.7 when replacing the functions I, I'_p and I'_q by their asymptotics as $\lambda \to \infty$ to the nonuniform asymptotic of the initial integral.

However, to determine the asymptotic $I(\lambda^{3/4}\alpha, \sqrt{\lambda}\beta)$ it is reduced, as seen from eqn. 3.2, to the solution of the cubic equation for stationary points of the phase function. The corresponding equations are cumbersome and are omitted here. Instead, equations are given of the leading terms when expanding the functions ψ_0, α, β and series P, Q and R in power series in a, b, i.e. the local asymptotic of $I(\lambda, \alpha, \beta)$.

Assume that $\varphi(x, a, b)$ is expanded into Taylor's series eqn. 3.6 where $a_4 > 0$ and the function $f(x)$ is expanded into a series

$$f(x) = f_0 + xf_1 + x^2 f_2 + x^3 f_3 + O(x^4) \tag{3.8}$$

The variable change $x = x(\xi)$ transforming $\varphi(x)$ into polynomial of the fourth degree has the form

$$\begin{aligned} x = c + a_4^{-1/4}\xi - [a_5/(4a_4^{3/2})]\,\xi^2 + (d\xi^3/a_4^{3/4}) + (e\xi^4/4) \\ + O(a^2 + b^2 + |a|\xi + |b|\xi + \xi^5) \end{aligned}$$

where

$$c = a_5 b/(8a_4) - ad/4, \qquad d = 7a_5^2/(32a_4^2) - a_6/(4a_4)$$

$$e = (-a_5^3/a_4^4) + 2(a_5 a_6/a_4^3) - a_7/\alpha_4^2$$

For ψ_0, α, β and the amplitudes P, Q, R,

$$\left.\begin{aligned} &\psi_0 = \varphi_0 + O(a^2 + b^2) \\ &\alpha = a a_4^{-1/4} + O(a^2 + b^2) \\ &\beta = (b/\sqrt{a_4}) - (a a_5/(4a_4^{3/2}) + O(a^2 + b^2) \\ &P = a_4^{-1/4}[f_0 + cf_1 - (a\psi/4) + O(\lambda^{-1} + a^2 + b^2)] \\ &Q = (i/\sqrt{a_4})[-a_5 f_0/(2a_4) - f_1 - 2cf_2 + a_5 c f_1/(2a_4) \\ &\qquad + (\beta\psi/2) + O(\lambda^{-1} + a^2 + b^2)] \\ &R = -i a_4^{-3/4}\{3df_0 - 3a_5 f_1/(4a_4) + f_2 + O[\lambda^{-1} + \sqrt{(a^2 + b^2)}]\} \end{aligned}\right\} \tag{3.9}$$

where

$$\psi = [(a_5^2/a_4^3) - a_6/a_4^2]f_1 - (a_5 f_2/a_4^2) + (f_3/a_4) + ef_4$$

Thus, the following theorem is obtained.

Theorem 3.1: Let the integral $I(\lambda, a, b)$ be defined by eqn. 3.1 where the phase function has three close stationary points in a vicinity of zero, its fourth derivative is bounded in magnitude from below and is expanded in a Taylor's series eqn. 3.6. Let g be a cutting function and f be an analytical function in a vicinity of supp *(g) expanded in a Taylor's series eqn. 3.8. Then for $a_4 > 0$ the asymptotic of $I(\lambda, a, b)$ $(\lambda \to \infty)$ uniform over a, b is expressed in terms of the Piercey integral and its derivatives by eqn. 3.7. The leading terms of the power expansions in a, b defining this asymptotic (the local asymptotic of I) are of form of eqn. 3.9. The case $a_4 < 0$ is reduced to that of $a_4 > 0$ by complex conjugation.*

3.2 Two stationary points near integration domain boundary

Consider an integral

$$I(\lambda, a, b) = \int_b^\infty f(x)g(x) \exp [i\lambda\varphi(x, a)]\, dx \tag{3.10}$$

where for $a > 0$ the phase function $\varphi(x, a)$ has two stationary points, x_1 and x_2, close to the endpoint of the integration interval, $x = b$, the points approaching each other as $a \to 0$, $g(x)$ is a cutting function. The functions $f(x)$ and $\varphi(x)$ are analytical and $\varphi'''_{xxx}(x, a) > 0$ in the region supp $[g(x)]$ (the case of $\varphi'''_{xxx}(x) < 0$ is reduced to that of $\varphi'''_{xxx}(x) > 0$ by complex conjugation).

Theorem 3.2 for the uniform asymptotic of $I(\lambda, a, b)$ is now obtained. In this case the model integral is

$$\begin{aligned} I(\lambda, \alpha, \beta) &= \int_\beta^\infty \exp \{i\lambda[x^3/3 - \alpha x]\}\, dx \\ &= 2\pi\lambda^{-1/3} V(-\lambda^{2/3}\alpha, \lambda^{1/3}\beta) \end{aligned} \tag{3.11}$$

where $V(p, q)$ is the incomplete Airy function (Section 6.8)

$$V(p, q) = (1/2\pi) \int_q^\infty \exp \{i[t^3/3 + tp]\}\, dt$$

This function satisfies the equation

$$\begin{aligned} V''_{pp} - pV &= (i/2\pi) \int_q^\infty d \exp \{i[t^3/3 + tp]\} \\ &= (-i/2\pi) \exp \{i[q^3/3 + qp]\} \\ &= iV'_q \end{aligned} \tag{3.12}$$

When introducing the cutting function $g(x)$ under the integral sign the exact equality (eqn. 3.11) is transformed into the asymptotic one

$$I_0 = \int_\beta^\infty \exp\{i\lambda[(x^3/3) - x\alpha]\}g(x)\,dx \simeq 2\pi\lambda^{-1/3}V(-\lambda^{2/3}\alpha, \lambda^{1/3}\beta)$$

Applying the operator $[(i/\lambda)\partial/\partial\alpha]^n$ to this equality gives

$$I_n = \int_\beta^\infty \exp\{i\lambda[(x^3/3) + xa]\}x^n g(x)\,dx = (i/\lambda)^n \frac{\partial^n}{\partial\alpha^n} I_0$$

$$\simeq 2\pi i^n \lambda^{-n-1/3} \frac{\partial^n}{\partial\alpha^n} V(-\lambda^{2/3}\alpha, \lambda^{1/3}\beta)$$

If $f(x)$ is a polynomial, using this relation and eqn. 3.12 obtains

$$\int_\beta^\infty g(x)f(x)\exp\{i\lambda[(x^3/3) - x\alpha]\}\,dx$$
$$\simeq 2\pi\lambda^{-1/3}\{AV(-\lambda^{2/3}\alpha, \lambda^{1/3}\beta) + iB\lambda^{-1/3}V_\beta'(-\lambda^{2/3}\alpha, \lambda^{1/3}\beta)$$
$$+ iC\lambda^{-2/3}\exp[i\lambda(\beta^3/3) - \alpha\beta]\} \qquad (3.13)$$

where A, B, C are polynomials in λ^{-1} the coefficients of which are regular functions of α, β. Since for any n each analytical function $f(x)$ can be approximated by polynomials in such a way that the integral of their difference is estimated by λ^{-n} uniformly over α, β, then eqn. 3.13 appears to be valid for any analytical function $f(x)$, but the coefficients A, B, C are an infinite asymptotic power series in λ^{-1}.

The general case of eqn. 3.10 is reduced to eqn. 3.13 by an analytical and reversible variable change $x = x(\tau, a)$ depending on a as a parameter and transforming the phase function $\varphi(x, a)$ into a cube polynomial (eqn. 2.35)

$$\varphi[x(\tau), a] = \varphi_0(a) + \tau^3/3 - \xi(a)\tau$$

As noted, the case of $\varphi''' < 0$ is reduced to that of $\varphi''' > 0$ by complex conjugation. This results in the following theorem [11]:

Theorem 3.2: Let $I(\lambda, a, b)$ be the integral eqn. 3.10 in which for $a > 0$ and $a \to 0$ the phase function $\varphi(x, a)$ has two merging stationary points x_1 and x_2, the function φ''' in a vicinity of these points being bounded in magnitude from below. Let $g(x)$ be a cutting function and $f(x)$ a function analytical in the region supp *(g). Then as $\lambda \to \infty$ the asymptotic of I uniform over a distance between the points x_1 and x_2 and the endpoint b of the integration interval is expressed in terms of the incomplete Airy function and its derivative as*

$$I(\lambda, a, b) = (2\pi/\lambda^{1/3})\exp[i\varphi_0(a)]\{A\Phi(-\Lambda^{2/3}\xi, \lambda^{1/3}q)$$
$$-(iB/\lambda^{1/3})\,\text{sign}\,\varphi'''\Phi_p'(-\lambda^{2/3}\xi, \lambda^{1/3}q)$$
$$+2\pi i(\lambda^{-2/3})\,\text{sign}\,\varphi''' C\exp[i\lambda\varphi(b, a)]\} \qquad (3.14)$$

Here, $\Phi(p, q)$ for $\varphi''' > 0$ coincides with the incomplete Airy function, while for $\varphi''' < 0$ with its complex conjugate

$$\Phi(p, q) = \begin{cases} V(p, q) & \text{for } \varphi''' > 0 \\ V^*(p, q) = \mathrm{Ai}\,(p) - V(p, -q) & \text{for } \varphi''' < 0 \end{cases}$$

Functions $\varphi_0(a)$, $\xi = \xi(a)$, $q = q(a, b)$ and the coefficients A_n, B_n, C_n of the asymptotic series $A = \Sigma(i/\lambda)^n A_n$, $B = \Sigma(i/\lambda)^n B_n$, $C = \Sigma(i/\lambda)^n C_n$ are regular functions of the parameters a, b. The functions $\varphi_0(a)$, $\xi(a)$, A_n, B_n, C_n are the same as for theorem 2.3, i.e. determined by eqn. 2.36. The function $q(a, b) = \tau(b)$ where $\tau(x)$ is an analytical and reversible variable change transforming the phase function into a cubic polynomial (eqn. 2.35). For $a < 0$ the function $q(a, b)$ is the only root of the equation

$$\varphi(b, a) = \varphi_0(a) + (q^3/3) - \xi(a)q \tag{3.15}$$

while for $a > 0$ it can be defined as an analytical continuation from the region $a < 0$.

Finally, the leading term C_0 of the asymptotic series C has the form

$$C_0 = [f(b)/\varphi_b'(a, b)] - (A_0 - B_0 q)/(q^2 - \xi) \tag{3.16}$$

Write local expansions for the functions ψ_0, $\xi = \xi(a)$, $q = q(a, b)$ and for the coefficients A_0, B_0, C_0, i.e. the first terms of their power expansions in a, b. Assume that functions $f(x)$ and $\varphi(x, a)$ are expanded into the series

$$f(x) = f_0 + xf_1 + x^2 f_2 + \cdots \tag{3.17a}$$

$$\varphi(x, a) = a_0 - ax + a_3 x^3/3 + a_4 x^4 + \cdots \tag{3.17b}$$

where $a_3 > 0$ (the case of $a_3 < 0$ is reduced to that of $a_3 > 0$ by the variable change $x \to -x$). Then

$$\varphi_0 = a_0 - a^2 a_4 a_3^{-2} + a^3\left(\tfrac{64}{3} a_4^3 a_3^{-5} - 10 a_5 a_4 a_3^{-4} + a_6 a_3^{-3}\right) + O(a^4) \tag{3.18a}$$

$$\xi = a a_3^{-1/3} + a^2\left(4 a_4^2 a_3^{-10/3} - a_5 a_3^{-7/3}\right) + O(a^3) \tag{3.18b}$$

$$\begin{aligned} \tau(x, a) = {} & a a_4 a_3^{-5/3} + a^2\left[\left(a_6 a_3^{-8/3} + \tfrac{53}{3} a_4^3 a_3^{-14/3}\right.\right. \\ & \left.\left. - 9 a_4 a_5 a_3^{-11/3}\right] + O(a^3)\right) \\ & + x\left[a_3^{1/3} - a\left(3 a_4^2 a_3^{-8/3} - a_5 a_3^{-5/3}\right) + O(a^2)\right] \\ & + x^2\left[a_4 a_3^{-2/3} - a\left(6 a_4 a_5 a_3^{-8/3} - a_6 a_3^{-5/3}\right.\right. \\ & \left.\left. - \tfrac{26}{3} a_4^3 a_3^{-11/3}\right) + O(a^2)\right] \\ & + x^3\left[a_5 a_3^{-2/3} - a_4^2 a_3^{-5/3} + O(a)\right] \\ & + x^4\left[a_6 a_3^{-2/3} - 2 a_4 a_5 a_3^{-5/3} + \tfrac{5}{3} a_4^3 a_3^{-8/3} + O(a)\right] + O(x^5) \end{aligned} \tag{3.18c}$$

$$\begin{aligned} A_0 = {} & f_0 a_3^{-1/3} + a\left[f_0\left(14 a_4^2 a_3^{-10/3} - 4 a_5 a_3^{-7/3}\right)\right. \\ & \left. - 4 f_1 a_4 a_3^{-7/3} + f_2 a_3^{-4/3}\right] + O(a^2) \end{aligned} \tag{3.18d}$$

$$\begin{aligned} B_0 = {} & f_0\left[2 a_4 a_3^{-5/3} + a\left(-52 a_4 a_5 a_3^{-11/3} + 100 a_4^3 a_3^{-14/3} + 6 a_6 a_3^{-8/3}\right)\right] \\ & + f_1\left[-a_3^{-2/3} + a\left(-26 a_4^2 a_3^{-11/3} + 6 a_5 a_3^{-8/3}\right)\right] \\ & + 6 f_2 a a_4 a_3^{-8/3} - f_3 a a_3^{-5/3} + O(a^2) \end{aligned} \tag{3.18e}$$

$$C_0 = f_0(9a_4^2a_3^{-3} - 3a_5a_3^{-2}) - 3f_1a_4a_3^{-2} + f_2a_3^{-1}$$
$$+ b[f_0(-\tfrac{140}{3}a_4^3a_3^{-4} + 28a_4a_5a_3^{-3} - 4a_6a_3^2)$$
$$+ f_1(14a_4^2a_3^{-3} - 4a_4a_3^{-2}) - 6f_2a_4a_3^{-2} + f_3a_3^{-1}]$$
$$+ O(a) + O(b^2) \qquad (3.18f)$$

3.3 Two stationary points near pole

Consider an integral

$$I(\lambda, a, b) = \int_{-\infty}^{\infty} \exp[i\lambda\varphi(x, a)] \frac{f(x)g(x)}{x - b + iO} dx \qquad (3.19)$$

where for $a > 0$ the phase function $\varphi(x, a)$ has two stationary points merging as $a \to 0$ and close to the pole $x = b$, $g(x)$ is a cutting function, $f(x)$ and $\varphi(x, a)$ are analytical and $|\varphi'''_{xxx}(x, a)| > C > 0$ uniformly over a in the region supp (g) and the pole $x = b$ is passed around from above.

The model integral is a relatively new special function, the Airy–Fresnel integral, $Af(\alpha, \beta)$ (Section 6.9)

$$I(\lambda, \alpha, \beta) = \int_{-\infty + i0}^{\infty + i0} \exp\left[i\lambda\left(\frac{\tau^3}{3} - \tau\alpha\right)\right] \frac{d\tau}{\tau - \beta}$$
$$= -2\pi i Af(-\lambda^{2/3}\alpha, \lambda^{1/3}\beta)$$

where

$$Af(\alpha, \beta) = (i/2\pi) \int_{-\infty}^{\infty} \exp\left[i\left(\frac{\tau^3}{3} + \alpha\tau\right)\right] \frac{d\tau}{\tau - \beta + iO}$$

Assume that $\varphi'''(x) > 0$. The case of $\varphi'''(x) < 0$ is reduced to that of $\varphi'''(x) > 0$ by the variable change $x \to -x$ or by complex conjugation. One should take into account that after each of these procedures the pole $x = b$ would be passed round in the lower halfplane.

To reduce the initial integral $I(\lambda, a, b)$ to the model one perform the variable change $x = x(\tau)$ transforming the function $\varphi(x, a)$ into a cubic polynomial (as in Section 2.3)

$$\varphi(x, a) = \varphi_0(a) + \tau^3/3 - \xi(a)\tau$$

Then $I(\lambda, a, b)$ is written as

$$I(\lambda, a, b) = \exp[i\lambda\varphi_0(a)] \int_{-\infty}^{\infty} \exp\left[i\lambda\left(\frac{\tau^3}{3} - \tau\xi\right)\right] \frac{h(\tau)}{\tau - \beta + iO} d\tau$$

where $\beta = \tau(b)$ and $\tau(x, a)$ is a function inverse to $x(\tau, a)$. The function $h(\tau)$ is defined by the expression

$$h(\tau) = \frac{\tau - \beta}{x(\tau) - b} f[x(\tau)]g[x(\tau)] \frac{dx(\tau)}{d\tau}$$

Since

$$\lim_{\tau\to\beta}\frac{x(\tau)-b}{\tau-\beta}=\lim_{\tau\to\beta}\frac{x(\tau)-x(\beta)}{\tau-\beta}=\frac{dx(\beta)}{d\tau}$$

the function $h(\tau)$ can be represented as $h(\tau)=h(\beta)+h_1(\tau)(\tau-\beta)$, where $h(\beta)=f[x(\beta)]=f(b)$. Therefore,

$$I(\lambda, a, b)=\exp[i\lambda\psi_0]f(b)\int_{-\infty}^{\infty}\exp\left[i\lambda\left(\frac{\tau^3}{3}-\tau\xi\right)\right]\frac{d\tau}{\tau-\beta+i0}$$
$$+\exp[i\lambda\psi_0]\int_{-\infty}^{\infty}\exp\left\{i\lambda\left[\frac{\tau^3}{3}-\tau\xi\right]\right\}h_1(\tau)\,d\tau$$

The first of these integrals is expressed in terms of the Airy–Fresnel integral, the second, according to the results of Section 2.3, is expressed in terms of the Airy function and its derivative. As a result, the following theorem is obtained.

Theorem 3.3: The uniform asymptotic of eqn. 3.19 is of the form

$$\begin{aligned}I(\lambda, a, b)=\exp(i\lambda\varphi_0)\{&-2\pi i f(b)Af(-\lambda^{2/3}\xi, \lambda^{1/3}\beta)\\&+\pi[A\lambda^{-1/3}Ai(-\lambda^{2/3})-Bi\lambda^{-2/3}Ai'(-\lambda^{2/3}\xi)]\}\end{aligned}\tag{3.20}$$

where $Af(p, q)$ is the Airy–Fresnel integral, $Ai(x)$ is the Airy function, x_1, x_2 are the stationary points of $\varphi(x, a)$ φ_0 and ξ are defined as

$$\varphi_0=[\varphi(x_1)+\varphi(x_2)]/2$$
$$\xi=\{3[\varphi(x_1, a)-\varphi(x_2, a)]/4\}^{2/3}$$

$\beta=\tau(b)$ where $\tau(x)$ is the variable change transforming the phase function into the cube polynomial eqn. 2.35, A and B are the asymptotic power series in λ^{-1} with the coefficients regular for small a, b.

The leading terms of these series are

$$\begin{aligned}A_0=&\left\{\frac{f(x_2)}{(x_2-b)}\sqrt{[2/|\varphi''(x_2)|]}+\frac{f(x_1)}{(x_1-b)}\sqrt{[2/|\varphi''(x_1)|]}\right\}\xi^{1/4}\\&-\frac{2\beta f(b)}{\xi-\beta^2}\\B_0=&\left\{\frac{f(x_2)}{(x_2-b)}\sqrt{[2/|\varphi''(x_2)|]}-\frac{f(x_1)}{(x_1-b)}\sqrt{[2/|\varphi''(x_1)|]}\right\}\xi^{-1/4}\\&-\frac{2f(b)}{\xi-\beta^2}\end{aligned}\tag{3.21a}$$

If the phase function (x, a) and the nonexponential factor $f(x)$ are expanded into the power series eqn. 3.17, then φ_0 and ξ have the local expansions of eqn. 3.18, while the local expansions of A_0 and B_0 are of the form

$$\begin{aligned}A_0 = a^{-1/3}\{ &f_0[-2a_4a_3^{-1} + a(60a_5a_4a_3^{-3} - 8a_6a_3^{-2} - 280a_4^3a_3^{-4}/3)\\ &+b(4a_4^2a_3^{-2} - 2a_5a_3^{-1}) + b^2(8a_4a_5a_3^{-2} - 2a_6a_3^{-1} - 28a_4^3a_3^{-3}/3)]\\ &+f_1[2 + a(28a_4^2a_3^{-3} - 8a_5a_3^{-2}) - 2ba_4a_3^{-1} + b^2(4a_4^2a_3^{-2} - 2a_5a_3^{-1})]\\ &+f_2(2b - 8aa_4a_3^{-2} - 2b^2a_4a_3^{-1}) + f_3(2b^2 + 2aa_3^{-1}) + O(a^2) + O(b^3)\}\\ B_0 = a^{-2/3}\{ &f_0[10a_4^2a_3^{-2} - 4a_5a_3^{-1} + b(20a_4a_5a_3^{-2} - 4a_6a_3^{-1} - 80a_4^3a_3^{-3}/3)]\\ &+f_1[-4a_4a_3^{-1} + b(10a_4^2a_3^{-2} - 4a_5a_3^{-1})]\\ &+f_2(2 - 4ba_4a_3^{-1}) + 2bf_3 + O(a) + O(b^2)\}\end{aligned} \tag{3.21b}$$

3.4 Pole and stationary point near integration domain boundary

Consider an integral

$$I(\lambda, a, b) = \int_a^\infty \exp[i\lambda\varphi(x)] \frac{f(x)g(x)}{(x-b)}\,dx \tag{3.22}$$

where $g(x)$ is the cutting function, the phase function $\varphi(x)$ has a single stationary point $x = x_0$, $|\varphi''(x_0)| \neq 0$. The functions $\varphi(x)$ and $f(x)$ are analytical in the region supp $[g(x)]$. For real $b > a$ the integral of eqn. 3.22 is understood in the sense of the principal Cauchy value, which simplifies the equations. In the case under consideration the model integral I_1 is also understood in a sense of the principal Cauchy value.

$$\begin{aligned}I_1(\sqrt{\lambda}\alpha, \sqrt{\lambda}\beta) &= \int_\alpha^\infty \exp(i\lambda\tau^2)\frac{d\tau}{\tau-\beta}\\ &= \int_{\sqrt{\lambda}\alpha}^\infty \exp(i\xi^2)\frac{d\xi}{\xi - \sqrt{\lambda}\beta}\end{aligned} \tag{3.23}$$

It is expressed in terms of the generalised Fresnel integral, integral sine and cosine by eqn. 3.27 and in terms of a new special function, the Ff-integral, by eqn. 3.28.

In this Section the following theorem is proved.

Theorem 3.4: For $\varphi''(x_0) > 0$ the uniform asymptotic of eqn. 3.22 has the form

$$\begin{aligned}I(\lambda, a, b) = &\exp(i\lambda\varphi_0)f(b)I_1(\sqrt{\lambda}\alpha, \sqrt{\lambda}\beta)\\ &+A\exp(i\lambda\varphi_0)\sqrt{(\pi i/\lambda)}F(-\sqrt{\lambda}\alpha)\\ &+(i/\lambda)B\exp[i\lambda\varphi(a)]\end{aligned} \tag{3.24}$$

where

$$\begin{aligned}\alpha &= \text{sign}(a - x_0)\sqrt{[\varphi(a) - \varphi(x_0)]}\\ \beta &= \text{sign}(b - x_0)\sqrt{[\varphi(b) - \varphi(x_0)]}\\ \varphi_0 &= \varphi(x_0)\end{aligned} \tag{3.25}$$

$F(\xi)$ is the Fresnel integral, A and B are asymptotic series in powers of λ^{-1} with the coefficients regular as $\alpha \to \beta$, $\alpha \to x$, $\beta \to x$. The leading terms of these series are defined by the expressions

$$A_0 = \frac{f(x_0)\sqrt{(2/\varphi_0'')}}{(x_0 - b)} + \frac{f(b)}{\beta}$$

$$= \frac{1}{(x_0 - b)} \left\{ f(x_0)\sqrt{(2/\varphi_0'')} - f(b)\sqrt{\left[\frac{(b - x_0)^2}{\varphi(b) - \varphi(x_0)}\right]} \right\}$$

$$B_0 = \frac{f(a)}{(a - b)\varphi'(a)} - \frac{f(b)}{2\alpha(\alpha - \beta)} - \frac{A_0}{2\alpha}$$

$$= \frac{f(a)}{(a - b)\varphi'(a)} - \frac{f(b)}{2\beta(\alpha - \beta)} - \frac{f(x_0)}{\alpha(x_0 - b)\sqrt{(2\varphi_0'')}} \tag{3.26}$$

The case $\varphi''(x_0) < 0$ is reduced to that considered by complex conjugation of all the quantities in eqn. 3.22.

To prove theorem 3.4 change the variable in eqn. 3.22, $\xi = \xi(x) = \operatorname{sign}(x - x_0)\sqrt{[\varphi(x) - \varphi(x_0)]}$. As a result,

$$I(\lambda, a, b) = \exp(i\lambda\varphi_0) \int_\alpha^\infty \exp(i\lambda\xi^2) \frac{h(\xi)\, d\xi}{(\xi - \beta + i0)}$$

$$= \exp(i\lambda\varphi_0) \left[h(\beta) \int_\alpha^\infty \exp(i\lambda\xi^2) \frac{d\xi}{(\xi - \beta + i0)} + \int_\alpha^\infty \frac{\exp(i\lambda\xi^2)[h(\xi) - h(\beta)]\, d\xi}{\xi - \beta} \right]$$

where

$$h(\xi) = f[x(\xi)]g[x(\xi)] \frac{\xi - \beta}{x(\xi) - b} \frac{dx(\xi)}{d\xi}$$

Since $h(\beta) = f(b)$, the first integral on the right-hand side of this expression coincides with $\exp(i\lambda\varphi_0)f(b)I_1(\lambda, a, b)$, i.e. with the first summand in eqn. 3.24. A nonexponential factor in the second integral is a regular function of ξ as $\xi \to \beta$. Therefore the uniform asymptotic of the integral written according to theorem 2.2 is expressed in terms of the Fresnel integral and gives the second and third summands in eqn. 3.24. Thus, theorem 3.4 is proved.

Express now the integral $I_1(\sqrt{\lambda}\alpha, \sqrt{\lambda}\beta)$ in terms of the known special functions.

$$I_1(\sqrt{\lambda}\alpha, \sqrt{\lambda}\beta) = \int_\alpha^\infty \frac{\exp(i\lambda\tau^2)}{\tau - \beta}\, d\tau$$

$$= \int_\alpha^\infty \frac{(\tau + \beta)\exp(i\lambda\tau^2)}{\tau^2 - \beta^2}\, d\tau$$

$$= \beta \int_{\alpha}^{\infty} \frac{\exp(i\lambda\tau^2)}{\tau^2 - \beta^2} d\tau + \int_{\alpha}^{\infty} \frac{\tau \exp(i\lambda\tau^2)}{\tau^2 - \beta^2} d\tau$$

$$= I_2 + I_3$$

The first of the integrals is expressed in terms of the generalised Fresnel integral $G(x, y)$

$$I_2 = (2\pi/i) \exp(i\lambda\beta^2) G(\sqrt{\lambda}\alpha, i\sqrt{\lambda}\beta)$$

where

$$G(x, y) = (y/2\pi) \int_{x}^{\infty} \frac{\exp[i(t^2 + y^2)]}{t^2 + y^2} dt$$

and for imaginary y and $x < y$ the integral in a vicinity of the poles $t = \pm|y|$ is understood in a sense of the principal Cauchy value. Taking into account that the integrand is an odd function transform the second integral I_3

$$I_3 = \int_{\alpha}^{\infty} \frac{\tau \exp(i\lambda\tau^2)}{\tau^2 - \beta^2} d\tau = (1/2) \int_{\alpha^2}^{\infty} \frac{\exp(i\lambda\xi)}{\xi - \beta^2} d\xi$$

$$= (1/2) \exp(i\lambda\beta^2) \int_{\lambda(\alpha^2 - \beta^2)}^{\infty} \frac{\cos t + i \sin t}{t} dt$$

$$= (1/2) \exp(i\lambda\beta^2) \left[\int_{\lambda(\alpha^2 - \beta^2)}^{\infty} \frac{\cos t}{t} dt + i \int_{\lambda(\alpha^2 - \beta^2)}^{\infty} \frac{\sin t}{t} dt \right]$$

$$= (1/2) \exp(i\lambda\beta^2) \left[-\mathrm{Ci}(\lambda|\alpha^2 - \beta^2|) + i \left\{ \frac{\pi}{2} - \mathrm{Si}[\lambda(\alpha^2 - \beta^2)] \right\} \right]$$

where $\mathrm{Ci}(z)$ and $\mathrm{Si}(z)$ are the integral sine and cosine (see Section 6.5):

$$\mathrm{Ci}(z) = -\int_{z}^{\infty} \frac{\cos t}{t} dt, \qquad \mathrm{Si}(z) = \int_{0}^{z} \frac{\sin t}{t} dt$$

The final expression of the integral I_1 is of the form

$$\begin{aligned} I_1(\sqrt{\lambda}\alpha, \sqrt{\lambda}\beta) = &-2\pi i \exp(i\lambda\beta^2) G(\sqrt{\lambda}\alpha, i\sqrt{\lambda}\beta) \\ &+ (1/2) \exp(i\lambda\beta^2)[-\mathrm{Ci}(\lambda|\alpha^2 - \beta^2|) \\ &+ i\{(\pi/2) - \mathrm{Si}[\lambda(\alpha^2 - \beta^2)]\}] \end{aligned} \tag{3.27}$$

The integral I_1 can be expressed in terms of a new special function, the $Ff(p, q)$ (Ff-integral)

$$\begin{aligned} I_1(\sqrt{\lambda}\alpha, \sqrt{\lambda}\beta) = &\, 2\pi Ff(\sqrt{\lambda}\alpha, \sqrt{\lambda}\beta) \\ &- i\pi\chi(-\alpha) \operatorname{sign}(\beta) \exp(i\lambda\beta^2) \end{aligned} \tag{3.28}$$

where

$$Ff(p, q) = \frac{1}{2\pi}\int_{-\infty}^{p} \frac{\exp(it^2)}{t-q+iO}\,dt$$

In what follows the function Ff will appear again, the uniform asymptotic of the two-fold integral with a stationary saddle point near the corner point of the integration domain boundary is expressed in terms of the double Fresnel integral.

Write local expansions of the coefficients A_0 and B_0 in theorem 3.4. Let $x_0 = 0$ and the functions $\varphi(x)$ and $f(x)$ be expanded into the series

$$\varphi(x) = \varphi_0 + \varphi_2 x^2 + \varphi_3 x^3 + \varphi_4 x^4 + \varphi_5 x^5 + \cdots$$
$$f(x) = f_0 + f_1 x + f_2 x^2 + f_3 x^3 + \cdots$$

Then

$$\begin{aligned} A_0 = {} & (1/\sqrt{\varphi_2})\{f_0[-\tfrac{1}{2}\varphi_3\varphi_2^{-1} + b(-\tfrac{1}{2}\varphi_4\varphi_2^{-1} + \tfrac{3}{8}\varphi_3^2\varphi_2^{-2}) \\ & + b^2(-\tfrac{1}{2}\varphi_5\varphi_2^{-1} + \tfrac{3}{4}\varphi_3\varphi_4\varphi_2^{-2} - \tfrac{5}{16}\varphi_3^3\varphi_2^{-3})] \\ & + f_1[1 - \tfrac{1}{2}b\,\varphi_3\varphi_2^{-1} + b^2(-\tfrac{1}{2}\varphi_4\varphi_2^{-1} + \tfrac{3}{8}\varphi_3^2\varphi_2^{-2})] \\ & + f_2 b(1 - \tfrac{1}{2}b\,\varphi_3\varphi_2^{-1}) + f_3 b^2 + O(b^3)\} \end{aligned} \tag{3.29a}$$

$$\begin{aligned} B_0 = {} & \frac{f}{2}[-\varphi_4\varphi_2^{-2} + \varphi_3^2\varphi_2^{-3} + a(-\tfrac{3}{2}\varphi_5\varphi_2^{-2} + \tfrac{15}{4}\varphi_3\varphi_4\varphi_2^{-3} - 2\varphi_3^3\varphi_4^{-4}) \\ & + b(-\varphi_5\varphi_2^{-2} + 2\varphi_3\varphi_4\varphi_2^{-3} - \varphi_3^3\varphi_2^{-4})] \\ & + \frac{f_1}{2}[-\varphi_3\varphi_2^{-2} + b(-\varphi_4\varphi_2^{-2} + \varphi_3^2\varphi_2^{-3}) \\ & + a(-\tfrac{3}{2}\varphi_4\varphi_2^{-2} + \tfrac{15}{8}\varphi_3^2\varphi_2^{-3})] \\ & + \frac{f_2}{2}\left[\varphi_2^{-1} - \left(\frac{3a}{2} + b\right)\varphi_3\varphi_2^{-2}\right] + \frac{f_3}{2}(a+b)\varphi_2^{-1} + O(a^2+b^2) \end{aligned} \tag{3.29b}$$

Chapter 4

Uniform asymptotics of two-fold integrals

This Chapter is devoted to asymptotics of two-fold integrals

$$I(\lambda)=\iint_{\Omega} \exp\,[i\lambda\varphi(x,y)]f(x,y)\,dx\,dy \tag{4.1}$$

Evidently, the number of qualitatively different situations is essentially greater than for one-dimensional integrals, the number of situations being concerned with the nature of stationary points for the functions $\varphi(x,y)$ and $f(x,y)$ as well as with the relative location of these points and the boundary Σ of the integration domain Ω. What follows is restricted to the simplest (and the most frequent) cases only.

4.1 Stationary phase method for two-fold integrals

Nonuniform asymptotics of the integral $I(\lambda)$ as $\lambda\to\infty$ for fixed relative location of critical points are considered.

4.1.1 Partition of unity for two-fold integrals

Assume that the phase function $\varphi(x,y)$ is analytical in the domain Ω, while the function $f(x)$ is infinitely differentiable and the boundary Σ of the integration domain Ω is a piecewise analytical curve, the phase function $\varphi(x,y)$ being not constant on any part of the boundary. If the length measured along the boundary Σ is denoted by s, the last-mentioned condition means that the function

$$\varphi'(s)=\frac{d\varphi[x(s),y(s)]}{ds}=\varphi'_x x'_s+\varphi'_y y'_s$$

does not vanish identically on any interval over s. Prove that under these conditions the asymptotic of the integral $I(\lambda)$ as $\lambda\to\infty$ is a sum of contributions due to critical points. These critical points are

—Stationary points of the phase function $\varphi(x,y)$
—Points on the boundary Σ where the derivative of the phase function $\varphi(s)$ along the boundary vanishes
—Corner points of the boundary Σ (points where the analyticity Σ is violated).

This statement is proved using partition of unity. Denoting the critical points of the integral $I(\lambda)$ by $A_1,\ldots,A_N$ and the cutting functions concentrated

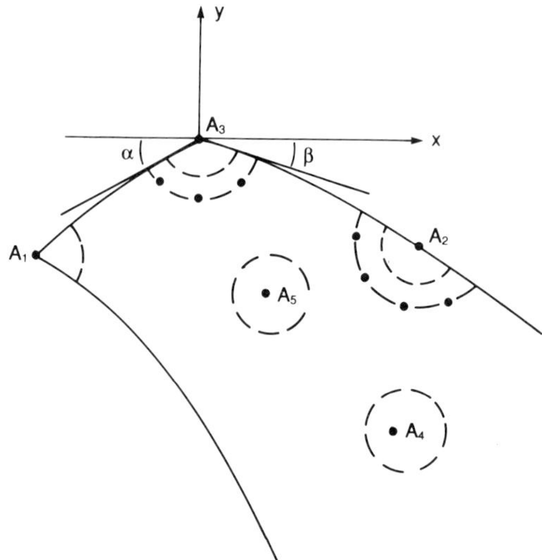

Figure 4.1 Critical points and decomposition of unity

near these points by $g_1(x, y), \ldots, g_N(x, y)$

$$I(\lambda) = \sum_{n=1}^{N} I_n + I_0 = \sum_{n=1}^{N} \iint_{\Omega} f g_n \exp(i\lambda\varphi)\, dx\, dy$$

$$+ \iint_{\Omega} fh \exp(i\lambda\varphi)\, dx\, dy$$

where the function $h = 1 - g_1, \ldots, g_N$ identically vanishes in a vicinity of each of these critical points. In Figure 4.1 points A_1, A_3 are corner points of the boundary, point A_2 is a point of the boundary where the longitudinal derivative of the phase function φ_3' vanishes and points A_4, A_5 are stationary points of $\varphi(x, y)$ inside Ω, vicinities of each of these critical points where h vanishes are marked by dotted lines. The asymptotics of the integrals I_n where $n = 1, 2, \ldots, N$ are expressed in terms of values of f, φ and their derivatives at A_n. When A_n lies on the boundary Σ the asymptotic is expressed in terms of the derivatives (at A_n) of the function determining the boundary as well. Appropriate equations follow.

Now prove that the integral I_0 falls off more rapidly than any power of λ^{-1}. To obtain such estimations in the one-dimensional case integration was by parts; here, the counterpart of such integration is the Gauss's formula

$$\iint_{\Omega} \operatorname{div} \bar{V}\, dx\, dy = \iint_{\Omega} \left[\frac{\partial V_x}{\partial x} + \frac{\partial V_y}{\partial y}\right] dx\, dy = \int_{\Sigma} V_n\, ds$$

where $\boldsymbol{V} = (V_x, V_y)$ is the vector field and V_n is a normal component (a flux) of the field on the boundary Σ. Set

$$V_x = (1/i\lambda) u \exp(i\lambda\varphi); \qquad V_y = (1/i\lambda) v \exp(i\lambda\varphi)$$

where

$$u = hf\varphi'_x/|\text{grad } \varphi|^2, \qquad v = hf\varphi'_y/|\text{grad } \varphi|^2 \tag{4.2}$$

Since $|\text{grad } \varphi|$ is bounded from below in the region supp (h) the functions u, v are bounded and infinitely differentiable. It is obvious that

$$\text{div } (V) = hf \exp (i\lambda\varphi) - (i/\lambda)(u'_x + v'_y) \exp (i\lambda\varphi)$$

Therefore using Gauss's formula,

$$\iint_\Omega hf \exp (i\lambda\varphi)\, dx\, dy = (i/\lambda) \iint_\Omega \left[\frac{\partial u}{\partial x} + \frac{\partial v}{\partial y}\right] \exp (i\lambda\varphi)\, dx\, dy$$
$$- (i/\lambda) \int_\Sigma hf (\text{grad } \varphi)^{-2} \frac{\partial \varphi}{\partial n} \exp (i\lambda\varphi)\, ds \tag{4.3}$$

Here the integral over the boundary Σ is a sum of integrals over the segments between the critical points. On each of the segments the boundary Σ is an analytical curve, the function φ is strictly monotonic and the nonexponential factor vanishes at the ends of the interval. Therefore the integral over the boundary falls off more rapidly than any power of λ^{-1} and it follows from eqn. 4.3 that

$$\iint_\Omega hf \exp (i\lambda\varphi)\, dx\, dy \simeq (i/\lambda) \iint_\Omega f_1 \exp (i\lambda\varphi)\, dx\, dy$$

where

$$f_1 = \frac{\partial}{\partial x}\left[\frac{hf\varphi'_x}{|\text{grad } \varphi|^2}\right] + \frac{\partial}{\partial y}\left[\frac{hf\varphi'_y}{|\text{grad } \varphi|^2}\right]$$

Repeating the procedure many times one obtains that the integral I_0 falls off more rapidly than any power of λ^{-1}. Thus, the determination of the asymptotic of eqn. 4.1 as $\lambda \to \infty$ is reduced to the determination of the asymptotics for integrals I_n, i.e. to the determination of the critical point contributions.

4.1.2 Asymptotic contribution of nondegenerate stationary point

The following theorem will be proved

Theorem 4.1.2: Let the functions $f(x, y)$ and $\varphi(x, y)$ in the integral

$$I(\lambda) = \iint_\Omega f(x, y) g(x, y) \exp [i\lambda\varphi(x, y)]\, dx\, dy \tag{4.4}$$

be infinitely differentiable, $\varphi(x, y)$ have the stationary point $x = x_0$, $y = y_0$ with the nondegenerate matrix of the second derivatives

$$A = \begin{bmatrix} \varphi''_{xx} & \varphi''_{xy} \\ \varphi''_{yx} & \varphi''_{yy} \end{bmatrix}$$

and let $g(x, y)$ be a cutting function that singles out a sufficiently small vicinity of the stationary point $x = x_0$, $y = y_0$. Then as $\lambda \to \infty$ the integral $I(\lambda)$ has the *asymptotic expansion*

$$I(\lambda) \simeq \sum_{n=0}^{\infty} a_n \lambda^{-1-n} \exp\,[i\lambda\varphi(x_0, y_0)] \tag{4.5a}$$

where

$$a_0 = \frac{2\pi f(x_0, y_0) \exp\,(i\pi\delta/2)}{\sqrt{|\varphi_{xx}\varphi_{yy} - \varphi_{xy}^2|}} \tag{4.5b}$$

and $\delta = 1$, if both eigenvalues of the matrix A are positive, $\delta = 0$, if they have opposite signs, and $\delta = -1$, if both are negative.

When $\varphi(x, y)$ is the square form

$$\varphi(x, y) = ax^2 + by^2 + \varphi_0 \tag{4.6}$$

and the matrix A is diagonal, then it is easy to derive eqn. 4.5 by repeating the asymptotic integration.

$$\int f(x, y) g(x, y) \exp\,[i\lambda\,(ax^2 + by^2)]\; dy$$

$$\simeq \sqrt{(\pi/\lambda|b|)} \exp\,[i\pi \text{ sign } (b)/4] \sum_{n=0}^{\infty} c_n (i/\lambda)^n \exp\,[i\lambda\,(ax^2 + \varphi_0)]$$

where $c_0 = f(x, 0)g(x, 0)$. Integrating this expression with respect to x and taking into account that $g(0, 0) = 1$ obtains eqn. 4.5.

To consider the general case, note that this formula is invariant under the integration variable change, $u = u(x, y)$, $v = v(x, y)$ which is analytical and reversible in a vicinity of the point $x = x_0$, $y = y_0$. When such a change occurs, the matrix A changes as well as the nonexponential factor fg. However, the values both of the integral $I(\lambda)$ and its asymptotic and, in particular, the coefficient a_0 of the asymptotic remain the same. Therefore it is sufficient to show that there exists an analytical and reversible (in a vicinity of $x = x_0$, $y = y_0$) variable change which transforms the phase function $\varphi(x, y)$ into a square one of the form of eqn. 4.6.

Now shift the origin to the point $x = x_0$, $y = y_0$, rotating the axes so that the function $\varphi''_{xy}(0, 0)$ vanishes with new variables. Then in the Taylor series of the function φ there is no term with xy, and the summands of the series independent of y and linearly dependent on y can be singled out:

$$\varphi(x, y) = \varphi_0 + x^2\psi_0(x) + x^2 y\psi_1(x) + y^2\psi_2(x, y) \tag{4.7}$$

where $\psi_0(x)$, $\psi_1(x)$, $\psi_2(x, y)$ are analytical functions. Since the matrix A (the elements of which are the second derivatives of φ) is supposed to be nondegenerate at $x = y = 0$ the values of $\psi_0(0) = a$ and $\psi_2(0, 0) = b$ are not equal to zero. Set

$$u = x\sqrt{\{[\psi_0(x) + y\psi_1(x)]/\psi_0(0)\}} = x + \cdots \tag{4.8}$$

$$v = y\sqrt{[\psi_2(x, y)/\psi_2(0, 0)]} = y + \cdots \tag{4.9}$$

(dots denote the quadratic and the subsequent terms of u, v expansions in powers of x, y). Obviously, this change is nondegenerate in a vicinity of zero. It follows from eqn. 4.7 that

$$\varphi = \varphi_0 + u^2\psi_3(0) + v^2\psi_2(0, 0)$$

i.e. that the change of eqn. 4.8 to eqn. 4.9 transforms φ into the square form eqn. 4.6.

Note that by a variable extension one can reduce the phase function $\varphi(x, y) = ax^2 + by^2 + \varphi_0$ to a sum (in the case of the extremum point of $\varphi(x, y)$) or to a difference (for the saddle point) of squares,

$$\varphi = \pm x^2 \pm y^2 + \varphi_0$$

4.1.3 Critical point on smooth boundary segment

Prove the following theorem:

Theorem 4.1.3: In the region supp *(g) let the integral $I(\lambda)$ defined by eqn. 4.4 have a single critical point A at a smooth part of the boundary Σ. The derivative of $\varphi(x, y)$ along Σ vanishes at this point and $g = 1$ in the vicinity of this point. Then $I(\lambda)$ is expanded into the asymptotic series*

$$I(\lambda) \simeq \sum_{n=0}^{\infty} (1/\lambda)^{n+3/2} b_n \exp\,[i\lambda\varphi(x_0, y_0)] \tag{4.10}$$

where (x_0, y_0) are the co-ordinates of the point A.

In the local co-ordinate system where the integration domain in the vicinity of A is defined by the equation $y > y(x)$ the coefficient b_0 is of the form

$$b_0 = \frac{i\sqrt{(2\pi)}f(x_0, y_0)\exp\,(i\pi\delta/4)}{\varphi'_y\sqrt{|\varphi''_{xx} + 2\varphi'_x\varphi'_y + \varphi''_{yy}y'^2 + \varphi'_y y''_{xx}|}} \tag{4.11a}$$

All the partial derivatives of the phase function $\varphi(x, y)$ and the derivatives of the function $y(x)$ specifying the boundary are taken at the point $x = x_0, y = y_0$. If A is a local minimum point of $\varphi[x, y(x)]$, then $\delta = 1$ and $\delta = -1$, if A is a local maximum point.

In the co-ordinate system where the x-axis is directed along a tangent to Σ at the point A, i.e. where $y'_x(x_0) = 0$, eqn. 4.11a is simplified and the coefficient b_0 takes the form

$$b_0 = \frac{i\sqrt{(2\pi)}f(x_0, y_0)\exp\,(i\pi\delta/4)}{\varphi'_y\sqrt{|\varphi''_{xx} + \varphi'_y y''_{xx}|}} \tag{4.11b}$$

Denoting a length measured along Σ by s obtains even more simple formula for b_0

$$b_0 = \frac{i\sqrt{(2\pi)}f(x_0, y_0)\exp\,(i\pi\delta/4)}{\varphi'_n\sqrt{|\varphi''_{ss}|}} \tag{4.11c}$$

where φ'_n is a derivative of φ along the normal to the boundary directed inside the domain Ω, the derivative being taken at the point A.

To prove theorem 4.1.3 it is sufficient to note that in the co-ordinate system where Σ is described by the equation $y = y(x)$ derivative φ'_y does not vanish.

Indeed, the derivative of φ along the boundary has the form

$$\frac{d\varphi\,[x, y(x)]}{dx} = \varphi'_x + \frac{dy}{dx}\,\varphi'_y = 0$$

If φ'_y were equal to zero at the point A it would follow from this equality that at A the derivative $\varphi'_x = 0$ as well and therefore in spite of the assumption A would be a stationary point of φ.

Since the derivative $\varphi'_y \neq 0$ at A it does not vanish in a vicinity of A and one can consider the function g to vanish identically outside this vicinity. Therefore after writing integral eqn. 4.3 in the form

$$I(\lambda) = \int_{-\infty}^{\infty} dx \int_{y(x)}^{\infty} f(x, y)g(x, y) \exp\,[i\lambda\varphi(x, y)]\; dx$$

one can perform internal integration with respect to y. It follows from theorem 1.1 that

$$\int_{y(x)}^{\infty} f(x, y)g(x, y) \exp\,[i\lambda\varphi(x, y)]\; dx\, dy$$

$$\simeq \sum_{n=0}^{\infty} (i/\lambda)^{1+n} a_n(x) \exp\,\{i\lambda\varphi[x, y(x)]\}$$

where

$$a_0(x) = \frac{f[x, y(x)]g[x, y(x)]}{\varphi'_y[x, y(x)]}$$

Integrating the expression term by term with respect to x and taking into account that $x = x_0$ (where $[x_0, y(x_0)]$ are the co-ordinates of the point A) is a stationary point of the function $\varphi[x, y(x)]$ and that in its vicinity the cutting function $g \equiv 1$, one obtains eqns. 4.10 and 4.11 and, thus, theorem 4.1.3 is proved.

4.1.4 Corner point of the integration domain boundary

The following theorem holds:

Theorem 4.1.4: Let the integral $I(\lambda)$ be defined by eqn. 4.4 in the region supp (g) *and let a corner point P of the boundary Σ be the only critical point of the integral, the one-sided derivatives of the phase function taken along the boundary being not vanished at this point. Then the asymptotic of $I(\lambda)$ as $\lambda \to \infty$ is of the form*

$$I(\lambda) = (1/\lambda)^2 \sum_{n=0}^{\infty} b_n(i/\lambda)^n \exp\,[i\lambda\varphi(A)] \tag{4.12a}$$

In the local co-ordinate system where $\varphi'_y \neq 0$ and the integration domain is defined by the equation $y < u(x)$ where $u(x)$ is a piecewise analytical function with a derivative discontinuity at $P=A_3$ (Figure 4.1), the coefficient b_0 has the form

$$b_0=\frac{f(P)}{\varphi_y'(P)}\left\{\left[\frac{d\varphi_+}{dx}\right]^{-1}-\left[\frac{d\varphi_-}{dx}\right]^{-1}\right\}$$

$$=\frac{(u_-'-u_+')f(P)}{(\varphi_x'+u_+'\varphi_y')(\varphi_x'+u_-'\varphi_y)} \tag{4.12b}$$

where

$$\frac{d\varphi_\pm}{dx}=\frac{d\varphi[x,\,u(x)]}{dx}=\varphi_x'+u_\pm'\varphi_y'$$

The right-side derivative of $u(x)$ is denoted by u_+', while u_-' relates to the left. The expression for b_0 can be written in a form that does not depend on the co-ordinate choice of the x, y-plane. Indeed,

$$u_-'=tg\ \alpha, \qquad u_+'=-tg\ \beta$$

$$\varphi_x'+u_+'\varphi_y'=\frac{1}{\cos\beta}\frac{d\varphi_+}{ds}, \qquad \varphi_x'+u_-'\varphi_y'=\frac{1}{\cos\alpha}\frac{d\varphi_-}{ds}$$

where the angles α, β are shown in Figure 4.1, while $d\varphi_+/ds$ and $d\varphi_-/ds$ are the right- and left-hand side derivatives of $\varphi(x, y)$ along the boundary at P and s is the length measured along the boundary. Using these expressions one gets

$$b_0=\frac{f(P)\sin\gamma}{\dfrac{d\varphi_+}{ds}\dfrac{d\varphi_-}{ds}} \tag{4.12c}$$

where $\gamma=\alpha+\beta$ is an angle between two tangents at the corner point P.

To prove the theorem perform the asymptotic integration with respect to y in integral eqn. 4.3

$$\int_{-\infty}^{u(x)} f(x, y)g(x, y)\exp\,[i\lambda\varphi(x, y)]\ dy$$

$$\simeq-(i/\lambda)\sum_{n=0}^{\infty}(i/\lambda)^n a_n\exp\,\{i\lambda\varphi[x,\,u(x)]\}$$

where

$$a_0=\frac{f[x,\,u(x)]g[x,\,u(x)]}{\varphi_y'(x,\,u(x))}$$

Integrating asymptotically with respect to x, the expressions obtained should take into account that the function $\varphi[x,\,u(x)]$ (as well as $u(x)$) has discontinuous derivatives at P starting from the first one. As a result eqn. 4.12*a* is obtained where b is of form of eqn. 4.12*b*.

Thus, the asymptotic contribution of the boundary corner point is of order of λ^{-2} and its leading term is determined in the invariant form by eqn. 4.12*c*.

If P is a point of the boundary curvature discontinuity, i.e. the magnitude of the boundary inflection at P is equal to zero, then the leading term of the asymptotic for the integral $I(\lambda)$ vanishes as $\lambda\to\infty$ and the next term determined by the curvature discontinuity at P becomes the leading one. Omitting appropriate mathematics, write the expression for the leading term of the asymptotic of $I(\lambda)$ as

$$I(\lambda)=-\frac{if(P)(\rho_{+}-\rho_{-})\exp[i\lambda\varphi(P)]}{\lambda^{3}[\varphi_{s}'(P)]^{3}} \tag{4.13}$$

where s is a length measured along the boundary, ρ_{+} and ρ_{-} are the left- and right-hand side limits of the boundary curvature at the point P.

4.2 Integrals with close critical points reduced to one-dimensional integrals

This Section considers integrals for which one can choose such local co-ordinates in the vicinity of merging critical points which give a possibility to integrate asymptotically with respect to one of the variables. The uniform asymptotic of the obtained one-dimensional integral is determined by the foregoing results.

4.2.1 Close stationary points of phase function

Assume the phase function in eqn. 4.1 depends on a parameter α and that two stationary points, $O_1=(x_1,y_1)$ and $O_2=(x_2,y_2)$ approach the common limit O as $\alpha\to0$. If this situation is stable, i.e. if there appear no other critical points for small perturbations of the phase function in the vicinity of O, one can always choose such co-ordinates x, y in a vicinity of O that the functions φ_{xx}'' and φ_{yyy}''' should be bounded in magnitude from below uniformly over α.

The qualitative behaviour of the phase function φ is shown in Figure 4.2. One of the stationary points (O_1) is always an extremum point, while the second point (O_2) is a saddle point.

Single out a vicinity of the close critical points O_1 and O_2 by means of the cutting function $g(x,y)$. One can consider the derivative φ_{xx}'' in the region supp (g) to be bounded in magnitude from below uniformly over α. Therefore one can perform the asymptotic integration with respect to x in the integral

$$I(\lambda)=\iint f(x,y)g(x,y)\exp[i\lambda\varphi(x,y)]\,dx\,dy \tag{4.14}$$

Since $|\varphi_{xx}''|\neq0$ the equation $\varphi_x'(x,y)=0$ has a single root $x=x(y)$. The curve L: $[x(y),y]$ is shown in Figure 4.2 by a broken curve. At points on the curve the tangent to the level lines of φ is horizontal. Obviously, the curve L passes through the stationary points O_1, O_2. By performing the asymptotic integration in $I(\lambda)$ with respect to x one gets

$$I(\lambda)=\int\Sigma a_n(y)\lambda^{-n-1/2}\exp\{i\lambda\varphi[x(y),y]\}\,dy \tag{4.15}$$

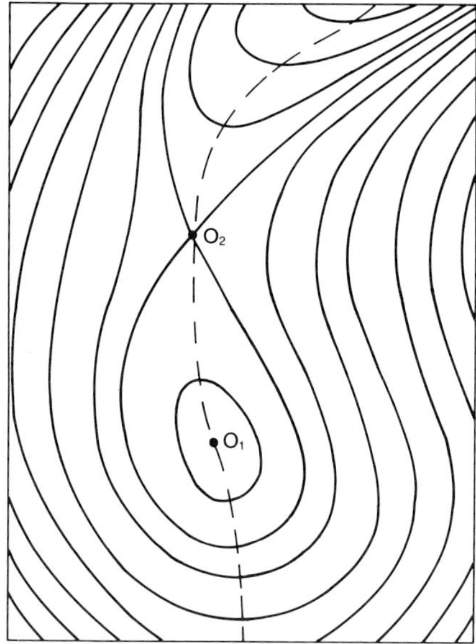

Figure 4.2 Level lines for close critical points

where under the integral sign there is an asymptotic series in powers of λ^{-1} and

$$a_0 = \frac{\sqrt{(2\pi)} f[x(y), y]}{\sqrt{|\varphi''_{xx}[x(y), y]|}} \exp(i\pi\delta/4)$$

where $\delta = \text{sign}(\varphi''_{xx})$.

The values $y = y_1$, $y = y_2$ corresponding to the points O_1 and O_2 and tending to the common limit y_0 as $\alpha \to 0$ are stationary points of the phase function in eqn. 4.15. Since φ'''_{yyy} is bounded in magnitude from below the asymptotic of the integral $I(\lambda)$ under these conditions is expressed according to theorem 2.3 in terms of the Airy function and its derivative. This results in the theorem:

Theorem 4.2.1: Let the phase function φ in eqn. 4.14 depend on a parameter α and for $\alpha > 0$ have two critical points, O_1, O_2, tending to a common limit O as $\alpha \to 0$ and $\varphi_1 = \varphi(O_1) > \varphi_2 = \varphi(O_2)$. Assume that in a vicinity of O one can introduce such co-ordinates x, y for which the derivatives φ''_{xx} and φ'''_{yyy} are bounded in magnitude from below uniformly over α and φ''_{yy} tends to zero as $\alpha \to 0$. Then the uniform asymptotic of eqn. 4.14 has the form

$$I(\lambda) \simeq \exp\{i[(\pi\delta_0/4) + \lambda\varphi_0]\}[A\lambda^{-5/6} Ai(-\lambda^{2/3}\xi) + B\lambda^{-7/6} Ai'(-\lambda^{2/3}\xi)] \tag{4.16}$$

where Ai and Ai' are the Airy function and its derivative, $\delta_0 = \text{sign}(\varphi''_{xx})$

$$\varphi_0 = (\varphi_1 + \varphi_2)/2; \qquad \xi = [3(\varphi_1 - \varphi_2)/4]^{2/3}$$

A and B are the asymptotic series in powers of λ^{-1} *with regular coefficients as* $\alpha \to 0$.

$$A = \sum_{n=0}^{\infty} a_n \lambda^{-n}; \qquad B = \sum_{n=0}^{\infty} b_n \lambda^{-n}$$

In particular, the coefficients a_0 and b_0 are

$$a_0 = 2\pi^{3/2}[(f_1/\sqrt{|g_1|}) + f_2/\sqrt{|g_2|}]\,\xi^{1/4}$$
$$b_0 = 2i\pi^{3/2}[(f_1/\sqrt{|g_1|}) - f_2/\sqrt{|g_2|}]\,\xi^{-1/4}$$

Here $f_1 = f(O_1)$, $f_2 = f(O_2)$ and g_1, g_2 are the values of the determinant

$$\begin{vmatrix} \varphi_{xx} & \varphi_{xy} \\ \varphi_{yx} & \varphi_{yy} \end{vmatrix}$$

at the points O_1 and O_2, respectively.

In a similar way the case of the phase function depending on two parameters α and β and having three close stationary points merging as $\alpha \to 0$, $\beta \to 0$ is considered. If such a situation is stable, one can also (in a similar way to the case of two merging stationary points) choose such co-ordinates x, y in which the function φ''_{xx} is bounded in magnitude from below in a vicinity of stationary points. Then one can integrate asymptotically with respect to x, the uniform asymptotic of the obtained one-dimensional integral being expressed (according to the results of Section 3.1) in terms of the Piercey integral and its derivatives.

4.2.2 Critical point on boundary near corner point

In the integral

$$I(\lambda) = \iint_{\Omega} f(x, y) g(x, y) \exp\,[i\lambda\varphi(x, y)]\; dx\, dy \tag{4.18}$$

let the boundary Σ of the integration domain Ω have a corner point P and in a vicinity of it consist of two analytical curves, PQ and PR. Assume that the gradient of φ does not vanish in a vicinity of P and the phase function φ along the arc PQ or its analytical continuation PQ' reaches the extremum at a point A near P (see Figure 4.3).

The cutting function g in eqn. 4.18 singles out a vicinity of P and the points P and A can be considered to be the only critical points in the region supp (g). Under these conditions the nonuniform asymptotic of the integral $I(\lambda)$ consists of two summands which are asymptotic contributions of the points P and A

$$I(\lambda) \simeq \sum_{n=0}^{\infty} b_n \lambda^{-n-2} \exp\,[i\lambda\varphi(P)] + \sum_{n=0}^{\infty} l_n \lambda^{-n-3/2} \exp\,[i\lambda\varphi(A)]\,\chi(A) \tag{4.19}$$

Here, the first summand is an asymptotic contribution of the corner point P, the second one is a contribution of the critical point A. If A lies on the arc PQ, then $\chi(A) = 1$ and $\chi(A) = 0$, if A lies on its extension PQ'. The leading terms b_0 and l_0 of the asymptotics are specified by eqns. 4.12 and 4.11, respectively.

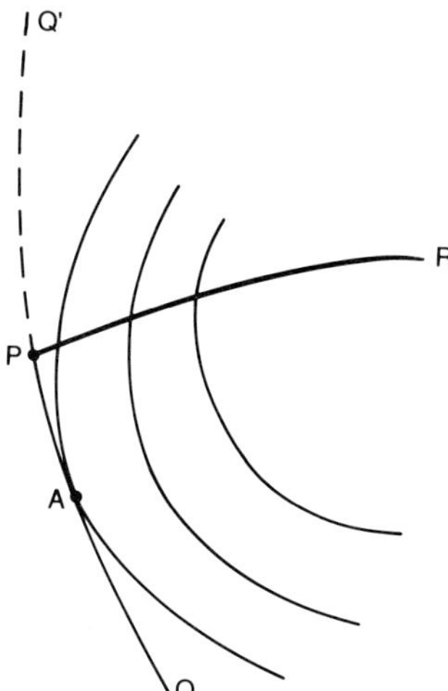

Figure 4.3 Critical point on boundary near corner

If critical point A approaches P, the derivative of the phase function at P along the arc PQ tends to zero and the coefficient b_0 (and the next terms of the asymptotic of $I(\lambda)$) tends to infinity according to eqn. 4.12. Therefore the asymptotic is inapplicable as A tends to P. It will be proven that the uniform asymptotic of $I(\lambda)$ is determined by the following theorem.

Theorem 4.2.2: The asymptotic of eqn. 4.18 uniform over a distance between the boundary corner point P and the nondegenerate critical point A on the boundary (i.e. the point where grad $\varphi \neq 0$, *the first derivative of the phase function along the boundary is equal to zero, while the second one is not) is expressed (for* $\varphi(A) < \varphi(P)$*) in terms of the Fresnel integral*

$$I(\lambda) = \sum_{n=0}^{\infty} a_n \lambda^{-n-3/2} \exp[i\lambda\varphi(A)] F\{\sqrt{\lambda}\sqrt{[\varphi(P) - \varphi(A)]}\}$$
$$+ \sum_{n=0}^{\infty} b_n^* \lambda^{-n-2} \exp[i\lambda\varphi(P)] \tag{4.20a}$$

Here, the value of $\sqrt{[\varphi(P) - \varphi(A)]}$ *is taken positive, if A lies on the arc PQ and negative, if A is in its extension PQ'. For* $\varphi(A) > \varphi(P)$ *the uniform asymptotic of eqn. 4.18 is expressed in terms of the conjugate Fresnel integral*

$$I(\lambda) = \sum_{n=0}^{\infty} a_n \lambda^{-n-3/2} \exp\{i\lambda\varphi(A)] F^*\{\sqrt{\lambda}\sqrt{[\varphi(A) - \varphi(P)]}\}$$
$$+ \sum_{n=0}^{\infty} b_n^* \lambda^{-n-2} \exp[i\lambda\varphi(P)] \tag{4.20b}$$

The coefficients a_n *in eqn. 4.20 are the same as that for nonuniform asymptotic eqn. 4.19, the coefficients* b_n^* *are determined by asymptotic matching when using nonuniform asymptotic eqn. 4.19.*

To prove the theorem introduce local co-ordinates x, y in such a way that the integration domain in a vicinity of P and the arc PQ is specified by the inequalities $u>0$, $v>0$ and $v=0$, $u>0$, respectively. This can be done, for instance, on specifying points inside Ω by their distances to the arcs PQ and PR. In the variables u, v the integral $I(\lambda)$ is written as

$$I(\lambda)=\int_0^\infty du \int_0^\infty f_1(u, v)g(u, v) \exp [i\lambda\varphi(u, v)]\, dy$$

where

$$f_1(u, v)=f[x(u, v), y(u, v)] \frac{\partial(x, y)}{\partial(u, v)}$$

and

$$\frac{\partial(x, y)}{\partial(u, v)}$$

is the Jacobian of the variable change to u, v.

Since the derivative φ_u' at P taken along the arc PQ tends to zero as the critical point A tends to P, while $|\text{grad}\, \varphi| \neq 0$, then the derivative φ_v' at P is bounded in magnitude from below. The region supp (g) can be considered to be so small for $|\varphi_v'|$ to be bounded from below in the whole region. Therefore one can integrate asymptotically in $I(\lambda)$ with respect to v to get

$$I(\lambda)=\int_0^\infty \sum_{n=0}^{\infty} c_n(u)\lambda^{-n-1} \exp [i\lambda\varphi(u, 0)]\, du$$

where

$$c_0=i\frac{f_1(u, 0)g(u, 0)}{\varphi_v'(u, 0)}$$

Here, integration is along the arc PQ on which the phase function $\varphi(u, 0)$ has a stationary point A close to the endpoint P of the integration interval. The conditions under which theorem 2.1 is applicable are satisfied. Evaluating the uniform asymptotic of the integral according to this theorem leads to theorem 4.2.2.

4.2.3 *Two close critical points on smooth part of boundary*

In eqn. 4.18 let there be two close critical points O_1 and O_2 on a smooth part of the boundary, i.e. two close points at which the phase function derivative along the boundary vanishes. Let us assume that $\varphi(O_1)>\varphi(O_2)$ and there are no other critical points near these points. Then at the points O_1, O_2 the phase function derivative along the normal to the boundary is bounded in magnitude from below. Therefore one can introduce such local co-ordinates x, y that the y-direction is close to the direction of the normal to the boundary and $d\varphi/dy$ is bounded from below in the vicinity of O_1, O_2. One can consider that $g=0$ outside the vicinity. Then one can asymptotically integrate with respect to y and eqn. 4.18 is reduced to the one-dimensional integral over the boundary Σ with two close stationary points of the phase function. According to theorem 2.3 the uniform asymptotic of such integral is expressed in terms of the Airy function and its derivative. This results in the theorem

Theorem 4.2.3: Let the phase function φ in eqn. 4.18 depend on a parameter α and there be two critical points on a smooth part of the boundary for $\alpha > 0$, the points being tended to common limit O as $\alpha \to 0$ and $\varphi_1 = \varphi(O_1) > \varphi_2 = \varphi(O_2)$. Assume the phase function third derivative along the boundary to be bounded in magnitude from below uniformly over α in a vicinity of O and there be no other critical points in the region supp *(g). Then the uniform asymptotic of eqn. 4.18 has the form*

$$I(\lambda) \simeq \exp\{i[(\pi/2) + \lambda\varphi_0]\} \times [A\lambda^{-4/3}Ai(-\lambda^{2/3}\xi) + B\lambda^{-5/3}Ai'(-\lambda^{2/3}\xi)] \tag{4.21}$$

where Ai and Ai' are the Airy function and its derivative

$$\varphi_0 = (\varphi_1 + \varphi_2)/2, \qquad \xi = [3(\varphi_1 - \varphi_2)/4]^{2/3}$$

A and B are asymptotic series in powers of λ^{-1} with regular coefficients as $\alpha \to 0$

$$A = \sum_{n=0}^{\infty} a_n \lambda^{-n}, \qquad B = \sum_{n=0}^{\infty} b_n \lambda^{-n}$$

In particular, the coefficients a_0 and b_0 are

$$a_0 = \sqrt{(2)}\pi \left[\frac{f_2(\partial^2\varphi_2/\partial s^2)^{-1/2}}{\partial\varphi_2/\partial n} + \frac{f_1|\partial^2\varphi_1/\partial s^2|^{-1/2}}{\partial\varphi_1/\partial n} \right] \xi^{1/4}$$

$$b_0 = \sqrt{(2)}\pi i \left[\frac{f_1|\partial^2\varphi_1/\partial s^2|^{-1/2}}{\partial\varphi_1/\partial n} - \frac{f_2(\partial^2\varphi_2/\partial s^2)^{-1/2}}{\partial\varphi_2/\partial n} \right] \xi^{-1/4}$$

where $f_1 = f(O_1)$, $f_2 = f(O_2)$, while $\partial\varphi_1/\partial n$ and $\partial\varphi_2/\partial n$ are the values of the phase function derivative along the normal to the boundary taken at O_1 and O_2, respectively, $\partial^2\varphi_1/\partial s^2$ and $\partial^2\varphi_2/\partial s^2$ are the second derivatives along the boundary; s is the length measured along the boundary.

4.2.4 Phase function stationary point near smooth part of boundary

Consider the integral $I(\lambda)$ defined by eqn. 4.18 assuming that a stationary point P of the phase function is close to a smooth part of the boundary Σ and there is a critical point A on this part of the boundary where the phase function second derivative along the boundary is bounded in magnitude from below. The cutting function g singles out a vicinity of close points A, P in such a way that integral (eqn. 4.18) has no other critical points in the region supp (g).

Given these conditions, the nonuniform asymptotic of $I(\lambda)$ is a sum of asymptotic contributions due to the points A, P.

$$I(\lambda) \simeq \sum_{n=0}^{\infty} \lambda^{-n-1} a_n \exp[i\lambda\varphi(P)]\chi(P) + \sum_{n=0}^{\infty} \lambda^{-n-3/2} b_n \exp[i\lambda\varphi(A)] \tag{4.22}$$

where $\chi(P)=1$, if the stationary point P is in the integration domain, and $\chi(P)=0$, if P is outside Ω. According to eqns. 4.5 and 4.11a the leading terms a_0, b_0 of the asymptotic are

$$a_0 = 2\pi \frac{f(P)\exp[i\pi\delta(P)/2]}{\sqrt{|g(P)|}}$$
$$b_0 = i\sqrt{(2\pi)}\frac{f(A)\exp[i\pi\delta(A)/4]}{\varphi'_n\sqrt{|\varphi''_{ss}|}} \tag{4.23}$$

where $\delta(P)=1$, 0 or -1 depending on the signs of the matrix eigenvalues, the matrix being composed of the second partial derivatives of φ at the point P (see Section 4.2.1); $g(P)$ is a determinant of the matrix. Here φ'_n and φ''_{ss} are the values of the normal derivative and the second derivative along s, respectively, the derivatives being taken at the point A, and $\delta(A)=\text{sign}\,[\varphi''_{ss}(A)]$.

Asymptotic eqn. 4.22 is nonuniform since when the point P approaches the boundary Σ the derivative φ'_n tends to zero, while the coefficient b_0 tends to infinity.

To find the uniform asymptotic change variables $u=u(x, y)$, $v=v(x, y)$ transforming the domain Ω into the halfplane $0<u<\infty$, $-\infty<v<\infty$. Since $\varphi''_{ss}\neq 0$ at A, one has (in new variables) $\varphi''_{vv}\neq 0$ at A and one can consider the function φ''_{vv} to be bounded in magnitude from below in the whole region supp (g). Therefore the integral over v can be asymptotically evaluated by the standard stationary phase method. Denote the stationary point of $\varphi(u, v)$ (as function of v for u fixed) by $v(u)$, it is a root of the equation $\varphi'_v=0$. The values of the functions φ, f and g at the point $[u, v(u)]$ are denoted by $\varphi(u), f(u)$ and $g(u)$. The integral $I(\lambda)$ can be written as

$$I(\lambda)\simeq \exp\left[i\frac{\pi}{4}\,\text{sign}\,\varphi''_{vv}\right]\cdot\int_0^\infty \sum_{n=0}^{\infty} c_n\lambda^{-n-1/2}\exp[i\lambda\varphi(u)]\,du \tag{4.24}$$

where

$$c_0=\sqrt{(2\pi)}\frac{f(u)g(u)}{\sqrt{|\varphi''_{vv}|}} \tag{4.24a}$$

Obviously, $\varphi(0)=\varphi(A)$ where A is the critical point at the boundary S, the extremum value of $\varphi(u)$ is reached at $u=u_0$ corresponding to the stationary point P, $u_0=u(P)$, and it coincides with the value of φ at this point. When P approaches Σ the value of u_0 tends to zero and the problem is reduced to the problem to find the uniform asymptotic of eqn. 4.24 when the stationary point $u=u_0$ of the phase function is close to the integration domain boundary $u=0$.

The uniform asymptotic of such an integral is determined by theorem 2.1 from which

Theorem 4.2.4: In eqn. 4.18 let the stationary point P of the phase function be close to a smooth part of the boundary Σ, the phase function second derivative along the boundary Σ being bounded in magnitude from below at the corresponding critical point A on the boundary. Then for $\varphi(A)>\varphi(P)$ the asymptotic of eqn. 4.18 uniform over a distance between the points

P and A is expressed in terms of the Fresnel integral

$$I(\lambda)=\sum_{n=0}^{\infty} a_n\lambda^{-n-1}\exp\,[i\lambda\varphi(P)]F\{\pm\sqrt{\lambda}\sqrt{[\varphi(A)-\varphi(P)]}\}$$
$$+\sum_{n=0}^{\infty} b_n^*\lambda^{-n-3/2}\exp\,[i\lambda\varphi(A)] \qquad (4.25a)$$

For $\varphi(A)<\varphi(P)$ the asymptotic is expressed in terms of the complex conjugate Fresnel integral

$$I(\lambda)=\sum_{n=0}^{\infty} a_n\lambda^{-n-1}\exp\,[i\lambda\varphi(O)]F^*\{\pm\sqrt{\lambda}\sqrt{[\varphi(P)-\varphi(A)]}\}$$
$$+\sum_{n=0}^{\infty} \lambda^{-n-3/2}b_n^*\exp\,[i\lambda\varphi(A)] \qquad (4.25b)$$

The coefficients a_n in eqn. 4.25 are the same as those in the nonuniform asymptotic (eqn. 4.22), the coefficients b_n are regular as $P\to\Sigma$ and are determined by asymptotic matching to the nonuniform asymptotic (eqn. 4.22). The signs of the radicals are chosen positive, if the stationary point is in the integration domain Ω, otherwise they are negative.

The uniform asymptotic of eqn. 4.25 is applicable only when the phase function second derivative along the boundary Σ is bounded in magnitude from below at the critical point A. If the stationary point is an extremum of the phase function, this condition is always satisfied for P sufficiently close to the boundary. If P is a saddle point the condition can be violated.

Let, for instance, the phase function be $\varphi=xy$ and the boundary Σ be a straight line at an angle ψ to the x-axis and let it be at a distance ρ from the origin. In parametric form, Σ is written as

$$x=x(s)=-\rho\sin\psi+s\cos\psi$$

$$y=y(s)=\rho\cos\psi+s\sin\psi$$

and the phase function on Σ is

$$\varphi(s)=x(s)y(s)=-(\rho^2/2)\sin 2\psi+\rho s\cos 2\psi+(s^2/2)\sin 2\psi$$

The function $\varphi''_{ss}=2\sin\psi\cos\psi$ tends to zero as $\psi\to 0,\ \pi/2$ and eqn. 4.25 is inapplicable when the direction of Σ is close to the x-, y-axes, i.e. to the phase function level lines outgoing from the stationary point O (see Figure 4.4). In a general case of a nondegenerate stationary saddle point the situation is similar. To apply eqn. 4.2 it is required (for O close to Σ) the direction of Σ in a vicinity of O to differ from the directions of the phase function level lines outgoing from the point O.

One can say that when a stationary saddle point O is close to a smooth part of the integration domain boundary Σ asymptotic (eqn. 4.25) is uniform over a distance from the boundary, but it is nonuniform over a direction of Σ. The asymptotic uniform over both a distance and boundary directions is obtained in the next Section.

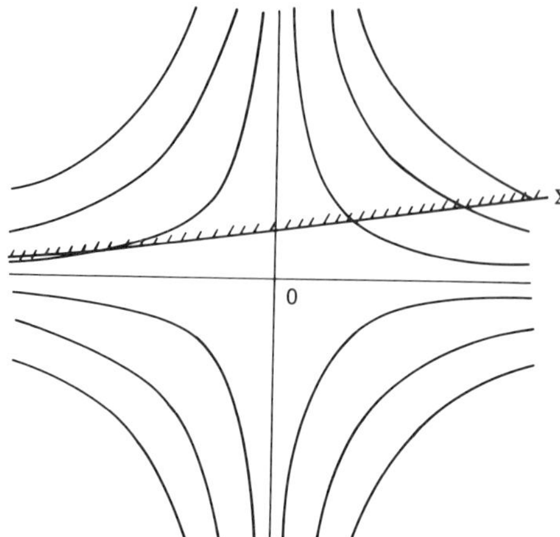

Figure 4.4 Stationary saddle point near boundary of integration domain

4.3 Stationary saddle point near boundary

Consider the integral $I(\lambda)$ defined by eqn. 4.18 assuming that the nondegenerate saddle stationary point O is close to the integration domain boundary Σ. As shown, if Σ has no corner points near O and the direction of Σ is not close to the phase-function level lines emanating from O (called separatrises), the asymptotic of $I(\lambda)$ uniform over a distance from O to the boundary Σ is expressed by eqn. 4.25 in terms of the Fresnel integral or the complex conjugate Fresnel integral. These restrictions are removed in what follows.

First, the boundary Σ consisting of two arcs AQ and QB will be considered when assuming that the corner point Q is near the point O and the directions of the arcs TQ and QR near O are not close to the directions of the separatrises outgoing from O. In this case, the uniform asymptotic of $I(\lambda)$ is expressed in terms of the Ff-integral introduced in Section 3.4.

Secondly, the case of the smooth boundary Σ near O will be considered, the direction of Σ being close to that of one of the separatrises, e.g. to the direction of the separatrix AOB (see Figure 4.5). In this situation the asymptotic of the integral $I(\lambda)$ specified by eqn. 4.18 can be nonlocal. For instance, if Σ coincides with the separatrix AOB, i.e. the phase function $\varphi(x,y)\equiv\varphi(O)$ on Σ, the asymptotic of $I(\lambda)$ as $\lambda\to\infty$ is expressed in terms of the integral of $f(x)g(x)$ taken with a weight over Σ (eqns. 4.28 and 4.29 below where $\varphi[x, h(x)] = \varphi(O)$ and $\exp(i\lambda\varphi)$ are factored out of the integral sign).

Therefore the requirement that the boundary curvature near O differs from the curvature of the separatrix AOB is introduced. This means that for any points $l\in AOB$ and $s\in\Sigma$ in the region supp (g) the curvature difference $\rho(l)-\rho(s)$ of the separatrix and the boundary, respectively, should be bounded in magnitude from below. The phase function third derivative along the boundary is bounded in magnitude from below provided this requirement is satisfied.

To obtain the uniform asymptotic of $I(\lambda)$ one should reduce the initial integral to that over a boundary having a pole on the integration contour.

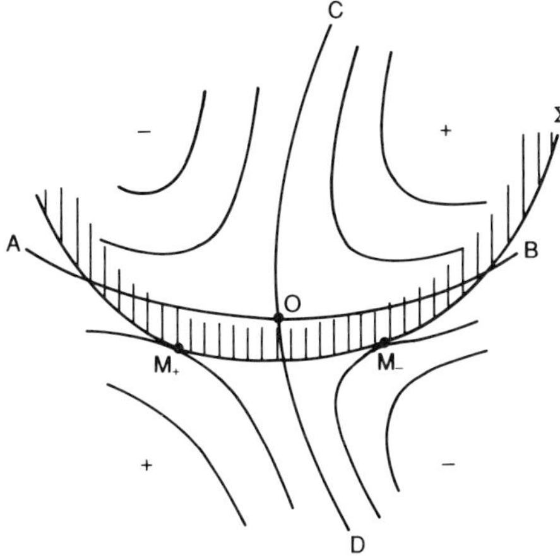

Figure 4.5 *Boundary Σ direction close to separatrix direction*

Choose co-ordinates x, y with O as the origin in such a way that the phase function is specified by the equation

$$\varphi = \varphi_0 + pxy + \cdots \tag{4.26}$$

where $p > 0$, the dots denote terms following the quadratic form and the integration domain is specified by the equation $y > h(x)$

$$I(\lambda) = \iint_{y > h(x)} \exp\left[i\lambda\varphi(x, y)\right] fg \, dx \, dy \tag{4.27}$$

It is now proven that there exists such asymptotic series

$$T(\lambda, x) = \sum_{n=0}^{\infty} \lambda^{-n} T_n(x)$$

with finite and infinitely differentiable coefficients $T_n(x)$ that

$$I(\lambda) \simeq I_0(\lambda) \tag{4.28}$$

where

$$I_0(\lambda) \simeq \int_{-\infty}^{\infty} \frac{\exp\{i\lambda\varphi[x, h(x)]\} T(\lambda, x)}{(x - x_0 + iO)} dx \tag{4.29}$$

and $[x_0, h(x_0)]$ is the point of the intersection of the boundary Σ and the separatrix COD.

Before proving eqn. 4.28 the equation is used to find the asymptotic of $I(\lambda)$ for the previously mentioned cases.

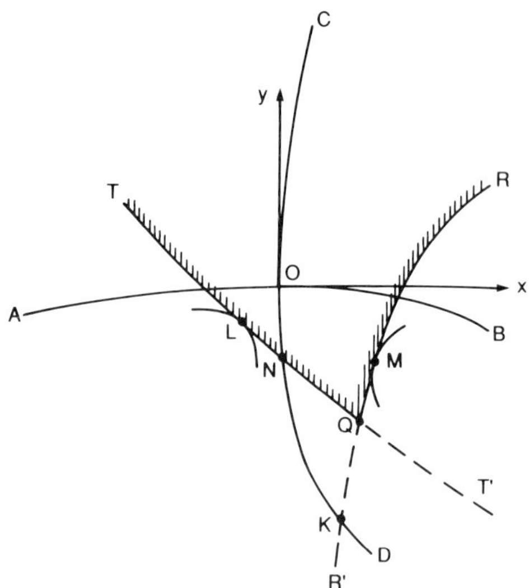

Figure 4.6 Stationary saddle point near corner point of boundary

4.3.1 Corner point of boundary near stationary point

Consider the relative location of a stationary point O and the integration domain boundary (Figure 4.6), the final equations are given for arbitrary relative location of O, Σ and the directions of the Σ sides.

First, let a corner point Q be in the fourth quadrant. Then the integral $I(\lambda)$ can be written as

$$I_0(\lambda) = I_1 + I_2 = \int_{-\infty}^{x_Q} \frac{\exp\{i\lambda\varphi[x, h_1(x)]\} T_1(\lambda, x)}{(x - x_N + iO)} dx + \int_{x_Q}^{\infty} \frac{\exp\{i\lambda\varphi[x, h_2(x)]\} T_2(\lambda, x)}{(x - x_K + iO)} dx \qquad (4.30)$$

where x_Q, x_K and x_N are the co-ordinates of the boundary corner point Q and the points at which the arcs TQ and QR (or their extensions) intersect the separatrix COD, the functions $y = h_1(x)$ and $y = h_2(x)$ being the equations of these arcs.

Since the directions of the arcs QT and QR are not close to those of the separatrises the functions $\varphi[x, h(x)]$ and $\varphi[x, h_2(x)]$ have the second derivatives bounded in magnitude from below and stationary points $x = x_L$ and $x = x_M$ where L and M are the tangency points of the arcs QT and QR (or their extensions) with the phase-function level lines. If the corner point Q is close to the stationary point O, then the critical points of the integrals in eqn. 4.30 are close (here the critical points are poles of the integrand $x = x_N$ or $x = x_K$ the boundary of the integration interval $x = x_Q$ and the stationary points $x = x_L$ or $x = x_M$).

The uniform asymptotic of such integrals has been considered in Section 3.4, it is expressed in terms of *Ff*-integral (see Section 6.11)

$$Ff(\alpha, \beta) = (1/2\pi) \int_{-\infty}^{\alpha} \frac{\exp(it^2)}{(t - \beta + iO)} dx$$

the Fresnel integral $F(\xi)$ and their complex conjugate functions. As a result, one obtains the asymptotic of $I(\lambda)$ when a corner point is in the fourth quadrant and the slopes of arcs QT and QR as shown in Figure 4.6 (the arc QT goes to the second quadrant, while QR to the first one)

$$\begin{aligned} I(\lambda) \simeq \frac{A(\lambda)}{\lambda} \Bigg[& \exp(i\lambda\varphi_L) \int_{-\infty}^{\sqrt{(\varphi_L - \varphi_Q)}} \frac{\exp(-i\lambda t^2)\,dt}{t - \sqrt{(\varphi_L - \varphi_0)} + iO} \\ & + \exp(i\lambda\varphi_M) \int_{\sqrt{(\varphi_M - \varphi_Q)}}^{\infty} \frac{\exp(-i\lambda t^2)\,dt}{t - \sqrt{(\varphi_M - \varphi_0)} + iO} \Bigg] \\ & + \lambda^{-3/2} [B(\lambda) \exp(i\lambda\varphi_L) F^*\{\sqrt{[\lambda(\varphi_L - \varphi_Q)]}\} \\ & + C(\lambda) \exp(i\lambda\varphi_M) F^*\{\sqrt{[\lambda(\varphi_M - \varphi_Q)]}\}] \\ & + \lambda^{-2} D(\lambda) \exp(i\lambda\varphi_Q) \end{aligned} \quad (4.31)$$

where $A(\lambda)$, $B(\lambda)$, $C(\lambda)$ are the asymptotic series in powers of λ^{-1}. Here φ_L, φ_M etc. are the values of the phase function at the corresponding points. When writing these expressions one takes into account that the points N and K are on the separatrix and therefore $\varphi_N = \varphi_K = \varphi_O$.

The signs of radicals in eqn. 4.31 are chosen so that critical points $t = 0$, $t = \sqrt{(\varphi_L - \varphi_0)}$ and $t = 0$, $t = -\sqrt{(\varphi_0 - \varphi_M)}$ are located, together with the integration domain boundaries $t = -\infty$, $t = \sqrt{(\varphi_L - \varphi_Q)}$ and $t = \infty$, $t = -\sqrt{(\varphi_Q - \varphi_M)}$, in the same sequence as the points L, N and K, M, together with the boundary points T, Q and Q, R of the arcs TQ and QR, respectively.

It follows that if Q is in the fourth quadrant, then one needs to set $\sqrt{(\varphi_L - \varphi_Q)} > 0$, $\sqrt{(\varphi_0 - \varphi_M)} > 0$. The radical $\sqrt{(\varphi_L - \varphi_0)}$ is chosen positive, if the arc QT is lower than the point O, otherwise it is negative. Similarly, $\sqrt{(\varphi_Q - \varphi_M)} > 0$, if Q is between K and the stationary point M, otherwise it is negative. So defined, radicals appear to be analytical functions of the position of the corner point Q. Obviously, the integrals in eqn. 4.31 are expressed in terms of the complex conjugate *Ff*-integral.

Let a distance between the point O and the corner point Q be bounded from below. Then according to the results of Section 4.2 the uniform asymptotic of $I(\lambda)$ is expressed in terms of the Fresnel integral $F\{\sqrt{[\lambda(\varphi_L - \varphi_0)]}\}$ and its complex conjugate. Eqn. 4.31 agrees with this result. Indeed, if the arc TQ approaches the stationary point O, the point N at which the arc intersects the separatrix approaches the point L, the tangency point of the arc and the phase-function level line, and in the first integral of eqn. 4.28 the pole $t = \sqrt{[(\varphi_L - \varphi_0)]} = \sqrt{[(\varphi_L - \varphi_N)]}$ tends to the phase function stationary point $t = 0$. Therefore, according to the results of Section 2.2, the asymptotic of the integral is also expressed in terms of $F\{\sqrt{\lambda[(\varphi_L - \varphi_0)]}\}$.

Other cases of the position of the corner point Q and the directions of the arcs QT and QR are considered in a similar way. The results can be formulated as follows.

Let N be an intersection of the arc QT or its extension QT' with the separatrix COD tangent to the y-axis at the point O and let L be a point where the arc TQT' is tangent to the level line of the phase function φ. Let φ_1 be a value of φ at the point L. Set

$$\alpha_1 = \pm\sqrt{|\varphi_1 - \varphi_Q|}, \qquad \beta_1 = \pm\sqrt{|\varphi_1 - \varphi_O|}$$

where φ_O and φ_Q are the values of the phase function at O and Q, respectively, and the radical signs are chosen in such a way that the numbers $-\infty$, 0, β_1, α_1 are located on the real axis in the same sequence as the points T, L, N, Q on the arc TQ. Introduce numbers

$$\xi_1 = \text{sign}\,(\varphi_0 - \varphi_1), \qquad \delta = \pm 1,\ \omega_1 = \pm 1$$

where $\delta_1 = 1$, if the integration domain is to the right from the arc QT when moving from Q to T, otherwise $\delta_1 = -1$, and $\omega_1 = 1$, if $x < x(Q)$ on the arc QT, otherwise $\omega_1 = -1$. Similarly, introduce numbers φ_2, α_2, β_2, ξ_2, $\delta_2 = -\delta_1$, ω_2 corresponding to the arc QR. The nonuniform asymptotic can be written as

$$\begin{aligned} I(\lambda, \alpha) \simeq (1/\lambda) A_0(\lambda) \exp\,(i\lambda\varphi_0)\chi_0 + \lambda^{-3/2} B_0(\lambda) \exp\,(i\lambda\varphi_1)\chi_1 \\ + \lambda^{-3/2} C_0(\lambda) \exp\,(i\lambda\varphi_2)\chi_2 + \lambda^{-2} D_0(\lambda) \exp\,(i\lambda\varphi_Q) \end{aligned} \tag{4.32}$$

where $\chi_0 = 1$, if the point O is inside the integration domain Ω, otherwise $\chi_0 = 0$, and $\chi_1 = 1$, if there is a phase function stationary point L on the arc QT, i.e. if $\alpha > 0$, otherwise $\chi_1 = 0$, and $\chi_2 = 1$, if there is a stationary point on the arc QR, otherwise $\chi_2 = 0$.

To write the uniform asymptotic of $I(\lambda)$ introduce the following functions expressed in terms of the Ff-integral

$$\Phi(\alpha, \beta, 1, 1) = iFf(\alpha, \beta) = \frac{-1}{2\pi i}\int_{-\infty}^{\alpha} \frac{\exp\,(is^2)}{s - \beta + i0}\, ds$$

$$\Phi(\alpha, \beta, -1, 1) = \frac{-1}{2\pi i}\int_{-\infty}^{\alpha} \frac{\exp\,(is^2)}{s - \beta - i0}\, ds = iFf(\alpha, \beta) + \begin{cases} 0 & \text{for } \beta > \alpha \\ \exp\,(i\beta^2) & \text{for } \beta < \alpha \end{cases}$$

$$\Phi(\alpha, \beta, 1, -1) = \Phi^*(\alpha, \beta, 1, 1) = \frac{1}{2\pi i}\int_{-\infty}^{\alpha} \frac{\exp\,(-is^2)}{s - \beta - i0}\, ds$$

$$\Phi(\alpha, \beta, -1, -1) = \Phi^*(\alpha, \beta, -1, 1) = \frac{1}{2\pi i}\int_{-\infty}^{\alpha} \frac{\exp\,(-is^2)}{s - \beta + i0}\, ds$$

Let $F(\alpha, 1) = F(\alpha)$ and $F(\alpha, -1) = F^*(\alpha)$ where $F(\alpha)$ and $F^*(\alpha)$ are the Fresnel integral and its complex conjugate, respectively. The following theorem holds.

Theorem 4.3.1: The asymptotic of eqn. 4.18, uniform with respect to the relative location of the saddle point O and the corner point on the boundary Σ of the integration domain Ω,

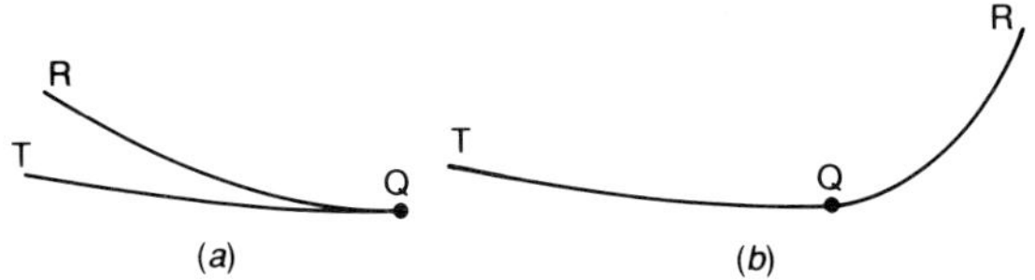

Figure 4.7 Boundary segments mutually tangent at point Q

is of the form

$$\begin{aligned} I(\lambda) \simeq \frac{1}{\lambda} A_0(\lambda)[&\delta_1\omega_1 \exp(i\lambda\varphi_1)\Phi(\sqrt{\lambda}\alpha_1, \sqrt{\lambda}\beta_1, \omega_1, \xi_1) \\ &+ \delta_2\omega_2 \exp(i\lambda\varphi_2)\Phi(\sqrt{\lambda}\alpha_2, \sqrt{\lambda}\beta_2, \omega_2, \xi_2)] \\ &+ \lambda^{-3/2}[\tilde{B}(\lambda) \exp(i\lambda\varphi_1)F(\sqrt{\lambda}\alpha_1, \xi_1) \\ &+ \tilde{C}(\lambda) \exp(i\lambda\varphi_2)F(\sqrt{\lambda}\alpha_2, \xi_2)] + \lambda^{-2}\tilde{D}(\lambda) \exp(i\lambda\varphi_Q) \end{aligned} \qquad (4.33)$$

if the boundary directions at the corner point differ from the directions of the phase-function level lines emanating from O. Here the numbers $\omega_{1,2}$, $\xi_{1,2}$, $\delta_{1,2}$, arguments $\alpha_{1,2}$ and $\beta_{1,2}$ of the Fresnel integrals and the phase functions φ_1, φ_2, φ_Q are as previously defined and A_0, $\tilde{B}$, $\tilde{C}$, $\tilde{D}$ are the asymptotic series in powers of λ^{-1}.

The series A_0 in eqn. 4.33 is the same as in eqn. 4.32, while the series $\tilde{B}$, $\tilde{C}$, $\tilde{D}$ are determined by the nonuniform asymptotic of eqn. 4.32 by asymptotic matching.

Note that eqn. 4.33 depends on the choice of the co-ordinates. If in the co-ordinate system with O as the origin in which the phase function is $\varphi(x, y) \equiv xy$ x and y are interchanged (i.e. perform the reflection with respect to the bisectrix $x = y$), then eqn. 4.33 would have different analytical form. However, one can show directly that the new expression coincides with the old one.

Note also that eqn. 4.33 is applicable even when the arcs TQ and QR are tangent to each other at O forming either a corner point (Figure 4.7*a*) or the boundary TQR with the curvature discontinuity point at Q (Figure 4.7*b*).

Consider the second case (see Reference 30).

4.3.2 *Smooth boundary Σ close to saddle stationary point O*

Assume that its slope near O is close to the direction of the separatrix going out from O and its curvature near O differs from the separatrix curvature.

Let x, y be the co-ordinates introduced in which the phase function is of the form of eqn. 4.26, while the integration domain is determined by the inequality $y > h(x)$. Then the parameters $\alpha = h(0)$ and $\beta = h'(0)$ are small and the first and second derivatives of the phase function $\tau(x) = \varphi[x, h(x)]$ in eqn. 4.29 in the zero vicinity are also small, while the third derivative is bounded in magnitude from below. The stationary points (x_1 and x_2) of the function $\tau(x)$ correspond to the points M_+ and M_- at which the boundary Σ and the level lines of the function $\varphi(x, y)$ (Figure 4.8) are tangential to each other, the points being close to a pole of the nonexponential factor. Therefore, the asymptotic of eqn. 4.29 is determined by theorem 3.3 and is expressed in terms of the Airy–Fresnel

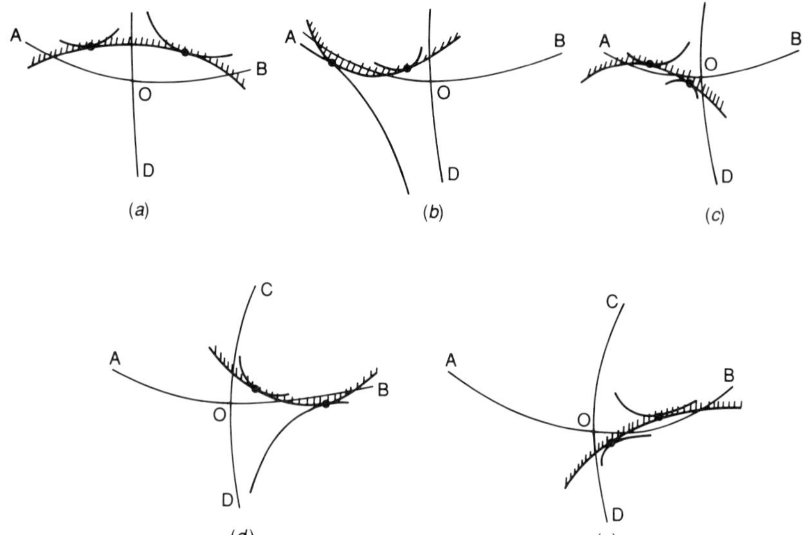

Figure 4.8 Mutual location of integration domain boundary and phase-function level lines

integral $Af(\eta, \xi)$, the Airy function $Ai(\xi)$ and its derivative. This results in the theorem:

Theorem 4.3.2: Eqn. 4.27 with the phase function eqn. 4.26, bounded in magnitude from below third derivative τ'''_{xxx} (where $\tau(x)=\varphi(x, h(x))$) and with small $\alpha=h(0)$ and $\beta=h'(0)$ has the uniform over α, β asymptotics as $\lambda\to\infty$. For $\tau'''_{xxx}>0$

$$I(\lambda)\simeq\frac{1}{\lambda}\exp(i\lambda\xi)[R_0(\lambda)Af(-\lambda^{2/3}\eta, \lambda^{1/3}\theta)$$
$$+\lambda^{-1/3}R_1(\lambda)Ai(-\lambda^{2/3}\eta)+\lambda^{-2/3}R_2(\lambda)Ai'(-\lambda^{2/3}\eta)] \tag{4.34a}$$

For $\tau'''_{xxx}<0$

$$I(\lambda)\simeq\frac{1}{\lambda}\exp(i\lambda\xi)\left\{R_0(\lambda)\left\{\exp\left[-i\lambda\left(\frac{\theta^3}{3}-\theta\eta\right)\right]\right.\right.$$
$$\left.-Af(-\lambda^{2/3}\eta, -\lambda^{1/3}\theta)\right\}$$
$$\left.+\lambda^{-1/3}R_1(\lambda)Ai(-\lambda^{2/3}\eta)+\lambda^{-2/3}R_2(\lambda)Ai'(-\lambda^{2/3}\eta)\right\} \tag{4.34b}$$

The functions ξ, η, θ and the coefficients of the asymptotic series R_0, R_1, R_2 in powers of λ^{-1} are analytical functions of α, β.

To find the functions ξ, η, θ and the asymptotic series use asymptotic matching. Determine the nonuniform asymptotic of the integral $I(\lambda)$. It is a sum of contributions due to the phase function stationary point O and the critical points

M_+ and M_- on the boundary Σ (i.e. the stationary points of the function $\tau[x, h(x)]$)

$$I(\lambda) \simeq (1/\lambda)P_0\chi_0 \exp(i\lambda\varphi_0) + \lambda^{-3/2}[P_+ \exp(i\lambda\varphi_+) + P_- \exp(i\lambda\varphi_-)] \tag{4.35a}$$

where $\chi_0 = 1$, if Ω contains the stationary point O, otherwise $\chi_0 = 0$.

$$\varphi_0 = \varphi(P), \qquad \varphi_\pm = \varphi(M_\pm), \qquad \varphi_- < \varphi_0 < \varphi_+$$

and P_0, P_+, P_- are asymptotic series in powers of λ^{-1}.

If there are no critical points on the boundary, then the asymptotic of $I(\lambda)$ includes only the contribution of the stationary point O

$$I(\lambda) \simeq (1/\lambda)P_0\chi_0 \exp(i\lambda\varphi_0) \tag{4.35b}$$

The integral $I(\lambda)$ does have critical points on the boundary if $\alpha\tau'''_{xxx} < 0$, i.e. if $\tau'''_{xxx} > 0$ and the integration domain contains the point O or if $\tau'''_{xxx} < 0$ and the integration domain does not contain the point O. On comparing eqn. 4.35*a* with the asymptotics obtained when replacing the functions Af, Ai, Ai' in eqn. 4.34 by their asymptotics one gets

$$\xi = (\varphi_+ + \varphi_-)/2, \qquad \eta = [3(\varphi_+ - \varphi_-)/4]^{2/3} \tag{4.36a}$$

These equations determine ξ and η for sign $\alpha = -\text{sign}(\tau'''_{xxx})$ when there exist stationary points x_1 and x_2 and the values of the phase function at the points $[x_1, h(x_1)]$, $[x_2, h(x_2)]$ are also defined. If sign $\alpha = \text{sign}(\tau'''_{xxx})$ and there are no stationary points the functions ξ, η are specified by their analytical continuations.

The function θ is defined as a root of the equation

$$\text{sign}\, \tau'''_{xxx} \cdot (\theta^3/3 - \theta\eta) + \xi = \varphi_0 \tag{4.36b}$$

For $\eta < 0$, i.e. when the phase function $\tau[x, h(x)]$ is strictly monotonic, the equation has a single root. For $\eta > 0$ the equation can have three roots, $\theta_1 < \theta_2 < \theta_3$, select the root which is an analytical continuation from the region $\eta < 0$. Depending on the signs of τ'''_{xxx}, $\alpha = h(0)$, $\beta = h'(0)$ this root is defined as

(i) For $\tau'''_{xxx} > 0$, $\alpha < 0$ (Figure 4.5) or for $\tau'''_{xxx} < 0$, $\alpha > 0$ (Figure 4.8*a*) the root $\theta = \theta_2$ is taken.

(ii) If for $\tau'''_{xxx} > 0$, $\alpha > 0$, $\beta > 0$ (Figure 4.8*b*) or for $\tau'''_{xxx} < 0$, $\alpha < 0$, $\beta < 0$ (Figure 4.8*c*) there are three roots, then the root $\theta = \theta_3$ is taken.

(iii) If for $\tau'''_{xxx} > 0$, $\alpha > 0$, $\beta < 0$ (Figure 4.8*d*) or for $\tau'''_{xxx} < 0$, $\alpha < 0$, $\beta > 0$ (Figure 4.8*e*) there are three roots, then the root $\theta = \theta_1$ is taken.

So defined, root θ depends analytically on the parameters α, β. The asymptotic series R_0 coincides with P_0, and for two leading terms $R_1^{(0)}$ and $R_2^{(0)}$ of the series R_1 and R_2 one gets

$$\begin{aligned} R_1^{(0)} &= \eta^{1/4}\sqrt{\pi}(P_+^{(0)}e^{\pi i/4} + P_-^{(0)}e^{-\pi i/4}) - \frac{iR_0^{(0)} \cdot \theta}{\eta - \theta^2} \\ R_2^{(0)} &= -\eta^{-1/4}\sqrt{\pi}(P_+^{(0)}e^{-\pi i/4} + P_-^{(0)}e^{\pi i/4}) + \frac{R_0^{(0)}\,\text{sign}\,\tau'''_{xxx}}{\eta - \theta^2} \end{aligned} \tag{4.36c}$$

where $R_0^{(0)}$, $P_\pm^{(0)}$ are the leading terms of the appropriate asymptotic series.

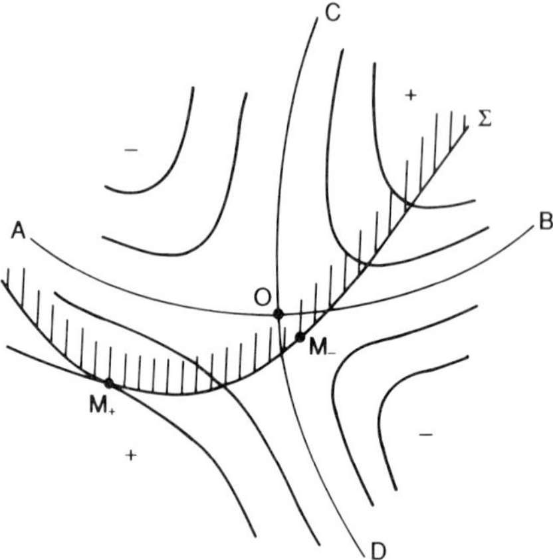

Figure 4.9 Integration domain boundary and critical points as $\alpha \to 0$ and $\beta > const > 0$

Consider the qualitative behaviour of the functions η, θ, $R_1^{(0)}$, $R_2^{(0)}$. Set $\tau''' > 0$ and fix $\beta > 0$ (Figure 4.6). Then $M_- \to O$ as $\alpha \to 0$. Besides $\varphi_- \to \varphi_0$ the difference $\varphi_+ - \varphi_-$, together with η, are bounded from below and $\tau \to \sqrt{\eta}$. Therefore in the integral representation of the Airy–Fresnel function $Af(-\lambda^{2/3}\eta, \lambda^{1/3}\theta)$ the pole $\tau = \lambda^{1/3}\theta$ appears to be near the phase-function stationary point and the asymptotic of Af is expressed in terms of the complex conjugate Fresnel integral, while eqn. 4.34*a* is transformed into eqn. 4.25*b*. When moving to the limit the term $P_-^{(0)}$ (and the subsequent coefficients of the series P_-) tends to infinity, but the sum

$$P_-^{(0)} + \frac{R_0^{(0)} \exp(-i\pi/4)}{2\sqrt{\pi}(\sqrt{\eta} - \theta)\sqrt{\eta}}$$

remains bounded. If $\alpha \to 0$ and $\beta \to 0$ simultaneously, then $P_\pm^{(0)}$ and $1/[\sqrt{\eta} \pm \theta]^{-1}\eta^{-1/4}$ tend to infinity, however, all the growing summands on the right-hand sides of eqns. 4.36*c* cancel.

4.3.3 Proof of eqn. 4.28

To finish the proof of theorems 4.31 and 4.32 it remains to prove eqn. 4.28. To do this, perform an analytical and reversible variable change transforming the phase function into $\varphi_0 + xy$. Then the equation is written as

$$I = \iint_{y > h(x)} \exp(i\lambda xy) fg\, dx\, dy$$

$$\simeq \int_{-\infty}^{\infty} \exp[i\lambda x h(x)]\, Q(\lambda, x) \frac{dx}{x + iO} \tag{4.37}$$

and after proving eqn. 4.37 one can return to eqn. 4.28 by means of the inverse variable change. Represent $f(x, y)$ as a sum

$$f(x,y) = \sum_{n,m=0}^{2N} a_{n,m} x^n y^m + x^N f_1(x,y) + y^N f_2(x,y)$$

Then

$$I = \sum_{n,m=0}^{2N} a_{n,m} I_{n,m} + I_N^{(1)} + I_N^{(2)}$$

where

$$I_{n,m} = \iint_\Omega \exp(i\lambda xy) x^n y^m g(x,y)\, dx\, dy$$

$$I_N^{(1)} = \iint_\Omega \exp(i\lambda xy) x^N f_1(x,y) g(x,y)\, dx\, dy$$

$$I_N^{(2)} = \iint_\Omega \exp(i\lambda xy) y^N f_2(x,y) g(x,y)\, dx\, dy$$

and the proof of eqn. 4.37 is reduced to the proof of the following relations

$$I_{n,m} \simeq \int_{-\infty+iO}^{\infty+iO} \exp[i\lambda h(x)]\, Q_{m,n}(\lambda, x) \frac{dx}{x} \tag{4.38a}$$

$$I_N^{(1,2)} = \int_{-\infty+iO}^{\infty+iO} \exp[i\lambda x h(x)]\, R_{1,2}(\lambda, x)\, dx + O(\lambda^{-N}) \tag{4.38b}$$

where $Q_{m,n}$ and $P_{m,n}$ are polynomials in λ^{-I} with finite infinitely differentiable coefficients. Indeed, on setting

$$Q_N = \Sigma a_{m,n} Q_{m,n} + x(R_1 + R_2)$$

one obtains from eqn. 4.38 that for any $N > 0$

$$I = \int \exp[i\lambda x h(x)] \frac{Q_N(\lambda, x)\, dx}{x + iO} + O(\lambda^{-N})$$

and eqn. 4.37, obviously, follows from the foregoing results.

Now prove eqn. 4.38*a*. In so doing the expression for $I_{n,m}$ is integrated by parts to get

$$I_{n,m} = (im/\lambda) I_{n-1,m-1} + (i/\lambda) \int_{y=h(x)}^{\infty} x^{n-1} y^m g(x,y) \exp(i\lambda xy)\, dx$$

$$+ \frac{i}{\lambda} \iint_\Omega y^m x^{n-1} \frac{\partial g}{\partial y} \exp(i\lambda xy)\, dx\, dy$$

$$I_{n,0} = (i/\lambda) \int_{y=h(x)}^{\infty} x^{n-1} g(x,y) \exp(i\lambda xy)\, dx$$

$$+ \frac{i}{\lambda} \iint_\Omega x^{n-1} \frac{\partial g}{\partial y} \exp(i\lambda xy)\, dx\, dy$$

$$I_{0,m} = -(i/\lambda) \int_{h(x)}^{\infty} y^{m-1} g(x, y) h'(x) \exp(i\lambda xy)\, dx$$
$$+ \frac{i}{\lambda} \iint_{\Omega} y^{m-1} \frac{\partial g}{\partial x} \exp(i\lambda xy)\, dx\, xy$$

Since

$$\frac{\partial g}{\partial x} = \frac{\partial g}{\partial y} = 0$$

in the vicinity of the stationary point, so in regions supp $(\partial g/\partial x)$ and supp $(\partial g/\partial y)$ the written two-fold integrals have no critical points and therefore fall off more rapidly than any power of λ^{-1}. Reducing the integrals $I_{n,m}$ to $I_{n-m,0}$ for $n > m$ and to $I_{0,m-n}$ for $n < m$ one obtains that for $n \neq m$

$$I_{n,m} \simeq \int_{y=h(x)}^{\infty} R_{n,m}(\lambda, x) \exp(i\lambda xy)\, dx$$

i.e. eqn. 4.38*a* holds for $Q_{m,n} = xR_{n,m}$ where $R_{n,m}$ is a finite series in powers of λ^{-1}.

To prove eqn. 4.38*a* for $n = m$, consider first $I_{0,0}$ and choose such $c > 0$ that the function $g(x)$ is identically equal to unity for sufficiently small x and $h(x) < y < c$. Then

$$I_{0,0} = \int_{c}^{\infty} dy \int_{-\infty}^{\infty} g(x, y) \exp(i\lambda xy)\, dx$$
$$+ \int_{-\infty}^{\infty} dx \int_{h(x)}^{c} g(x, y) \exp(i\lambda xy)\, dy$$

As $\lambda \to \infty$ the first integral tends to zero more rapidly than any power of λ^{-1}. When integrating by parts the second integral can be reduced to the form

$$I_{11} + I_{12} = \int_{-\infty}^{\infty} \frac{\exp(i\lambda cx) g(x, c) - \exp[i\lambda x h(x)] g[x, h(x)]}{i\lambda x}\, dx$$
$$+ \frac{i}{\lambda} \iint_{h(x) < y < c} \frac{\partial g}{\partial x} \exp(i\lambda xy) \frac{dx\, dy}{x}$$

Since $(\partial g/\partial x) \equiv 0$ for small x and $h(x) < y < c$ the integral $I_{1,2}$ falls off more rapidly than any power of λ^{-1}. Shift the integration contour for the integral $I_{1,1}$ into the upper complex halfplane. Since the integral

$$\int_{-\infty}^{\infty} \exp(i\lambda cx) g(x, c) \frac{dx}{x + iO}$$

falls off more rapidly than any power of λ^{-1} one gets

$$I_{0,0} \simeq \iint_{y>h(x)} \exp(i\lambda xy) g(x, y)\, dx\, dy$$
$$\simeq \frac{i}{\lambda} \int_{y=h(x)}^{\infty} \exp(i\lambda xy) h(x, y) \frac{dx}{x+iO}$$

Differentiating this relation N times with respect to λ obtains eqn. 4.38*a* for $n=m$.

To estimate $I_N^{(1)}$ integrate the integral N times by parts with respect to y each time integrating $\exp(i\lambda xy)$. Summing the integrals arising in each partial integration over the boundary and estimating (in magnitude) the two-fold integral obtained after integrating N times gives the eqn. 4.38*b* for $I_N^{(1)}$. The integral $I_N^{(2)}$ is estimated in a similar way; note that the integral over the boundary arising when integrating by parts with respect to x

$$\int_\Sigma \exp(i\lambda xy) R(\lambda, x, y)\, dy$$

can be written (after the variable change and taking into account that $y=h(x)$ and $dy=h'(x)\, dx$ on Σ) as follows

$$\int_\Sigma \exp([i\lambda x h(x)] R[\lambda, x, h'(x)] h(x)\, dx$$

from which estimation eqn. 4.38*b* follows. Thus, eqn. 4.28 is proven.

4.4 Extremum point of phase function near corner point of boundary

Let the integral $I(\lambda)$ be specified by eqn. 4.18 where the phase function $\varphi(x, y)$ has a nondegenerate extremal stationary point O near a corner point Q on the boundary of the integration domain Ω. This Section obtains the expression for the asymptotic of $I(\lambda)$ uniform over a relative location of the points Q and O.

The nonuniform asymptotic of such integral has the form

$$\begin{aligned} I(\lambda) \simeq \lambda^{-1} R \exp(i\lambda\varphi_0)\chi_0 + \lambda^{-3/2} P_A \exp(i\lambda\varphi_A)\chi_A \\ + \lambda^{-3/2} P_B \exp(i\lambda\varphi_B)\chi_B + \lambda^{-2} D \exp[i\lambda\varphi_Q] \end{aligned} \tag{4.39}$$

Here φ_0 and φ_Q are the values of the phase function at the stationary point O and the corner point Q, respectively (see Figure 4.10), $\varphi_A = \varphi(M_A)$ is the extreme value of $\varphi(x, y)$ on the arc AQA' (QA' is an analytical extension of the arc AQ), and M_A is a point where the extremum is reached, the values φ_B and M_B are defined in a similar way. The function $\chi_0 = 1$, if the point O is in the integration domain Ω, otherwise $\chi_0 = 0$. The function $\chi_A = 1$, if M_A is on the arc AQ, and $\chi_A = 0$, if M_A is on the arc QA'. Similarly, $\chi_B = 1$, if M_B belongs to BQ and $\chi_B = 0$, if M_B belongs to the arc QB'.

The functions R, P_A, P_B, D are asymptotic series in powers of λ^{-1}, their leading terms are determined by the equations of Section 4.1.

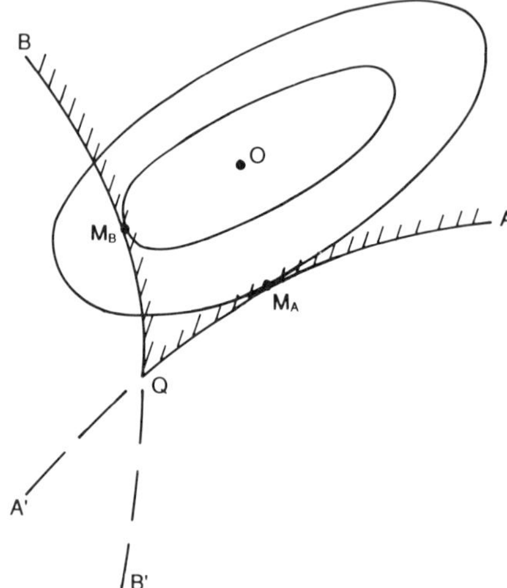

Figure 4.10 Extremal stationary point near corner point of boundary

If the point O approaches the arc AQ or BQ being far from Q, coefficients of the series P_A or P_B in eqn. 4.39 tend to infinity. If M_A or M_B tend to Q, the coefficients of the series D tend to infinity. As shown in Section 4.1.2, in these cases the uniform asymptotic of $I(\lambda)$ is expressed in terms of the Fresnel integral, if the point O is a minimum point of the phase function, and in terms of the complex conjugate Fresnel integral, if O is a point of maximum. If the point O approaches Q, the coefficients of all three series P_A, P_B and D tend to infinity.

To find a special function in terms of which the uniform asymptotic of the integral $I(\lambda)$ is expressed consider the particular case of the quadratic phase function

$$\varphi(x, y) = (x - \alpha)^2 + (y - \beta)^2$$

and the region Ω which is the internal part of an angle BQA (Figure 4.11).

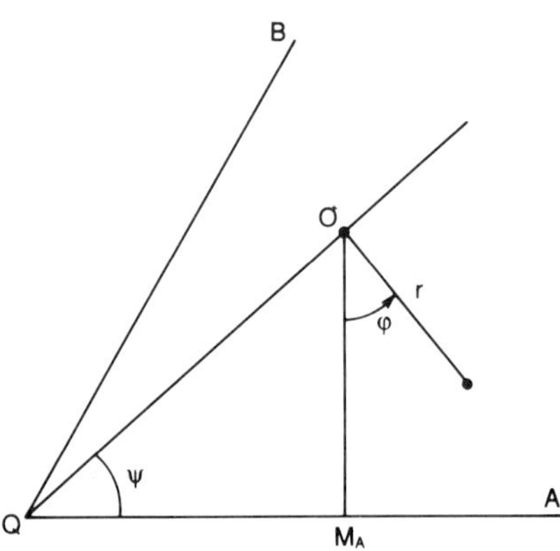

Figure 4.11 Geometrical parameters of model problem

First, let the non-exponential factor in the integral be equal to unity and the point O be inside Ω. Then $I(\lambda)$ is

$$I(\lambda)=I_0=\iint_\Omega \exp\{i\lambda[(x-\alpha)^2+(y-\beta)^2]\}\,dx\,dy=I_A+I_B$$

where I_A is the integral over the angle region OQA, while I_B is the integral over the region $O\,QB$, the co-ordinates of the point O being α, β.

To find I_A, introduce the polar co-ordinates $r=\sqrt{[(x-\alpha)^2+(y-\beta)^2]}$, φ as in Figure 4.11 and integrating with respect to r (for the integral to converge as $r\to\infty$ set $\mathrm{Im}\,(\lambda)=\varepsilon>0$) gives

$$I_A=\frac{1}{2i\lambda}\int_{\varphi-\pi/2}^{\pi/2}\exp\,[i\lambda r^2(\varphi)]\,d\varphi-\frac{\pi}{2i\lambda}$$

where

$$r(\varphi)=\rho_A(\cos\varphi)^{-1}=|OM_A|(\cos\varphi)^{-1}$$

After the variable change $\xi=\rho_A tg\varphi$ in the integral one obtains

$$\begin{aligned}I_A&=\frac{1}{2i\lambda}\int_{-|QM_A|}^{\infty}\exp\,[i\lambda\,(\rho_A^2+\xi^2)]\,\frac{\rho_A\,d\xi}{\rho_A^2+\xi^2}-\frac{\pi}{2i\lambda}\\&=-(\pi i/\lambda)G(-\sqrt{\lambda}\rho_{AQ},\sqrt{\lambda}\rho_A)+(\pi i/2\lambda)\end{aligned}$$

where $G(x,y)$ is the generalised Fresnel integral (see Section 6.10)

$$G(x,y)=(y/2\pi)\int_x^\infty \exp\,[i(t^2+y^2)]\,\frac{dt}{t^2+y^2}$$

and $\rho_{AQ}=|QM_A|$.

The integral $I_B(\lambda)$ is evaluated in a similar way. Considering $I(\lambda)$ for various versions of the point O position relative to sides of the angle region Ω, one obtains the unified formula

$$I_0=(\pi i/\lambda)[\chi_0-G(-\sqrt{\lambda}\rho_{AQ},\sqrt{\lambda}\rho_A)-G(-\sqrt{\lambda}\rho_{BQ},\sqrt{\lambda}\rho_B)] \tag{4.40}$$

Here, $\chi_0=1$, if the point O is inside Ω, otherwise $\chi_0=0$. The functions ρ_{AQ}, ρ_A, ρ_{BQ}, ρ_B are of the form

$$\begin{aligned}\rho_{AQ}&=\pm\sqrt{|\varphi(Q)-\varphi(M_A)|}, &\quad \rho_{BQ}&=\pm\sqrt{|\varphi(Q)-\varphi(M_B)|}\\ \rho_A&=\pm\sqrt{|\varphi(M_A)-\varphi(O)|}, &\quad \rho_B&=\pm\sqrt{|\varphi(M_B)-\varphi(O)|}\end{aligned} \tag{4.41}$$

The signs of ρ_A, ρ_{AQ}, ρ_B, ρ_{BQ} are chosen as follows. Set $\rho_A>0$, if the point O is at the same side from AQA' as Ω (QA' is an extension of AQ), otherwise $\rho_A<0$. If M_A (the minimum point of the phase function on AQA') is between the points Q and A, the quantity ρ_{AQ} is positive, otherwise $\rho_{AQ}<0$. The signs of ρ_B and ρ_{BQ} are defined in the same way.

Note that though χ_0 is a discontinuous function of the position of the point O, i.e. of the variables α, β, the right-hand side of eqn. 4.40 is a regular function of α, β, since the discontinuities of the function χ_0 are compensated by the discontinuities of the second and third summands. For instance, when the point O intersects the side AQ both χ_0 and $G(-\sqrt{\lambda}\rho_{AQ}, \sqrt{\lambda}\rho_A)$ become discontinuous simultaneously.

When the nonexponential factor under the integral sign is a polynomial in $(x-\alpha)$ and $(y-\beta)$ the expression of the integral $I(\lambda)$ is obtained when differentiating with respect to α and β many times. In this case one needs to differentiate the generalised Fresnel integral $G(p, q)$ with respect to its arguments p, q. According to Section 6.10 the derivatives are expressed in terms of the exponentials and the Fresnel integral

$$\frac{\partial G(p, q)}{\partial p} = -\frac{q}{2\pi}\,\frac{\exp\,[i(p^2+q^2)]}{p^2+q^2}$$

$$\frac{\partial G(p, q)}{\partial q} = \frac{ip}{2\pi}\,\frac{\exp\,[i(p^2+q^2)}{p^2+q^2} - \frac{F(-p)\exp\,(iq^2)}{\sqrt{(\pi i)}}$$

Therefore in this particular case the asymptotic of $I(\lambda)$ is expressed in terms of the Fresnel integral, the generalised Fresnel integral and exponentials

$$\begin{aligned} I(\lambda) \simeq {} & (1/\lambda)R\exp\,[i\lambda\varphi(O)]\,[\chi_0 - G(-\sqrt{\lambda}\rho_{AQ}, \sqrt{\lambda}\rho_A) \\ & - G(-\sqrt{\lambda}\rho_{BQ}, \sqrt{\lambda}\rho_B)] + \lambda^{-3/2}P_A^*\exp\,[i\lambda\varphi(A)]\,F(\sqrt{\lambda}\rho_{AQ}) \\ & + \lambda^{-3/2}P_B^*\exp\,[i\lambda\varphi(B)]\,F(\sqrt{\lambda}\rho_{BQ}) \\ & + \lambda^{-2}D^*\exp\,[i\lambda\varphi(Q)] \end{aligned} \tag{4.42}$$

It appears that in the general case the uniform asymptotic of $I(\lambda)$ is of the same form, i.e. the following theorem holds [31, 32]

Theorem 4.4: In the integral

$$I(\lambda) = \iint_\Omega f(x, y)g(x, y)\exp\,[i\lambda\varphi(x, y)]\;dx\,dy$$

let the phase function $\varphi(x, y)$ have a nondegenerate stationary point O (a minimum point) near a corner point Q on the boundary of the integration domain Ω. The cutting function $g(x, y)$ singles out a vicinity of the close stationary point O and the corner point Q. Then the uniform asymptotic of $I(\lambda)$ is of the form of eqn. 4.42 where G and F are the generalised and standard Fresnel integrals. The functions ρ_{AQ}, ρ_A, ρ_{BQ}, ρ_B are expressed by eqn. 4.41. The asymptotic series R in eqn. 4.42 is the same as in the nonuniform asymptotic eqn. 4.39. The asymptotic series P_A^*, Q_B^*, D^* have regular coefficients and are determined by the series P_A, P_B, D in the nonuniform asymptotic (eqn. 4.39) by means of asymptotic matching

$$\begin{aligned} & P_A^* = P_A + \sqrt{\lambda}RT(\sqrt{\lambda}\rho_A), \qquad P_B^* = P_B + \sqrt{\lambda}RT(\sqrt{\lambda}\rho_B) \\ & D^* = D + \sqrt{\lambda}[P_A^*T(\sqrt{\lambda}\rho_{AQ}) + P_B^*T(\sqrt{\lambda}\rho_{BQ})] \\ & \qquad + \lambda R[M(-\sqrt{\lambda}\rho_{AQ}, \sqrt{\lambda}\rho_A) + M(-\sqrt{\lambda}\rho_{BQ}, \sqrt{\lambda}\rho_B)] \end{aligned} \tag{4.43}$$

where $T(x)$ and $M(x, y)$ are asymptotic series (eqns. 6.6 and 6.42) in terms of which the asymptotic of $F(x)$ and of the generalised Fresnel integral $G(x, y)$ is expressed as $|x|, |y| \to \infty$.

If the phase function $\varphi(x, y)$ reaches its maximum at O the uniform asymptotic $I(\lambda)$ is expressed in terms of the conjugate generalised and standard Fresnel integrals

$$\begin{aligned} I(\lambda) \simeq \lambda^{-1} R \exp[i\lambda\varphi(0)][\chi_0 - G^*(-\sqrt{\lambda}\rho_{AQ}, \sqrt{\lambda}\rho_A) \\ - G^*(-\sqrt{\lambda}\rho_{BQ}, \sqrt{\lambda}\rho_B)] + \lambda^{-3/2} P_A^* \exp[i\lambda\varphi(A)] F^*(\sqrt{\lambda}\rho_{AQ}) \\ + \lambda^{-3/2} P_B^* \exp[i\lambda\varphi(B)] F^*(\sqrt{\lambda}\rho_{BQ}) \\ + \lambda^{-2} D^* \exp[i\lambda\varphi(Q)] \end{aligned} \quad (4.44)$$

In eqns. 4.43 for P_A^*, P_B^* and D^* one should replace the asymptotic series T and M by their complex conjugate.

The regularity of the coefficients of the series P_A^*, P_B^*, D^* means that poles of the expressions on the right-hand sides of eqn. 4.43 should cancel. Note further that eqns. 4.42 and 4.44 are applicable also in those cases when the function $f(x, y)$ vanishes at critical points of the integral $I(\lambda)$ or when at the corner point Q the arcs AQ and BQ are tangential to each other. In these cases the leading terms of the corresponding series in the equations vanish and one of the next terms becomes to be the leading one.

The proof of theorem 4.4 is laborious and therefore it is given separately. To proceed from the case when f reaches its maximum at a stationary point to the one when φ reaches a minimum, the complex conjugation in the integral $I(\lambda)$ is used. Hence, it is sufficient to be restricted to the proof of eqn. 4.42.

4.5 Proof of eqn. 4.42

Suppose the nondegenerate stationary point O of the phase function $\varphi(x, y)$ to be a minimum point and $\varphi(O) = 0$. Then, as shown in Section 4.1.2, there exists an analytical and reversible variable change transforming the phase function $\varphi(x, y)$ to a sum of squares

$$\varphi = (x - \alpha)^2 + (y - \beta)^2 \quad (4.45)$$

where (α, β) are the co-ordinates of O. It is sufficient to prove eqn. 4.42 for this particular case.

If φ is of the form of eqn. 4.45, the critical points M_A and M_B are projections of the point O to the arcs AQA' and BQB' and ρ_A, ρ_B (taken with the appropriate signs) are distances from O to M_A and M_B, $\rho_{AQ} = \pm\sqrt{|d^2 - \rho_A^2|}$, $\rho_{BQ} = \pm\sqrt{|d^2 - \rho_B^2|}$, where $d = |OQ|$. Eqn. 4.42 is written as

$$\begin{aligned} \iint_\Omega f(x, y) g(x, y) \exp\{i\lambda[(x-\alpha)^2 + (y-\beta)^2]\}\, dx\, dy \\ = (1/\lambda) R[\chi_0 - G(-\sqrt{\lambda}\rho_{AQ}, \sqrt{\lambda}\rho_A) - G(-\sqrt{\lambda}\rho_{BQ}, \sqrt{\lambda}\rho_B)] \\ + \lambda^{-3/2} P_A^* \exp(i\lambda\rho_A^2) F(\sqrt{\lambda}\rho_{AQ}) \\ + \lambda^{-3/2} P_B^* \exp(i\lambda\rho_B^2) F(\sqrt{\lambda}\rho_{BQ}) + \lambda^{-2} D^* \exp(i\lambda d^2) \end{aligned} \quad (4.46)$$

Eqn. 4.46 is proven assuming that the cutting function g is chosen in such a way that eqn. 4.46 has no critical points in the region supp $|\text{grad}(g)|$. First, eqn. 4.46 is proven for the particular case of $f(x, y) \equiv 1$. Then differentiating the expression obtained many times with respect to parameters α, β eqn. 4.46 is proven when $f(x, y)$ is a polynomial in x, y. Finally, it is shown that any function $f(x, y)$ can be approximated by a polynomial f_N for the difference $f(x, y) - f_N$ to fall off more rapidly than λ^{-N} uniformly over a distance between the point O and the boundary of the domain Ω.

Choose the points A and B on the sides of the boundary so far from Q and O for the cutting function g to vanish identically on the arc AB between these points. Then consider the integration in eqn. 4.46 to be performed over a curvilinear triangle ABQ (Figure 4.12).

To prove eqn. 4.46 for $f(x, y) = 1$, introduce polar co-ordinates with the point O as the origin. Let φ_A, φ_B, φ_Q be directions to A, B, Q, while $r_A(\varphi)$, $r_B(\varphi)$ and $R(\varphi)$ be equations of the arcs AQ, BQ and AB, respectively. Then

$$\iint_\Omega g(x, y) \exp \{i\lambda[(x-\alpha)^2 + (y-\beta)^2]\}\, dx\, dy$$
$$= \int_{\varphi_A}^{\varphi_B} d\varphi \int_0^{R(\varphi)} \exp(i\lambda r^2) r g(r, \varphi)\, dr$$
$$+ \int_{\varphi_B}^{\varphi_Q} d\varphi \int_0^{r_B(\varphi)} \exp(i\lambda r^2) r g(r, \varphi)\, dr$$
$$= \int_{\varphi_Q}^{\varphi_A} d\varphi \int_0^{r_A(\varphi)} \exp(i\lambda r^2) r g(x, \varphi)\, dr$$

Integrating each of the written integrals by parts (integrating $r \exp(i\lambda r^2)$), differentiating $g(x, y)$ and taking into account that the cutting function $g = 0$ on the arc AB and $g = 1$ in a vicinity of the point Q, gives

$$\iint_\Omega g(x, y) \exp \{i\lambda[(x-\alpha)^2 + (y-\beta)^2]\, dx\, dy$$
$$= \frac{i}{2\lambda} \iint_\Omega \exp(i\lambda r^2) \frac{\partial g(r, \varphi)}{\partial r}\, dr\, d\varphi$$
$$- \frac{i}{2\lambda} \int_{\varphi_B}^{\varphi_Q} \exp[i\lambda r_B^2(\varphi)] g[\varphi, r_B(\varphi)]\, d\varphi$$
$$- \frac{i}{2\lambda} \int_{\varphi_Q}^{\varphi_A} \exp[i\lambda r_B^2(\varphi)] g[\varphi, r(\varphi)]\, d\varphi + (\pi i \chi_0/\lambda) \tag{4.47}$$

Here, $\chi_0 = 1$, if the stationary point O is inside the curvilinear triangle $ABQ = \Omega$, otherwise $\chi_0 = 0$. Since the two-fold integral in eqn. 4.46 has no critical points in a region supp $|\text{grad}\ \varphi|$ it falls off more rapidly than any power of λ^{-1}. Therefore to prove eqn. 4.46 when $f(x, y) \equiv 1$ it is sufficient to prove that equation

$$I_B = \int_{\varphi_B}^{\varphi_Q} \exp[i\lambda r_B^2(\varphi)] g[\varphi, r_B(\varphi)]\, d\varphi$$
$$= 2\pi G(-\sqrt{\lambda}\rho_{BQ}, \sqrt{\lambda}\rho_B) + \lambda^{-1/2} P_0^* F(\sqrt{\lambda}\rho_{BQ}) \exp(i\lambda\rho_B)$$
$$+ (1/\lambda) D_0^* \exp(i\lambda d) \tag{4.48}$$

and a similar equation for the third integral in eqn. 4.47.

To prove eqn. 4.48 introduce a new independent variable $\xi = \pm\sqrt{[r_B^2(\varphi) - \rho_B^2]}$ on the arc QB where $\xi > 0$ between the points M_B and B and $\xi < 0$ between M_B and B' (Figure 4.10). Then $r_B^2(\varphi) = \xi^2 + \rho_B^2$ and eqn. 4.48 is written as

$$I_B = \int_{-\rho_{BQ}}^{\infty} \exp\,[i\lambda\,(\xi^2 + \rho_B^2)]g(\xi)h(\xi)\,d\xi \tag{4.49}$$

where $h(\xi) = (d\xi/d\varphi)^{-1}$ and the cutting function $g(\xi)$ vanishes outside a vicinity of the points Q, M_B, i.e. outside the vicinity of the points $\xi = 0$, $\xi = -\rho_{BQ}$.

Eqn. 4.48 follows from the relation

$$h(\xi) = (d\xi/d\varphi)^{-1} = [\rho_B/(\xi^2 + \rho_B^2)] + q(\xi) \tag{4.50}$$

where $q(\xi)$ is analytical and regular for small ρ_B, ξ. Indeed, substituting this expression into eqn. 4.49 gives

$$I_B = \rho_B \int_{\rho_{BQ}}^{\infty} \frac{\exp\,[i\lambda\,(\xi^2 + \rho_B^2)]}{\xi^2 + \rho_B^2}\, g(\xi)\,d\xi + \int_{\rho_{BQ}}^{\infty} \exp\,[i\lambda\,(\xi^2 + \rho_B^2)g(\xi)q(\xi)]\,d\xi$$

The first of these integrals is asymptotically equivalent to the generalised Fresnel integral, i.e. to the first summand in eqn. 4.48, the second integral is equivalent to the second and third summands. Therefore the proof of eqn. 4.48 is reduced to the proof that the function $g(\xi)$ in eqn. 4.50 is analytical. To prove this, introduce the co-ordinates u, v with the point M_B as the origin, the v-axis being directed along the tangent to the arc BQ, while the u-axis is normal to it and directed to the point O (Figure 4.12). Since $OM_B = \rho_B$ the variables φ, ξ and u, v are related on the arc QB by the equations

$$r_B = \sqrt{[(\rho_B - u)^2 + v^2]}, \qquad \varphi = \varphi_B + \operatorname{arc\,tg}[v/(\rho_B - u)]$$

$$\xi = \sqrt{(r_B^2 - \rho_B^2)} = \sqrt{(u^2 + v^2 - 2\rho_B u)}$$

where $u = u(v) = v^2 t(v)$ is the equation of the arc BQ, the function $t(v)$ being analytically dependent on the co-ordinates of the point O. Here, φ_B is the direction from O to the point M_B.

Therefore for sufficiently small v, $\rho = \rho_B$ the function

$$\xi = \sqrt{(u^2 + v^2 - 2\rho u)} = v\sqrt{[1 - 2\rho t(v) + v^2 t^2(v)]}$$

is an analytical and bounded function of v. Now,

$$\frac{\partial\varphi}{\partial v} = \frac{\rho - u + u'v}{(\rho - u)^2 + v^2}, \qquad \left[\frac{\partial\xi}{\partial v}\right]^{-1} = \frac{\sqrt{(u^2 + v^2 - 2\rho u)}}{v + uu' - \rho u'}$$

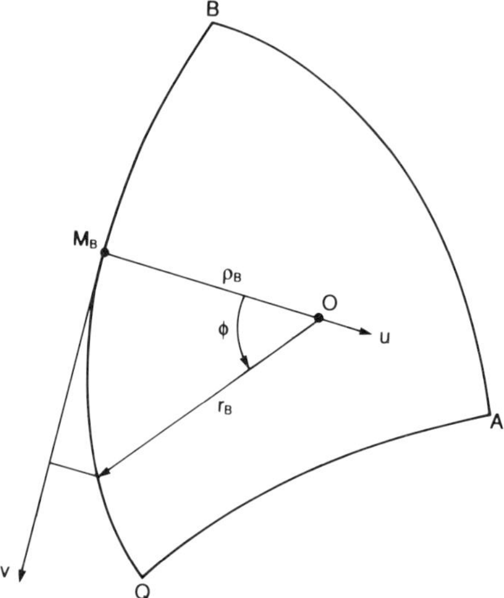

Figure 4.12 Co-ordinates u, v and ρ_B, φ

and therefore

$$q(\xi)=\left[\frac{\partial\xi}{\partial\varphi}\right]^{-1}-\frac{\rho}{\rho^2+\xi^2}=\left[\frac{\partial\xi}{\partial v}\right]^{-1}\frac{\partial\varphi}{\partial v}-\frac{\rho}{\rho^2+\xi^2}$$
$$=\frac{(\rho-u+vu')\sqrt{(u^2+v^2-2\rho u)}-\rho(v+uu'-\rho u')}{[(\rho-u)^2+v^2](v+uu'-\rho u')}$$

where

$$u'=\frac{du\,(v)}{dv}=\frac{d}{dv}\,[v^2t(v)]$$

To prove that $q(\xi)$ is regular it is sufficient to prove that the zeros of the numerator and denominator coincide. Since

$$v+uu'-\rho u'=v[1-2\rho t(v)+O(v)]$$
$$\sqrt{(u^2+v^2-2\rho u)}=v\sqrt{[1-2\rho t(v)+t^2(v)]}$$

it is sufficient to verify that the numerator vanishes for $(\rho-u)^2+v^2=0$, i.e. for $\rho=u\pm iv$. The statement is verified directly if one takes into account that for $\rho=u\pm iv$

$$\xi=\sqrt{(u^2+v^2-2\rho u)}=v\mp iu$$

Thus, the regularity of the function $q(\xi)$ in eqn. 4.50 is proved and thereby eqn. 4.46 for $f(x,y)\equiv 1$.

To proceed to the case of an arbitrary analytical function f it is proved that for any m, n

$$\iint_{\Omega} g(x, y) \exp \{i\lambda[(x-\alpha)^2 + (y-\beta)^2]\} x^m y^n \, dx \, dy$$
$$= \lambda^{-1} R_{m,n}\{\chi_0 - G[-\sqrt{\lambda}\rho_{AQ}, -\sqrt{\lambda}\rho_A]$$
$$-G[-\sqrt{\lambda}\rho_{BQ}, \sqrt{\lambda}\rho_B] + \lambda^{-3/2} P^*_{A,m,n} \exp(i\lambda\rho_A) F[\sqrt{\lambda}\rho_{AQ}]$$
$$+ \lambda^{-3/2} P^*_{B,m,n} \exp(i\lambda\rho_B F[\sqrt{\lambda}\rho_B] + \lambda^{-2} D_{m,n} \exp(i\lambda d) \qquad (4.51)$$

where the coefficients of the asymptotic series $R_{m,n}$, $P^*_{A,m,n}$, $P^*_{B,m,n}$, $D_{m,n}$ in powers of λ^{-1} are regular functions of α, β.

Eqn. 4.51 has been proved for $m = n = 0$. For other m, n it is proved by induction. On going from m to $m+1$ one should apply an operator to eqn. 4.51

$$\frac{i}{\lambda} \frac{d}{d\alpha} - \alpha$$

while to go from n to $(n+1)$ one should use the operator

$$\frac{i}{\lambda} \frac{d}{d\beta} + \beta$$

Therefore, to prove eqn. 4.51 it is sufficient to verify that after differentiating the right-hand side of this expression with respect to α or β one obtains again the expression with regular series R, P^*_A, P^*_B, D. For definiteness, in what follows differentiation is with respect to α.

Since ρ_A, ρ_B, ρ_{AQ} are analytical functions of α, and

$$\frac{\partial F[\sqrt{(\lambda)}\rho_{AQ}]}{\partial \alpha} = \sqrt{(\lambda/\pi i)} \frac{d\rho_{AQ}}{d\alpha} \exp(i\lambda\rho_{AQ})$$

$$\frac{\partial F[\sqrt{(\lambda)}\rho_{BQ}]}{\partial \alpha} = \sqrt{(\lambda/\pi i)} \frac{d\rho_{BQ}}{d\alpha} \exp(i\lambda\rho_{BQ})$$

the regularity of eqn. 4.51 can be violated only when differentiating the generalised Fresnel integral $G(x, y)$ with respect to α.

Using eqn. 6.40*a* for the derivatives of $G(x, y)$ one obtains that, if u, v are the functions of α, then

$$\frac{d}{d\alpha} G(u, v) = \frac{1}{\sqrt{(\pi i)}} F(-u) \frac{dv}{d\alpha} - \frac{i \exp[i(u^2 + v^2)]}{2\pi(u^2 + v^2)} \left[v \frac{du}{d\alpha} - u \frac{dv}{d\alpha} \right]$$

Therefore

$$\frac{d}{d\alpha} [G[-\sqrt{(\lambda)}\rho_{AQ}, \sqrt{(\lambda)}\rho_A] + G[-\sqrt{(\lambda)}\rho_{BQ}, \sqrt{(\lambda)}\rho_B]]$$

$$= \sqrt{(\lambda/\pi i)} F[\sqrt{(\lambda)}\rho_{AQ}] \frac{d\rho_A}{d\alpha} - \sqrt{(\lambda/\pi i)} F[\sqrt{(\lambda)}\rho_{BQ}] \frac{d\rho_B}{d\alpha}$$

$$+ \frac{i}{2\pi} H \frac{\exp(i\lambda d^2)}{d^2} \qquad (4.52)$$

where

$$H=\rho_A\frac{d\rho_{AQ}}{d\alpha}+\rho_B\frac{d\rho_{BQ}}{d\alpha}-\rho_{AQ}\frac{d\rho_A}{d\alpha}-\rho_{BQ}\frac{d\rho_B}{d\alpha}$$

and the relation $d^2=\rho_A^2+\rho_{AQ}^2=\rho_B^2+\rho_{BQ}^2$ is taken into account. Since ρ_A, ρ_B, ρ_{AQ}, ρ_{BQ} are analytical functions of α, β the regularity of the expression can be violated only as $d\rightarrow 0$, i.e. when the point O tends to the vertex Q. Therefore one needs to prove that Hd^{-2} remains an analytical function of α, β as $O\rightarrow Q$. This function can be written as

$$\begin{aligned}\frac{H}{d^2}&=\frac{d}{d\alpha}\,[\pi-\arcsin(\rho_A/d)-\arcsin(\rho_B/d)]\\&=\frac{d(\varphi_1+\varphi_2)}{d\alpha}+\frac{d\delta_1}{d\alpha}+\frac{d\delta_2}{d\alpha}\end{aligned}$$

where the angles φ_1, φ_2 are shown in Figure 4.13.

$$\delta_1=(\pi/2)-\arcsin(\rho_A/d)-\varphi_1$$

$$\delta_2=(\pi/2)-\arcsin(\rho_B/d)-\varphi_2$$

The sum $\varphi_1+\varphi_2$ is an analytical function of the point O position since it is equal to the angle between the normals to the arcs AQ and QB going out from point O. To prove that δ_2 is an analytical function let v, $u(v)$ be co-ordinates of the vertex Q in the coordinate system u, v shown in Figure 4.13. It is obvious that v, u are analytical functions of α, β the co-ordinates of the stationary point O:

$$d=\sqrt{[(\rho_B-u)^2+v^2]},\qquad \varphi_2=\arccos[(\rho_B-u/d)]$$

$$\begin{aligned}\sin(\delta_2)&=\sqrt{[(d^2-\rho_B^2)/d^2]}\cos\varphi_2-(\rho_B/d)\sin\varphi_2\\&=\frac{(\rho_B-u)\sqrt{(u^2+v^2-2\rho_B u)}-\rho_B v}{(\rho_B-u)^2+v^2}\end{aligned}$$

where $u=u(v)=v^2t(v)$. The direct calculations show that the numerator of this expression vanishes at the same $\rho_B=u\pm iv$ as the denominator does, therefore $\sin(\delta_2)$ is an analytical function of α, β. Similarly, one can verify that $\cos(\delta_2)$ is an analytical function. Hence, δ_2 and $d\delta_2/d\alpha$ are analytical functions of α, β as well as $d\delta_1/d\alpha$.

Thus, eqn. 4.52 is regular and, hence, eqn. 4.51 holds for any integers m, n. It follows from this that the uniform asymptotic eqn. 4.46 is valid if $f(x, y)$ is a polynomial in x, y.

To prove eqn. 4.46 for arbitrary $f(x, y)$ one needs to approximate it by such polynomials f_N that the difference $\delta f=f(x,y)-f_N$ has zeros of the sufficiently high order at all the critical points of eqn. 4.46, i.e. at the stationary point O, at its projections M_A and M_B on the arcs AQA' and BQB' and at the

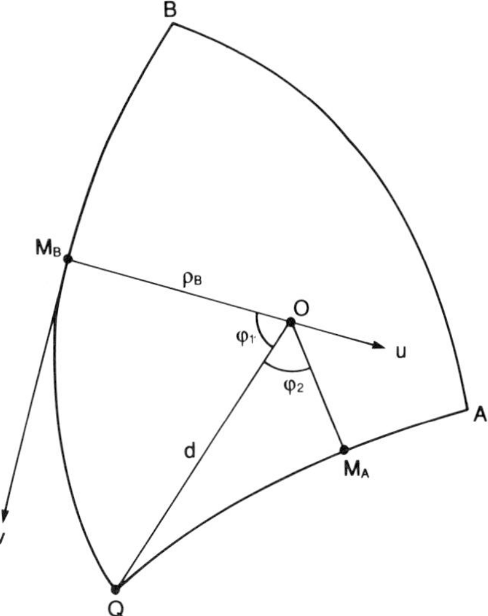

Figure 4.13 To the proof of eqn. 4.42

vertex Q:

$$I = \iint_\Omega g(x, y) f(x, y) \exp \{i\lambda[(x-\alpha)^2 + (y-\beta)^2]\} \, dx \, dy = I_1 + I_2$$

$$= \iint_\Omega g(x, y) f_N(x, y) \exp \{i\lambda[(x-\alpha)^2 + (y-\beta)^2]\} \, dx \, dy$$

$$+ \iint_\Omega g(x, y) \delta f(x, y) \exp \{i\lambda[(x-\alpha)^2 + (y-\beta)^2]\} \, dx \, dy$$

For the integral I_1 in the expression eqn. 4.46 holds. Therefore it is sufficient to show that the integral I_2 is estimated by const (λ^{-N}) uniformly over α, β, then eqn. 4.46 would be valid accurate up to λ^{-N} for the integral I as well.

Let δf have zeros of the order of $2N$ at the points O, M_A, M_B, Q. Then integrate

$$\iint_\Omega \exp (i\lambda r^2) \delta f(x, y) g(x, y) \, dx \, dy$$

N times by parts (the point O is an origin). If at the origin the function $\psi_n(x, y)$ has a zero of the order of $2N - 2n$, then using eqns. 4.2 and 4.3 one obtains

$$(i/\lambda)^n \iint_\Omega \exp (i\lambda r^2) \psi_n(x, y) \, dx \, dy$$

$$= (i/\lambda)^{n+1} \iint_\Omega \exp (i\lambda r^2) \psi_{n+1}(x, y) \, dx \, dy$$

$$- (i/\lambda)^{n+1} \int_\Sigma \exp (i\lambda r^2) \psi_n(x, y) \frac{\partial r}{\partial n} \frac{1}{r} \, ds$$

where s is the length measured along boundary Σ and

$$\psi_{n+1} = \frac{1}{r^2}\left[x\frac{\partial \psi_n}{\partial x} + y\frac{\partial \psi_n}{\partial y}\right]$$

has a zero of the order of $2N - 2(n+1)$ at the origin. Taking $\psi_0(x, y) = \delta f(x, y)g(x, y)$ and repeating the procedure N times one obtains a two-fold integral

$$\left[\frac{i}{\lambda}\right]^N \iint_\Omega \varphi_N(x, y) \exp\,(i\lambda r^2)\,dx\,dy$$

which can be estimated in magnitude. One-dimensional integrals along the boundary arising in this procedure are also estimated when integrating by parts. If one takes into account that at the phase function corner point (i.e. at the corner point Q of the boundary Σ) and at the stationary points M_A, M_B on the arcs BQ and AQ the functions ψ_n/r and $\partial\psi_n/\partial r$ have zeros of the order of $2(N-n)$, then one obtains that each of these integrals is also estimated by λ^{-N}.

4.6 Influence domain for nondegenerate stationary point

To estimate the applicability range for the nonuniform asymptotic of the integral $I(\lambda)$ defined by eqn. 4.1 in the case of approaching critical points, it is sufficient to find the conditions under which one can replace special functions in the uniform asymptotics obtained in Sections 4.2–4.4 by their asymptotic expansions applicable for large values of the arguments. The analysis of the conditions makes it possible to formulate a simple geometric criterion for the nonuniform asymptotic applicability, the criterion being based on a concept of the influence domain for a nondegenerate stationary point.

Let P be a critical point on a smooth part of the boundary Σ, i.e. it is a nondegenerate stationary point of the phase function $\varphi(s)$ (s is a length measured along the boundary). The influence domain Σ of this point is defined as an interval of the boundary Σ containing P on which

$$\lambda|\varphi(P) - \varphi(Q)| < \pi \tag{4.53}$$

i.e. an interval on which the phase (the exponential argument) of the integrand in $I(\lambda)$ differs from the phase at P not more than by π.

For the extremal nondegenerate critical point P inside the region Ω the influence domain Ω_P is specified as a vicinity of P defined by eqn. 4.53, i.e. as a vicinity of P bounded by the level line

$$\varphi(Q) = \varphi(P) \pm \pi/\lambda \tag{4.54}$$

where the minus sign is taken, if P is a local maximum, and plus if P is a local minimum. For $\lambda \gg 1$ the boundary of Ω_P is close to an ellipse with the halfaxis lengths being of the order of $\lambda^{-1/2}$.

In both cases the criterion for the nonuniform asymptotic to be applicable is the requirement for the phase function to have no other critical points in the influence domain. For the point P inside Ω this means, in particular, that the influence domain Ω_P should not intersect the boundary Σ of the integration domain Ω.

For a stationary saddle point definition eqn. 4.54 is not valid because the level lines of eqn. 4.54 are close to hyperbolas and extend far away from P. To define the influence domain for a stationary saddle point introduce such local co-ordinates x, y with P as the origin that $\varphi''_{xy}(P) = 0$ i.e. the term with x, y is absent in the expansion of this function into the Taylor series

$$\varphi = \varphi(P) + (x^2 \varphi''_{xx}/2) + (y^2 \varphi''_{yy}/2) + \cdots$$

Then the influence domain is defined by the inequality

$$\lambda[(x^2 |\varphi''_{xx}|/2) + y^2 |\varphi''_{yy}|/2] < \pi \tag{4.55}$$

For the nonuniform asymptotic to be applicable in the case of a stationary saddle point it is required, first [33], that the phase function has no other critical points inside the influence domain and, secondly, that the domain does not intersect the boundary Σ of the integration domain.

Note that the definition of eqn. 4.55 of the influence domain is also applicable for an extremal stationary point inside Ω, since the domain specified by this condition is close to the domain of eqn. 4.54; the sizes of these domains are of the order of $\lambda^{-1/2}$, while their boundaries are at a distance of the order of $\lambda^{-3/2}$ from each other.

Chapter 5

Some problems of diffraction and wave propagation

To solve many problems of the short-wave diffraction theory and wave propagation one manages to find their exact or approximate solutions as integrals of rapidly oscillating functions. The asymptotic analysis of these integrals is of interest from two complementary points of view.

On the one hand, diffraction problems are examples of how to apply the methods presented to find the uniform asymptotic. Moreover, in many cases the analysis of these problems shows that it is necessary to generalise the stationary phase method, to find the uniform asymptotic for new types of integrals and to introduce proper special functions. For instance, the Airy function was introduced as long ago as the first half of the previous century to describe the wavefield in a vicinity of caustics (the envelope of geometrical optics rays) [10].

On the other hand, the asymptotic for the integral describing wavefields makes it possible to get a vivid physical interpretation of the results obtained and also to give a qualitative description of the wavefields without any computing.

The nonuniform asymptotic of the integrals describing the sought-for field corresponds to the representation of this field in the form of a sum of a few ray fields (the primary field, edge field, etc.). Each of these fields, represented by the leading terms of their asymptotic, satisfies the standard laws of geometrical optics and propagates in a space region enclosed by the appropriate shadow–light boundaries. The uniform asymptotic corresponds to the field description in transient zones, i.e. in vicinities of the shadow–light boundaries. The wavefield in each type of transient zone is described by the proper special function.

This chapter obtains asymptotical solutions for a number of diffraction problems. I try to give a self-sufficient description so that a reader need not consult other books to verify and substantiate the facts and statements used.

5.1 Cylindrical functions and their asymptotics

The solutions $f(z)$ to the differential equation (the Bessel equation)

$$Lf = f'' + (f'/z) + (1 - \nu^2/z^2)f = 0 \tag{5.1}$$

are called the cylindrical functions. This equation arises in many problems of mathematical physics, in particular, if one looks for a solution to the two-dimensional Helmholtz equation

$$\Delta U + k^2 U = U_{xx} + U_{yy} + k^2 U = 0$$

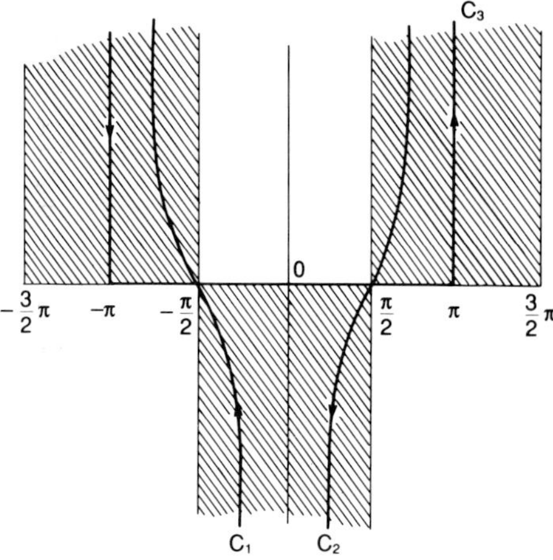

Figure 5.1 Integration contours for cylindrical functions

in the form $U=f(r)\exp(i\nu\varphi)$ where r, φ are the polar co-ordinates, then for $f(kr)$ one obtains eqn. 5.1. It is shown that the integral

$$f(z)=\int_C \exp(-iz\sin\theta+i\nu\theta)\,d\theta$$

is a solution to this equation. Here the contour C on the complex plane goes to infinity in the zones where the integrand falls off exponentially. Indeed, substituting this expression in eqn. 5.1 and differentiating under the integral sign gives

$$Lf=\int_C \exp(-iz\sin\theta+i\nu\theta)[\cos^2\theta+(i/z)\sin\theta-\nu^2/z^2]\,d\theta$$

$$=(i/z^2)\int_C d[(z\cos\theta+\nu)\exp(-iz\sin\theta+i\nu\theta)]=0$$

Depending on the choice of the contour C one obtains various solutions to eqn. 5.1. In Figure 5.1 the sectors are shadowed where the integrand magnitude is less than unity for $z>0$ and tends to zero exponentially as $|\theta|\to\infty$. Choosing the contours C_1 and C_2 (Figure 5.1) as C gives two linearly independent solutions to the Bessel equation, called the Hankel functions of the first and second kind

$$H_\nu^{(1)}(z)=-(1/\pi)\int_{C_1}\exp(-iz\sin\theta+i\nu\theta)\,d\theta$$

$$H_\nu^{(2)}(z)=(1/\pi)\int_{C_2}\exp(-iz\sin\theta+i\nu\theta)\,d\theta \tag{5.2}$$

For $z>0$ these functions are complex conjugate. To be convinced in this it is sufficient to transform the contours C_1, C_2 into the straight lines $\theta=\mp(\pi/2)+i\xi$, respectively, and to write $H_\nu^{(1,2)}(z)$ as integrals over ξ

$$H_\nu^{(1)}(z)=\frac{\exp(-i\nu\pi/2)}{i\pi}\int_{-\infty}^{\infty}\exp(iz\,\mathrm{ch}\,\xi-\nu\xi)\,d\xi$$
$$H_\nu^{(2)}(z)=\frac{i\exp(i\pi\nu/2)}{\pi}\int_{-\infty}^{\infty}\exp(-iz\,\mathrm{ch}\,\xi-\nu\xi)\,d\xi \tag{5.3}$$

The real parts of the functions $H_\nu^{(1)}$, $H_\nu^{(2)}$ are called the Bessel functions

$$I_\nu(z)=(1/2)[H_\nu^{(1)}(z)+H_\nu^{(2)}(z)]$$
$$=(1/2\pi)\int_{C_3}\exp(-iz\sin\theta+i\nu\theta)\,d\theta \tag{5.4}$$

where the contour C_3 is also shown in Figure 5.1. When the index $\nu=n$ is an integer the integrand is a periodic function, period 2π, the integrals along the vertical rays $\theta=\pm\pi+i\xi$ $(0<\xi<\infty)$ cancel and

$$I_n(z)=(1/2\pi)\int_{-\pi}^{\pi}\exp(-iz\sin\theta+in\theta)\,d\theta$$

Now find the asymptotic of $H_\nu^{(1)}(z)$ for $z\gg 1$ and ν fixed. Eqn. 5.3 is used where the phase function is $\varphi(\xi)=\mathrm{ch}\,\xi$, while the nonexponential factor is $f(\xi)=\exp(-\nu\xi)$. Evaluating the asymptotic of $H_\nu^{(1)}(z)$ according to theorem 1.2 and taking into account that the stationary point of the phase function $\xi=0$ is the only critical point of eqn. 5.3

$$H_\nu^{(1)}(z)\simeq(1/\pi)\exp\{i[-\nu(\pi/2)-(\pi/2)+(\pi/4)+z]\}\sqrt{(2\pi/z)}$$
$$=\sqrt{(2/\pi z)}\exp\{i[z-(\pi\nu/2)-\pi/4]\} \tag{5.5a}$$

Since $H_\nu^{(2)}$ is the complex conjugate to $H_\nu^{(1)}$ and $I_\nu=\mathrm{Re}\,(H_\nu^{(1)})$,

$$H_\nu^{(2)}(z)\simeq\sqrt{(2/\pi z)}\exp\{i[-z+\nu(\pi/2)+\pi/4]\}$$
$$I_\nu(z)\simeq\sqrt{(2/\pi z)}\cos[z-\nu(\pi/2)-\pi/4] \tag{5.5b}$$

The solution to the equation

$$\Delta G+k^2G=-\delta(x,y) \tag{5.6}$$

which satisfies the Sommerfeld radiation condition as $r=\sqrt{(x^2+y^2)}\to\infty$

$$G(kr)\sim\text{const.}\exp(ikr)/\sqrt{r} \tag{5.7}$$

is termed Green's function. To prove that $G(kr)$ is of the form

$$G(kr)=(i/4)H_0^{(1)}(kr) \tag{5.8}$$

The function $H_0^{(1)}(kr)$ satisfies the Helmholtz eqn. 5.6 for $r \neq 0$. It follows from eqn. 5.5*a* that

$$G(kr) \simeq \frac{\sqrt{i}}{2\sqrt{(2\pi kr)}} \exp(ikr)$$

as $r \to \infty$, i.e. the function G satisfies the radiation condition. Therefore it is sufficient to verify that as $r \to 0$ the function $H_0^{(1)}$ has a logarithmic singularity which provides that after applying the Laplace operator the δ function arises on the right-hand side of eqn. 5.6, i.e. to verify that

$$H_0^{(1)}(z) \simeq (2i/\pi) \ln(z) \tag{5.9}$$

as $z \to 0$. Change the integration variable sh $\xi = s$ in eqn. 5.3 and then change to the variable $g = sz$ (for $\nu = 0$)

$$\begin{aligned} H_0^{(1)}(z) &= (-i/\pi) \int_{-\infty}^{\infty} \exp[iz\sqrt{(1+s^2)}] \frac{ds}{\sqrt{(1+s^2)}} \\ &= (-2i/\pi) \int_0^{\infty} \exp[i\sqrt{(z^2+g^2)}] \frac{dg}{\sqrt{(z^2+g^2)}} \\ &= -(2i/\pi) \left[\int_1^{\infty} \exp[i\sqrt{(z^2+g^2)}] \frac{dg}{\sqrt{(z^2+g^2)}} + \int_0^1 \frac{dg}{\sqrt{(z^2+g^2)}} \right] \\ &\quad - \frac{2i}{\pi} \int_0^1 \frac{\exp[i\sqrt{(z^2+g^2)}] - 1}{\sqrt{(z^2+g^2)}} dg \end{aligned}$$

The first and third integrals remain bounded as $z \to 0$, while the second is evaluated analytically

$$\frac{-2i}{\pi} \int_0^1 \frac{dg}{\sqrt{(z^2+g^2)}} = (2i/\pi)\{\ln z - \ln[1 + (1+z^2)]\}$$

and this results in eqn. 5.9.

The stationary phase method can be used not only to obtain the asymptotic (i.e. approximate) equation. For example, to prove the following equality which has been used in Section 2.7 (Reference 26, eqn. 8.421)

$$I(\lambda, \alpha) = i\alpha^{\mu/2}\pi \exp(i\pi\mu/2) H_\mu^{(1)}(2\lambda\sqrt{\alpha}) \tag{5.10}$$

where

$$I(\lambda, \alpha) = \int_0^{\infty} \exp[i\lambda(\alpha x + 1/x)] x^{-\mu-1} \, dx \tag{5.11}$$

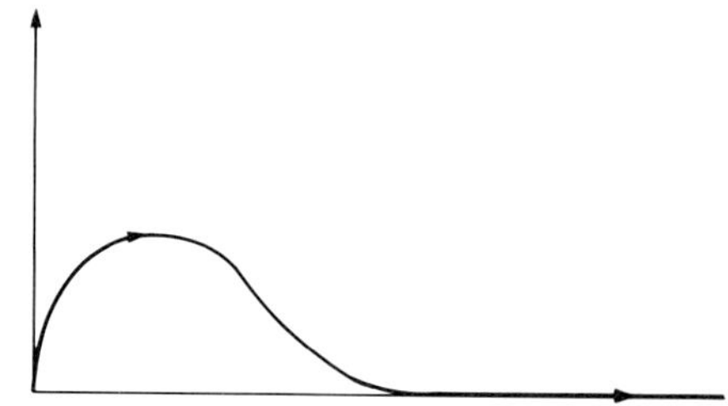

Figure 5.2 Shift of integration contour in eqn. 5.11

The integral converges at zero for $\mu < 1$. However, integrating along the contour shown in Figure 5.2, the exponential tends to zero as $|x| \to 0$ and $|x| \to \infty$ as $\exp(-\lambda/|x|)$ and $\exp(-\lambda\alpha|x|)$, respectively, i.e. the integral converges for any μ. To evaluate $I(\lambda, \alpha)$, i.e. to prove eqn. 5.10, reduce $I(\lambda, \alpha)$ to a function of one variable. Derive an equation for this function and solve it, and to obtain the integration constants apply the stationary phase method to the integral $I(\lambda, \alpha)$. Setting $y = \lambda\alpha x$ gives

$$I(\lambda, \alpha) = (\lambda\alpha)^{\mu} F(\lambda^2 \alpha)$$

where

$$F(\xi) = \int_0^{\infty} \exp\,[i(y + \xi/y)]y^{-\mu-1}\,dy$$

Now derive a differential equation for $F(\xi)$. In so doing integrate $F(\xi)$ by parts

$$\begin{aligned} F(\xi) &= -i\int_0^{\infty} \exp\,(i\xi/y)y^{-\mu-1}d[\exp\,(iy)] \\ &= i\int_0^{\infty}\left[-\frac{i\xi}{y^2} - \frac{1+\mu}{y}\right]\exp\,[i(y+\xi/y)y^{-\mu-1}\,dy \\ &= -\xi F''_{\xi\xi} - (\mu+1)F'_{\xi} \end{aligned}$$

The equation obtained

$$\xi F''_{\xi\xi} + (\mu+1)F'_{\xi} + F = 0$$

is reduced by the variable change $F = \xi^{-\mu/2} G(2\sqrt{\xi})$ to the Bessel equation

$$G''_{tt} + (G'_t/t) + (1 - \mu^2/t^2)G = 0$$

where $t = 2\sqrt{\xi}$. Therefore G is a linear combination of two Bessel functions of index μ for which one can take the Hankel functions of the first and second kind $H_{\mu}^{(1,2)}(2\sqrt{\xi})$ functions. Returning to $I(\lambda, \alpha)$ obtains

$$I(\lambda, \alpha) = \alpha^{\mu/2}[AH_{\mu}^{(1)}(2\lambda\sqrt{\alpha}) + BH_{\mu}^{(2)}(2\lambda\sqrt{\alpha})] \tag{5.12}$$

where A and B are constants (depending on parameter μ) to be determined. To find them set $\lambda \to \infty$ in eqn. 5.11. Evaluating the asymptotic of $I(\lambda, \alpha)$ by the stationary phase method yields

$$I(\lambda, \alpha) = \sqrt{(\pi/\lambda)}\alpha^{(2\mu-1)/4} \exp(i\pi/4) \exp(2i\lambda\sqrt{\alpha})$$

On the other hand, as $\lambda \to \infty$ the asymptotic of the right-hand side of eqn. 5.12 is of the form (see eqn. 5.5)

$$\alpha^{(2\mu-1)/4}\sqrt{(\pi\lambda)}\{A \exp[-(i\pi/4) - (i\pi\mu/2) + 2i\lambda\sqrt{\alpha}]$$
$$+B \exp[+(i\pi/4) + (i\pi\mu/4) - 2i\lambda\sqrt{\alpha}]\}$$

Equating these two expressions obtains $B = 0$, $A = i\pi \exp(i\pi\mu/2)$ i.e. eqn. 5.10.

5.2 Reflection from smooth surface in Kirchhoff approximation

Let a point source Q and a surface Σ be given. The reflected field $U_r(P)$ satisfying the Helmholtz equation and the radiation condition is termed the solution of the diffraction problems for the field $U_i(P) = G(P-Q)$ incident on Σ from the source Q if the total field $U(P) = U_i(P) + U_r(P)$ satisfies the boundary conditions on Σ.

The uniqueness of the solution is provided by the Sommerfeld radiation condition which in the two-dimensional case is of the form

$$U_r(r, \varphi) \simeq \frac{f(\varphi) \exp(ikr)}{\sqrt{r}} \tag{5.13}$$

as $r \to \infty$. Here the function $f(\varphi)$ is termed the pattern (or the radiation pattern) of the field $U_r(r, \varphi)$. This Section finds an approximate solution of the problem called the Kirchhoff approximation, or the physical optics approximation. This approximation is of quadrature form and its short-wave asymptotic is analysed using the methods developed.

Consider a two-dimensional problem in the domain Ω external to a smooth curve Σ on which the Dirichlét or Neumann boundary conditions are posed. Then according to eqn. 5.8 the incident field $U_i(P)$ is

$$U_i(P) = G(P-Q) = (i/4)H_0^{(1)}(k|PQ|)$$

where $|PQ|$ is the distance between the points P and Q.

Write the equation for the scattered field $U_r(M)$ at a point M in terms of the total field $U(P)$ and its normal derivative on Σ. In so doing consider an integral

$$-\iint_{\Omega_R} [U_r(P)L_pG(P-M) - G(P-M)L_pU_r(P)]\, dx\, dy$$
$$= \iint_{\Omega_R} \delta(P-M)U_r(P)\, dx\, dy$$
$$= U_r(M)$$

taken over a part Ω_R of the domain Ω inside a circle $r < R$ ((r, φ) are polar coordinates) and $L_p = \Delta_p + k^2$ is the Helmholtz operator, the index p shows with respect to which variable the function $G(P-M)$ is differentiated. It was taken into account that $L_p U_r(P)=0$ and $L_p G=\delta(P-M)$.

Applying Green's function to the written integral and taking into account that the integral over the boundary Ω_R on which $r=R$ vanishes as $R\to\infty$ due to the Sommerfeld radiation condition, one obtains

$$U_r(M)=\int_\Sigma\left[G(P-M)\frac{\partial U_r(P)}{\partial n}-U_r(P)\frac{\partial G(P-M)}{\partial n}\right]dP \qquad (5.14)$$

where $\partial/\partial n$ denotes the differentiation along the inward-directed normal to Σ. A similar integral in which the field $U(P)$ is replaced by $U_i=G(P-Q)$ is equal to zero

$$\int_\Sigma\left[G(P, M)\frac{\partial U_i(P)}{\partial n}-U_i(P)\frac{\partial G(P, M)}{\partial n}\right]dP$$

$$=\iint_\Omega [G(P, Q)L_pG(P, M)-G(P, M)L_pG(P, Q)]\ dx\ dy$$

$$=G(M, Q)-G(Q, M)=0$$

Therefore, in eqn. 5.14 the scattered field $U_r(P)$ can be replaced by the total field $U=U_i+U_r$

$$U_r(M)=\int_\Sigma\left[G(P, M)\frac{\partial U(P)}{\partial n}-U(P)\frac{\partial G(P, M)}{\partial n}\right]dP \qquad (5.15)$$

This equation is exact and reduces the problem to evaluate a reflected field in the domain Ω to the problem to determine the total field and its normal derivative on the boundary. Using various approximate expressions for $U(P)$ one obtains the corresponding approximations for the field $U_r(M)$.

The physical optics approximation, or the Kirchhoff approximation, is based on the assumption that for $\lambda=(2\pi/k)\ll L$ (L is a typical size of the problem, for instance, the typical radius of surface curvature), i.e. for $kL\gg 1$, one can use the approximation of geometrical optics for the total field $U(P)$ near the surface

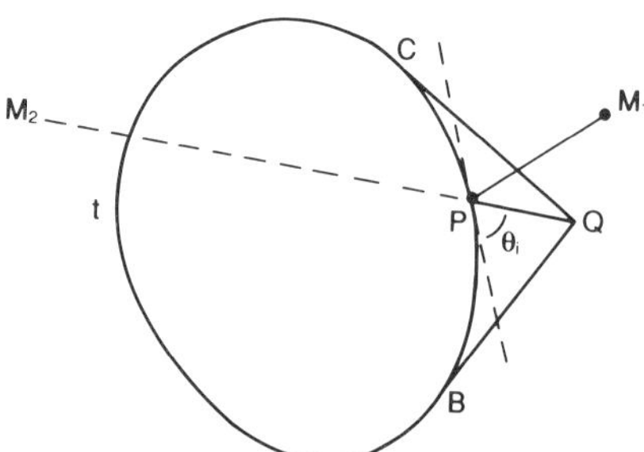

Figure 5.3 Evaluation of scattered field

Σ. In the shadow region (the arc BtC in Figure 5.3) set

$$U(P)=\frac{\partial U(P)}{\partial n}=0$$

When evaluating $U(P)$ and $\partial U(P)/\partial n$ in the illuminated region replace the boundary Σ by the tangent (dotted line in Figure 5.3). Then the total field is equal to a sum of the incident and specularly reflected fields for the Neumann condition or to a difference of these fields for the Dirichlét condition. Therefore, on Σ

$$U(P)=2U_i(P),\ \frac{\partial U(P)}{\partial n}=0 \qquad \text{for the Neumann condition}$$

$$U(P)=0,\ \frac{\partial U(P)}{\partial n}=2\frac{\partial U(P)}{\partial n} \qquad \text{for the Dirichlét condition}$$

Thus, for the reflected field and the Dirichlét condition

$$U_r(M)=2\int_{\Sigma'} G(P,M)\frac{\partial U_i(P)}{\partial n}\,dP \tag{5.16a}$$

and

$$U_r(M)=-2\int_{\Sigma'} U_i(P)\frac{\partial G(P,M)}{\partial n}\,dP \tag{5.16b}$$

for the Neumann condition. Here, Σ' is a part of the surface Σ illuminated by the source. The scattered field $U_r(M)$ determined by eqn. 5.16 is called the field in the physical optics approximation or the Kirchhoff approximation (KA), or the current method approximation. The last-mentioned name is explained by the fact that in electromagnetic problems the quantities $2U_i(P)$ or $2\partial U_i(P)/\partial n$ correspond to currents (evaluated according to the KA) induced by the field incident on the surface Σ.

Consider a high-frequency asymptotic of the scattered field $U_r(M)$ (in the KA). Replace the functions $U_i(P)$, $G(M, P)$ and their derivatives by the leading terms of their asymptotics:

$$\begin{aligned} &G(M,P)\simeq\sqrt{[i/(8\pi k|MP|)]}\ \exp(ik|MP|)\\ &\frac{\partial G(M,P)}{\partial n}\simeq\sin\theta_r\sqrt{[k/(8i\pi|MP|]}\ \exp(ik|MP|) \end{aligned} \tag{5.17}$$

where θ_r is the angle (taken with the sign) between the ray PM and the surface Σ. In Figure 5.3 the angle $\theta_r>0$ for $M=M_1$ and $\theta_r<0$ for $M=M_2$. Similarly,

$$U_i(P)\simeq A|PQ|^{-1/2}\exp(ik|PQ|)$$

$$\frac{\partial U_i}{\partial n}\simeq -ikA\sin\theta|PQ|^{-1/2}\exp(ik|PQ|)$$

where θ_i is the angle between the line PQ and the surface Σ and the pattern of the cylindrical wave incident on Σ is denoted by $A=\sqrt{(i/2\pi k)}/2$. Substituting this expression in eqn. 5.16*b* obtains for the Neumann condition

$$U_r(M)=-A\sqrt{(k/2\pi i)}\int_B^C \frac{\sin\theta_r \exp\left[ik(|MP|+|PQ|)\right]}{\sqrt{(|MP|\cdot|PQ|)}}\,ds \tag{5.18}$$

where s is the length element along the arc BC. In the case of the Dirichlét boundary condition the expression for $U(M)$ is obtained by replacing $\sin\theta_r$ by $-\sin\theta_i$.

Now find the asymptotic of eqn. 5.18. It is a sum of the contributions of the integration domain endpoints (the points B and C) as well as contributions of the phase function stationary points. The contributions of the points B and C are cylindrical waves emanating from B and C. These waves are absent in the exact solution; they arise in the KA since the exact field $U(P)$ in eqn. 5.15 is replaced by a discontinuous function equal to zero in the shadow region (on the arc BtC in Figure 5.3) and to the double primary field in the illuminated region (on the arc BC in Figure 5.3).

Consider the stationary points of eqn. 5.18. These points are determined from the condition for the length $|QP|+|PM|$ of the polygon QMP (as a function of the point P position on the boundary Σ) to be extremal, i.e. to be determined by the Fermat principal. This condition is analytically written as

$$\frac{d}{ds}\left[|QP|+|PM|\right]=\cos\theta_i-\cos\theta_r=0$$

from which follows that

$$\theta_r=\pm\theta_i$$

The second derivative of the phase function is expressed in terms of derivatives of the angles θ_r and θ_i. When evaluating these derivatives one should take into account that the angle θ_i, for example, changes when moving along Σ, first, owing to the change of the ray QP direction

$$\frac{d\psi}{ds}=\frac{-\sin\theta_i}{d_i}$$

where $d_i=|PQ|$ and, secondly, owing to the change of the tangent direction

$$\frac{d\omega}{ds}=\frac{1}{\rho}$$

where ρ is the curvature radius of Σ at the point P and $\rho>0$ for the convex Σ and $\rho<0$ for the concave Σ. Therefore,

$$\frac{d\theta_i}{ds}=-\frac{\sin\theta_i}{d_i}-\frac{1}{\rho}$$

In the same way, $d\theta_r/ds$ is evaluated. The intermediate manipulations are omitted and the ultimate results given.

Let a point M be in the shadow region ($M=M_2$ in Figure 5.3). Then $\theta_r=-\theta_i$. Since in the shadow region behind a smooth body the total field $U=$

$U_i + U_r$ tends to zero (as $k \to \infty$) like $\exp(-\text{const.}\, k^{1/3})$ (see, e.g. Reference 34), then the scattered field U_r should compensate the incident field U_i in the shadow region (within the accuracy of the exponentially small summands). Evaluating the asymptotic of eqn. 5.18 (or, more exactly, the contributions into the equation due to the stationary point P) shows that in KA the field U_r is

$$U_r \simeq -A|QM_1|^{-1/2} \exp(ik|QM_1|)$$

i.e. in KA the compensation occurs in the leading term of the asymptotic. The second term of the asymptotic of U_r does not compensate any longer the appropriate summands in U_i.

If the point M is in the illuminated region ($M = M_2$), then $\theta_r = \theta_i$, i.e. the incident field ray specularly reflected from Σ at P hits the point M. Evaluating U_r yields

$$U_r \simeq \varepsilon(A/\sqrt{d_i})\sqrt{[d_r/(d_r + PM_1)]} \exp[ik(|QP| + |PM_1|)] \qquad (5.19)$$

where $\varepsilon = 1$ for the Neumann boundary condition and $\varepsilon = -1$ for the Dirichlét boundary condition, d_i is the distance between the source Q and the point P or the curvature radius of the incident wavefront at P, d_r is the distance between P and the point R which is an intersection of two reflected rays outgoing from the infinitely close points P and P' (Figure 5.4), i.e. the curvature radius of the reflected wavefront at P. The quantities d_i and d_r are related by

$$\frac{1}{d_r} = \frac{1}{d_i} + \frac{2}{\rho \cos \theta_i}$$

where ρ is the curvature radius of the surface Σ at the reflection point P.

Eqn. 5.19, i.e. the leading term of the asymptotic (in the KA) for the field U_r coincides with the exact equation for the leading term of the reflected field asymptotic, i.e. with the geometrical optics approximation. The next term of the asymptotic U_r is not correct in the KA.

If $\rho > 0$ and the rays of the reflected wave form a diverging congruence, the amplitude of the field being decreased when the observation point moves along the reflected ray. If the surface Σ is concave, i.e. $\rho < 0$, then it may appear that

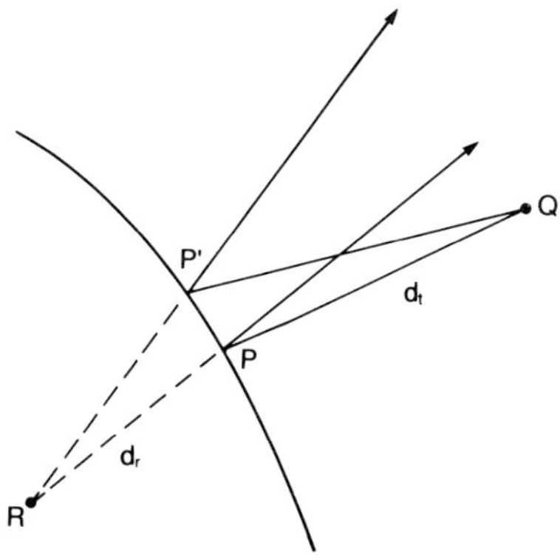

Figure 5.4 Rays of reflected wave

$d_r < 0$. Then the congruence of the reflected rays is convergent and the rays have an envelope, a caustic. When approaching the caustic the amplitude of the reflected field U_r increases. In the caustic vicinity both the approximation of geometrical optics and its equivalent evaluation of the leading term of the asymptotic for eqn. 5.18 (obtained by the stationary phase method) are inapplicable. Analytically, it manifests in the fact that the second derivative of the phase function in eqn. 5.8 proportional to $d_r + PM_1$ becomes small near the caustic. However, the Kirchhoff approximation, described by eqn. 5.18 remains applicable; its error does not depend on a distance from the caustic. This makes it possible to use the Kirchhoff approximation to describe the uniform asymptotic of the field in the caustic vicinity as well as in more complicated transient zones (caustic cusps, the intersection points of the caustics, light–shadow boundaries, etc.). Examples of such asymptotics are given in what follows.

5.3 Diffraction by wedge: asymptotic of exact solution

This Section uses the stationary phase method to find a short-wave asymptotic of Green's function for the problem of the diffraction by a wedge, i.e. to solve the problem of diffraction by a wedge for the incident wave $U_i(P)$ where

$$\begin{aligned} U_i(P) &= (i/4)H_0^{(1)}(k|PQ|) \\ &= (i/4)H_0^{(1)}\{k\sqrt{[r^2 + r_0^2 - 2rr_0\cos(\varphi - \varphi_0)]}\} \end{aligned} \tag{5.20}$$

Here, r_0, φ_0 are the polar co-ordinates of the source Q and r, φ are the polar co-ordinates of observation point P. Set the wedge apex to be at the origin and $\varphi = 0$, Φ $(0 < \Phi < 2\pi)$ to be the equations of its sides. So that the solution of such a problem be unique it is necessary to require (in addition to the Sommerfeld radiation condition) the total field $U(P)$ to satisfy the regularity condition (Meixner condition) at the apex of the wedge Σ as $r \to 0$. This means at as $r \to 0$ the total field $U(P)$ should remain bounded and the derivative $\partial U/\partial n$ should not grow more rapidly than $1/r$, the limit of $r\partial U/\partial n$ should be equal to zero.

The solution of the problem is of the Sommerfeld integral form (see, for example, Reference 35)

$$\begin{aligned} &U(k, r, \varphi, r_0, \varphi_0) \\ &= U_i + U_r \\ &= \int_\gamma H_{(0)}^{(1)}[k\sqrt{(r^2 + r_0^2 - 2rr_0\cos\alpha)}H(\alpha + \varphi, \varphi_0, \Phi)\, d\alpha \end{aligned} \tag{5.21}$$

The contour γ consists of two loops and passes round the branch points $A_{1,2}$: $\alpha = \pm i\,\mathrm{arcch}\,[(r^2 + r_0^2)/2rr_0]$, as shown in Figure 5.5 where the regions in which $\mathrm{Im}\cos\alpha < 0$ are shadowed, the function $H(\alpha + \varphi, \varphi, \Phi)$ is the Sommerfeld

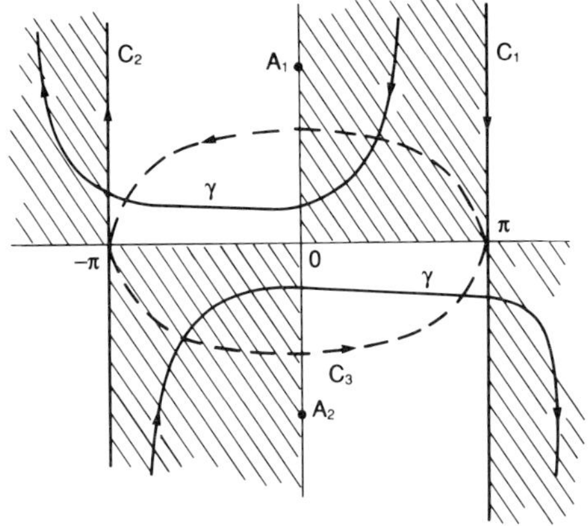

Figure 5.5 Integration contours in Sommerfeld integral

kernel

$$H(\alpha+\varphi,\varphi_0,\Phi)=\frac{\pi}{2\Phi}\left\{\operatorname{ctg}\left[\frac{\pi(\alpha+\varphi-\varphi_0)}{2\Phi}\right]\right.$$
$$\left.+\varepsilon\operatorname{ctg}\left[\frac{\pi(\alpha+\varphi+\varphi_0}{2\Phi}\right]\right\} \tag{5.22}$$

where $\varepsilon=1$ for the Neumann boundary condition and $\varepsilon=-1$ for the Dirichlét condition. To prove eqn. 5.22 one needs, first, to verify whether the written integral is a sum of the incident wave U_i and the regular solution of the Helmholtz equation, and secondly, to check, whether it satisfies the boundary conditions for $\varphi=0$, Φ, and thirdly, whether it satisfies the regularity conditions as $r\to 0$ and the Sommerfeld radiation conditions as $r\to\infty$.

To verify that the boundary conditions are satisfied note that the function $H(\alpha+\varphi,\varphi_0,\Phi)$ for $\varepsilon=-1$ and $\partial H(\alpha+\varphi,\varphi_0,\Phi)/\partial\varphi$ for $\varepsilon=1$ are even functions of α for $\varphi=0$, Φ and the integral of any even function taken over γ vanishes.

To prove that eqn. 5.21 satisfies the Helmholtz equation at $r\neq r_0$ it is sufficient to change variable $\beta=\alpha+\varphi$ and to note that both the function

$$(i/4)H_0^{(1)}\{k\sqrt{[r^2+r_0^2-rr_0\cos(\beta-\varphi)]}\}$$

and the integral of it with respect to β with the weight $H(\beta,\varphi,\Phi)$ are solutions of the Helmholtz equation.

As $r\to r_0$ the branch points of the radical

$$\sqrt{(r^2+r_0^2-2rr_0\cos\alpha)}$$

tend to the origin, while the integrand in eqn. 5.21 tends to infinity on the contour γ passing between these points. To remove the integration contour from these points and to make it pass through the phase function stationary points $\alpha=\pm\pi$ replace the contour γ by the sum of three contours, one involving the interval $(-\pi,\pi)$ of the closed contour C_3 (broken contour in Figure 5.5) and two straight lines, C_1: Im $\alpha = \pi$ and C_2: Im $\alpha = -\pi$. The integral over C_3 is expressed in terms of the sum of the integrand residues at the poles of the Sommerfeld kernel H which are between $-\pi$ and π. It appears to be equal to the geometrical optics (GO) approximation U_{GO} for the solution of the problem in question, i.e. to the sum of the incident wave (eqn. 5.20) and its reflections from the wedge faces, each wave being multiplied by the cutting function χ equal to unity in the region illuminated by this wave and to zero in a shadow region.

For example, if $\varphi > \pi$ and $\varphi_0 < \Phi - \pi$, then for $0 < \varphi < \pi - \varphi_0$ the integral over C_3 is a sum of the incident wave (eqn. 5.20) and its reflection from the face $\varphi = 0$

$$\varepsilon H_0^{(1)}\{k\sqrt{[r^2 + r_0^2 - rr_0 \cos(\varphi + \varphi_0)]}\}$$

For $\pi - \varphi_0 < \varphi < \pi + \varphi_0$ it coincides with the incident wave, while for $\varphi > \pi + \varphi_0$ it vanishes.

The integrals over C_1 and C_2 can be reduced by the variable change $\alpha = \pm\pi i + \beta$ to the integral over the real variable β and

$$\begin{aligned} U(k, r, \varphi, r_0, \varphi_0) &= U_{GO} + U_e \\ &= U_{GO} + \frac{i}{2\pi}\int_{-\infty}^{\infty} H_0^{(1)}\,[k\sqrt{(r^2 + r_0^2 + 2rr_0 \operatorname{ch}\beta)}] \\ &\quad \times H_1(\varphi + i\beta, \varphi_0, \Phi)\, d\beta \end{aligned} \tag{5.23}$$

where $H_1(\varphi + i\beta, \varphi_0, \Phi) = H(-\pi + i\beta + \varphi, \varphi_0, \Phi) - H(\pi + i\beta + \varphi, \varphi_0, \Phi)$. It follows that near the source (at $r \simeq r_0$, $\varphi \simeq \varphi_0$) the total field U is a sum of the incident wave U_i (in the integral for U_{GO} the residue of the pole $\alpha = \varphi_0 - \varphi$ of the Sommerfeld kernel corresponds to the incident wave, the pole being close to the origin for $\varphi \simeq \varphi_0$) and the regular solution to the Helmholtz equation. Besides, eqn. 5.23 is convenient to evaluate asymptotics of U as $kr \to \infty$, $kr_0 \to \infty$. Under these conditions the argument of the Hankel function under the radical sign is large and the function can be replaced by asymptotic eqn. 5.5.

Evaluating the integral obtained by the stationary phase method yields

$$U \simeq U_{GO} + U_e$$

where the second summand is an edge wave diverging from the apex of Σ

$$U_e \simeq \frac{\exp(ikr)}{\sqrt{r}} f(\varphi) \tag{5.24}$$

with a pattern

$$\begin{aligned} f(\varphi) &= U_i(O)D(\varphi, \varphi_0) \\ &= \frac{\exp(ikr_0 + \pi i/4)}{2\sqrt{(2\pi kr_0)}}\,\frac{\exp(i\pi/4)}{\sqrt{k}}\,V(\varphi, \varphi_0) \end{aligned}$$

Here, the first cofactor is the value of the incident wave $U_i(r, \varphi)$ (more exactly, the leading terms of its asymptotic for $kr \gg 1$) at the apex of Σ. In papers on the geometric diffraction theory the second cofactor is called the Keller diffraction coefficient. The function

$$V(\varphi, \varphi_0)$$

$$= \frac{1}{\sqrt{(2\pi)}}[H(-\pi+\varphi, \varphi_0, \Phi) - H(\pi+\varphi, \varphi_0, \Phi)]$$

$$= \frac{\sqrt{\pi}}{2\sqrt{2\Phi}}\left\{\operatorname{ctg}\left[\frac{\pi}{2\Phi}(\varphi-\varphi_0-\pi)\right] - \operatorname{ctg}\left[\frac{\pi}{2\Phi}(\varphi-\varphi_0+\pi)\right]\right.$$

$$\left. + \varepsilon \operatorname{ctg}\left[\frac{\pi}{2\Phi}(\varphi+\varphi_0-\pi)\right] - \varepsilon \operatorname{ctg}\left[\frac{\pi}{2\Phi}(\varphi+\varphi_0+\pi)\right]\right\} \tag{5.25}$$

is called the angular part of the diffraction coefficient. To prove that eqn. 5.21 gives the solution of the problem for the diffraction of the cylindrical wave (eqn. 5.21) by the wedge it remains to verify whether the regularity conditions on the wedge are satisfied. Appropriate mathematics are omitted.

Continuing with the asymptotic of eqn. 5.21 as $kr \to \infty$, $kr_0 \to \infty$, treat the case of the direction of the observation angle φ approaching the light–shadow boundary for the incident or reflected wave. Therefore, in this integral (more exactly, in the integral obtained when replacing the function $H_0^{(1)}[k\sqrt{(r^2+r_0^2-2rr_0 \operatorname{ch}\beta)}]$ by its asymptotic) the poles of the Sommerfeld kernel approach the phase function stationary point on the integration contour. Under these conditions the stationary phase method is inapplicable and asymptotic (eqn. 5.24) of the edge wave becomes invalid since $f(\varphi)$ tends to infinity.

As shown in Section 2.2, the uniform asymptotic of the integral in which the pole is close to a phase function stationary point is expressed in terms of the Fresnel integral. Therefore in the problem the field near the light–shadow boundary is also expressed in terms of the Fresnel integral. Let, for instance, $\Phi > \pi$ and $\varphi_0 < \theta - \pi$. Then there are two light–shadow boundaries, $\varphi = \pi + \varphi_0$ for the incident wave and $\varphi = \pi - \varphi_0$ for the reflected wave.

There is no reflected wave in the vicinity of the light–shadow boundary of the incident wave and the uniform asymptotic is of the form

$$U(k, r, \varphi, r_0, \varphi_0) = U_i F\{\sqrt{[k(s_e - s_i)]}\} + U_e^* \tag{5.26a}$$

Here, the incident wave is $U_i = (i/4)H_0^{(1)}(ks)$, the eikonals of incident (s_i) and edge (s_e) waves are equal to $s_i = \sqrt{[r^2 + r_0^2 - 2rr_0 \cos(\varphi - \varphi_0)]}$ and $s_e = r + r_0$, respectively. On the light–shadow boundary $\varphi = \pi + \varphi_0$ the difference $s_e - s_i$ has a second-order zero, the quantity $\sqrt{(s_e - s_i)}$ being understood as a positive value of the radical in the illuminated region and a negative value in the shadow region.

The asymptotic of the correction summand U_e^* is determined by means of asymptotic matching using nonuniform asymptotic eqn. 5.24 and is of the same form with the angular function $V(\varphi, \varphi_0)$ replaced by

$$V^*(\varphi, \varphi_0)$$

$$= V(\varphi, \varphi_0)$$

$$+ \frac{\sqrt{\{r + r_0 + \sqrt{[r^2 + r_0^2 - 2rr_0 \cos(\varphi - \varphi_0)]}\}}}{4\sqrt{\pi}[r^2 + r_0^2 - 2rr_0 \cos(\varphi - \varphi_0)]^{1/4} \sin[(\pi + \varphi_0 - \varphi)/2]} \tag{5.27a}$$

For $\varphi \to \pi + \varphi$ the poles on the right-hand side of this expression cancel and $V^*(\varphi, \varphi_0)$ remains to be regular. As $r_0 \to \infty$, i.e. proceeding to the plane-wave diffraction problem, the expression is essentially simplified

$$V^*(\varphi, \varphi_0) = V(\varphi, \varphi_0) + \frac{1}{2\sqrt{(2\pi)}\sin[(\pi + \varphi_0 - \varphi)/2]} \tag{5.28}$$

Similarly, in a vicinity of the light–shadow boundary of the reflected wave ($\varphi = \pi - \varphi_0$) the uniform asymptotic is

$$U(k, r, \varphi, r_0, \varphi_0) = U_i + U_r F\{\sqrt{[k(s_e - s_r)]}\} + U_e^{**} \tag{5.26b}$$

where $s_r = \sqrt{[r^2 + r_0^2 - 2rr_0 \cos(\varphi + \varphi_0)]}$. One has $\sqrt{[k(s_e - s_r)]} > 0$ for $\varphi < \pi - \varphi_0$, i.e. in a region where the reflected waves propagate, and $\sqrt{[k(s_e - s_i)]} < 0$ for $\varphi > \pi\varphi_0$. Here, U_i^{**} is a correction edge wave with asymptotic eqn. 5.24 where $V(\varphi, \varphi_0)$ is replaced by

$$\begin{aligned} V^{**}(\varphi, \varphi_0) &= V(\varphi, \varphi_0) \\ &+ \frac{\sqrt{\{r + r_0 + \sqrt{[r^2 + r_0^2 - 2rr_0\cos(\varphi + \varphi_0)]}\}}}{4\sqrt{\pi}[r^2 + r_0^2 - 2rr_0\cos(\varphi + \varphi_0)]^{1/4}\sin[(\pi - \varphi_0 - \varphi)/2]} \end{aligned} \tag{5.27b}$$

5.4 Diffraction by wedge and slit: physical optics approximation

Section 5.3 considered the exact solution for the problem of field diffraction by a wedge AOB for the isotropic source Q (Figure 5.6). In what follows the solution is obtained of the same problem in the physical optics approximation (the Kirchhoff approximation, KA) and find the uniform asymptotic of this solution. Comparing this asymptotic with that of the exact solution makes it possible to

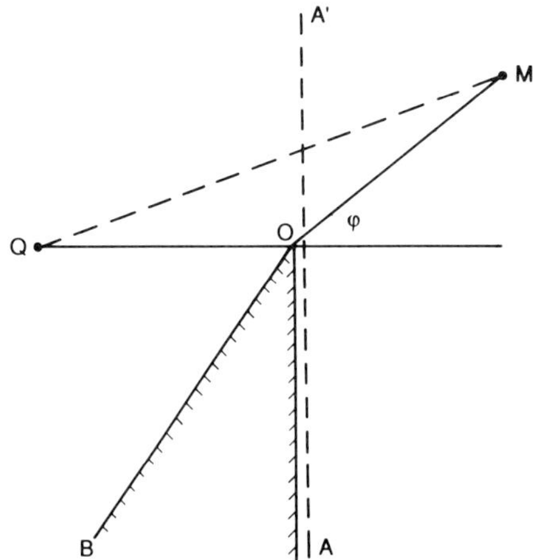

Figure 5.6 Diffraction by wedge

estimate the limits within which the Kirchhoff approximation is applicable, the error of the approximation, and this in turn makes it possible to use the KA in those situations where an exact solution is not known.

Let $U_i(P) = G(P, Q)$ be a field excited by a source at the point Q (Figure 5.6) and $U(P)$ be a solution for the problem of diffraction of the field U_i by the wedge AOB. Denote the continuation of the shadowed face AO by OA' and assume that observation point is to the right from the straight line AA'. Then for the field $U(M)$ one can write Green's function as

$$U(M) = \int_{AA'} \left[G(P, M) \frac{\partial U(P)}{\partial n} - U(P) \frac{\partial G(P, M)}{\partial n} \right] ds \tag{5.29}$$

where $\partial/\partial n$ denotes the differentiation along the normal to AA' directed to the point M. As with eqn. 5.15, eqn. 5.29 is exact. However, to use it one should find the field $U(P)$ and $\partial U/\partial n$ on the straight line AA'. Here, the physical optics approximation is reduced to the assumption that since the short-wave asymptotic is considered as $\lambda \to 0$, $k = 2\pi/\lambda \to \infty$ $U(P)$ and $\partial U/\partial n$ are replaced by their values according to geometrical optics, i.e. by zero in the shadow region (on the ray AO) and by the incident field U_i in the illuminated region (on the ray OA'), assuming the perturbations to the geometrical optics approximation to be small. In the other words, the field $U(M)$ in the physical optics approximation (KA) is

$$U(M) = \int_{OA'} \left[G(P, M) \frac{\partial U_i(P)}{\partial n} - U_i(P) \frac{\partial G(P, M)}{\partial n} \right] ds \tag{5.30}$$

Find the asymptotic of the integral as $k \to \infty$. Using the asymptotic of $G(P, M)$, U_i, $\partial U_i/\partial n$, $\partial G/\partial n$ in eqn. 5.30 it is easy to obtain that the integrand phase is $ik(|P-Q| + |P-M|)$, the stationary point is a minimum point of the phase and this point is at $AA' - MQ$ intersection.

The nonuniform asymptotic of the field $U(M)$ is determined, first, by the contribution of this stationary point, and secondly, by the contribution of the integration domain boundary (the point O). To find the stationary point contribution use eqn. 5.29, valid for the case when there is no wedge i.e. when the field $U(M)$ coincides with the primary field $U_i(M)$. Then eqn. 5.29 is of the form

$$U_i(M) = \int_{AA'} \left[G(P, M) \frac{\partial U_i(P)}{\partial n} - U_i(P) \frac{\partial G(P, M)}{\partial n} \right] ds$$

This integral differs from eqn. 5.30 in the integration domain only, it contains a single critical point which is the same as that in eqn. 5.30, i.e. the phase function stationary point. Therefore the contributions to these integrals are equal and they are equal to the incident field. Now,

$$U(M) \simeq U_i(M)\chi + U_e(M) = (i/4)H_0^{(1)}(k|M-P|) + U_e(M) \tag{5.31a}$$

where $\chi = 1$ when the point M is in the illuminated region and $\chi = 0$, if it is shadowed by the wedge. The second summand in eqn. 5.31a is a contribution of the integration domain boundary, it is evaluated according to theorem 1.1

and is of the form

$$U_e \simeq \frac{-i \exp\,[ik(r+r_0)]}{4\pi k\sqrt{(rr_0)}2\,\mathrm{tg}\,(\psi/2)} \tag{5.31b}$$

where the angle $\psi = \pi + \varphi_0 - \varphi$ is shown in Figure 5.6. If the point M is in a region illuminated by the incident wave then $\psi > 0$, otherwise $\psi < 0$. If the point M approaches the light–shadow boundary asymptotic eqn. 5.31 becomes inapplicable since in eqn. 5.30 the stationary point approaches the integration domain boundary.

According to theorem 2.1 the uniform asymptotic of this integral is expressed in terms of the Fresnel integral

$$U(M) \simeq U_i(M) F\{\sqrt{[k(s_e - s_i)]}\} + U_e^{**} \tag{5.32}$$

Here, $s_e = r + r_0$ and $s_i = \sqrt{[r^2 + r_0^2 - 2rr_0 \cos(\varphi - \varphi_0)]}$ are the same functions as in eqn. 5.2, the difference is only in the summand U^{**} regular on the light–shadow boundary.

The field in the Kirchhoff approximation is an exact solution to the Helmholtz equation, its approximate character is due to the fact that the boundary conditions are not satisfied accurately. This approximation gives the leading term (the approximation of geometrical optics, GO) of the solution which is equal to the incident wave in the illuminated region and is zero in the shadow region. It also gives the correct description of the field in a transient zone which is a vicinity of the light–shadow boundary for the incident wave. As well, the Kirchhoff approximation gives the correct phase of the second summand, i.e. the edge wave with the wedge apex as a centre. This wave amplitude is determined in the KA incorrectly. However, comparing eqn. 5.26 for the uniform asymptotic of the exact solution and eqn. 5.32 for the asymptotic in the KA one can see that near the light–shadow boundary the exact solution $U(M)$ can be written as

$$U(M) = U_k(M) + U_e^{***}$$

where $U_k(M)$ is the solution in the KA and $U_e^{***} \simeq U_e^{*} - U_e^{**}$ is the correction summand with the edge-wave phase which is regular in a vicinity of the light–shadow boundary.

Consider the diffraction by the aperture in the plane perfectly conducting screen (Figure 5.7). To evaluate the field at the point M behind the screen we again use eqn. 5.29 where the values of the field $U_i(P)$ and its derivative $\partial U_i(P)/\partial n$ on the aperture BB' and zeros on the lines BA and $B'A'$ are taken as the values of the field $U(P)$ and its derivative on the straight line AA'. Then, for the field $U(M)$,

$$U(M) = \int_{BB'} \left[G(P, M) \frac{\partial U_i(P)}{\partial n} - U_i(P) \frac{\partial G(P, M)}{\partial n} \right] dp$$

As previously, the leading term of the asymptotic corresponds to geometrical optics; it coincides with the incident wave in the illuminated region and vanishes in the shadow, and appears to be due to the phase-function stationary point contribution. The field asymptotic in the KA includes the integration interval endpoint contributions (the summands with the edge-wave phases).

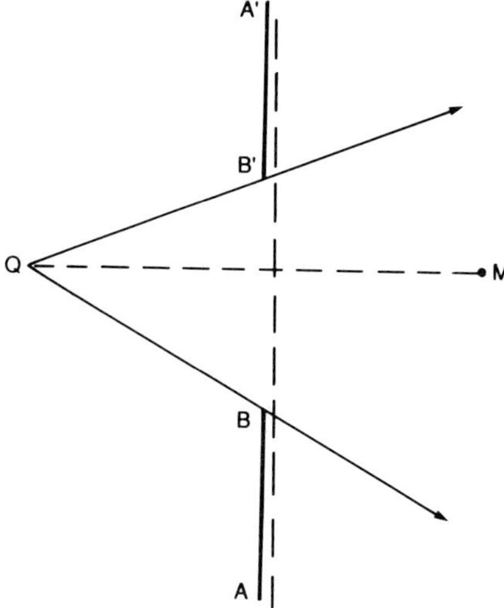

Figure 5.7 Diffraction by aperture

Compare this asymptotic with the exact asymptotic for the solution of the problem under consideration (see, for example, Reference 35). This consists of the GO summand, the two primary edge waves excited at the edges B and B' by the incident field, two secondary waves excited at opposite edges when each primary wave diffracts, and the waves of all subsequent diffractions. The Kirchhoff approximation does not describe waves of the subsequent diffractions; it describes correctly only the GO term of the field, the field in the transient zones (vicinity of the light–shadow boundary for the incident field) and the primary edge waves from edges B and B'.

5.5 Cylindrical wave reflection from plane interface of two media

Let the plane $y=0$ be an interface of two homogeneous fluids. The medium is characterised by the densities ρ, ρ_1 and the sound velocities c, c_1 in the upper ($y>0$) and lower regions, respectively. At the interface $y=0$ the pressure and normal to the boundary velocity component should be continuous.

Consider the harmonic oscillations at a frequency ω. Denote the velocity potentials in the upper and lower media by U and U_1, then obtain the equations

$$\Delta U + k^2 U = 0, \qquad y>0 \tag{5.33a}$$

$$\Delta U_1 + k_1^2 U_1 = 0, \qquad y<0 \tag{5.33b}$$

$$U = mU_1, \qquad \frac{\partial U}{\partial y} = \frac{\partial U_1}{\partial y}, \qquad y=0 \tag{5.33c}$$

where $k=\omega/c$, $k_1=\omega/c_1$ and $m=\rho/\rho_1$.

The two-dimensional electrodynamic problem is reduced to the same equations, the media being characterised by the dielectric constant and magnetic permeability ε and μ (the upper medium) and by ε_1, μ_1 (the lower medium), respectively. All the fields do not depend on z. The continuity of the longitudinal components of the fields E and H on the interface plays the role of the boundary conditions. For the E-polarised field where E_z, H_x, H_y are not equal to zero the functions

$$U = E_z \qquad \text{for } y > 0,$$

$$U_1 = (1/m)E_z \quad \text{for } y < 0$$

satisfy the system of eqn. 3.53 where

$$m = \mu/\mu_1, \qquad k = (\omega/c)\sqrt{(\varepsilon\mu)}, \qquad k_1 = (\varepsilon/c)\sqrt{(\varepsilon_1\mu_1)}$$

For the H-polarised field one should set

$$U = H_z \qquad \text{for } y > 0,$$

$$U_1 = (1/m)H_z \quad \text{for } y < 0$$

with the same k and k_1.

The spherical wave incident on a plane was treated in Reference 36. Here, the simpler problem of the cylindrical wave reflection is discussed. The field in the upper medium is sought in the form

$$U = U_i + U_r$$

where

$$U_i = H_0^{(1)}\{k\sqrt{[x^2 + (y - y_0)^2]}\} \tag{5.34}$$

is the field of the source located at the point $x = 0$, $y = y_0 > 0$, and U_r satisfies the homogeneous Helmholtz equation

$$\Delta U_r + k^2 U_r = 0$$

The field U_1 in the lower medium, i.e. the refracted field, satisfies eqn. 5.33*b*, the boundary conditions (eqn. 5.33*c*) being satisfied on the boundary $y = 0$. If the incident wave were plane

$$U_i = A \exp[ik(x \sin\theta + y \cos\theta)]$$

one could look for both reflected and refracted waves in the form of the plane waves

$$\begin{aligned} U_r &= AH_r \exp[ik(x \sin\theta + y \cos\theta)] \\ U_1 &= AH_1 \exp[ik_1(x \sin\theta_1 - y \cos\theta_1)] \end{aligned} \tag{5.35}$$

where the reflection coefficient H_r, the refraction coefficient H_1 and the direction of the refracted-wave propagation θ_1 are determined from the boundary conditions (eqn. 5.33*c*). Therefore to solve the problem for the wave incident on the interface one should represent the wave in the form of the integral over the plane waves and then for each of these waves to find the reflected and refracted waves. Consider the case $k_1 < k$ when for some angle θ there is total internal reflection and the reflected field contains so-called head waves (see later).

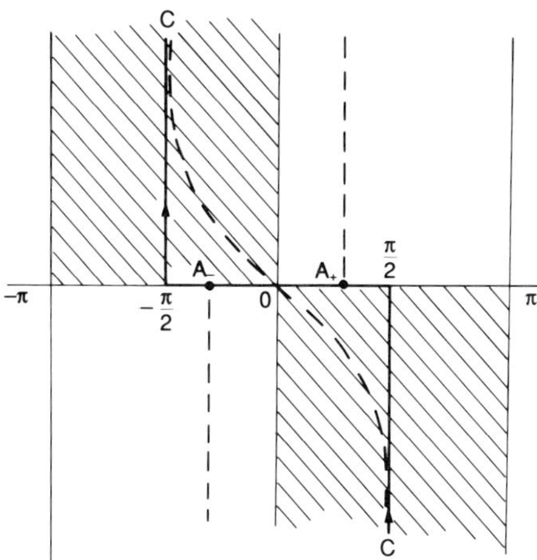

Figure 5.8 Integration contours in eqns. 5.36 and 5.37

Prove that for $y>0$ the following equation is valid

$$H_0^{(1)}[k\surd(x^2+y^2)] = -(1/\pi)\int_C \exp\,[ik\,(x\sin\theta+y\cos\theta]\,d\theta \qquad (5.36)$$

where the contour C is shown in Figure 5.8, the regions with Im $(\cos\theta)>0$ are shadowed. Indeed, for $y>0$ the integral along this contour converges. Setting $x=r\sin\psi$, $y=r\cos\psi$, where $-(\pi/2)<\psi<(\pi/2)$ one obtains that eqn. 5.36 is equivalent to

$$H_0^{(1)}(kr) = -(1/\pi)\int_C \exp\,[ikr\cos(\theta-\psi)]\,d\theta$$

The integration variable change $\theta_1=\theta-\psi-\pi/2$ reduces the integral to eqn. 5.2. For $y<y_0$, from eqn. 5.36,

$$H_0^{(1)}\{k\surd[x^2+(y-y_0)^2]\}$$
$$=-(1/\pi)\int_C \exp\,\{ik[x\sin\theta+(y_0-y)\cos\theta]\}\,d\theta$$

For the incident wave $U_i=A\exp\,[ik\,(x\sin\theta-y\cos\theta)]$ where $A=\exp\,(iky_0\cos\theta)$ look for the reflected and refracted waves in the form of eqn. 5.35 and obtain from the boundary condition $k\sin\theta=k_1\sin\theta_1$ from which Snell's law is obtained for the refraction angle θ_1: $\sin\theta_1=(k/k_1)\sin\theta=(1/n)\sin\theta$ where n is the refraction index. In particular, $k_1\cos\theta_1=k\surd(n^2-\sin^2\theta)$ and for reflected (U_r) and refracted (U_1) waves

$$U_r=\frac{m\cos\theta-\surd(n^2-\sin^2\theta)}{m\cos\theta+\surd(n^2-\sin^2\theta)}$$
$$\times\exp\,\{ik[x\sin\theta+(y+y_0)\cos\theta]\}$$

$$U_1 = \frac{2m \cos\theta}{m \cos\theta + \sqrt{(n^2 - \sin^2\theta)}}$$
$$\times \exp\{ik[x \sin\theta + y_0 \cos\theta - y\sqrt{(n^2 - \sin^2\theta)}]\}$$

For $|\sin\theta| > n$, i.e. for the incidence angles $\theta > \theta_r = \arcsin(n)$ for which the total internal reflection occurs, the refracted wave should decay when moving from the boundary, i.e. as $y \to -\infty$. Therefore one needs to set $\mathrm{Im}\,[\sqrt{(n^2 - \sin^2\theta)}] \geq 0$.

In the case of the incident wave (eqn. 5.34) the reflected and refracted waves obtained are

$$U_r = -\frac{1}{\pi}\int_C \frac{m \cos\theta - \sqrt{(n^2 - \sin^2\theta)}}{m \cos\theta + \sqrt{(n^2 - \sin^2\theta)}} \times \exp\{ik[x \sin\theta + (y + y_0)\cos\theta]\}\, d\theta$$
$$U_1 = -\frac{2}{\pi}\int_C \frac{\cos\theta}{m \cos\theta + \sqrt{(n^2 - \sin^2\theta)}} \times \exp\{ik[x \sin\theta + y_0 \cos\theta - y\sqrt{(n^2 - \sin^2\theta)}]\}\, d\theta \tag{5.37}$$

The condition $\mathrm{Im}\,[\sqrt{(n^2 - \sin^2\theta)}] > 0$ for $\sin\theta > n$ means that on the complex plane θ there are cuttings of the function $\sqrt{(n^2 - \sin^2\theta)}$, as shown by dotted lines in Figure 5.8.

Consider the asymptotic U_1 and U_r for fixed x, y, y_0 and $k \to \infty$. The phase function Φ for U_1 is of the form

$$\Phi = x \sin\theta + y_0 \cos\theta - y\sqrt{(n^2 - \sin^2\theta)} \tag{5.38}$$

and the equation for the stationary point $\theta = \theta_0$ is written

$$\Phi'_\theta = x \cos\theta - y_0 \sin\theta + \frac{y \sin\theta \cos\theta}{\sqrt{(n^2 - \sin^2\theta)}}$$

from which setting $\theta_1 = \arcsin(\sin\theta / n)$ gives

$$x = y_0 \operatorname{tg}\theta + |y| \operatorname{tg}\theta_1 \tag{5.39}$$

where it is taken into account that $y < 0$ and $|y| = -y$. The relation has a simple physical meaning, namely, that the angle θ_0 is an angle of the ray (in geometrical optics) emanating from the source Q and refracting on the interface according to Snell's law to the observation point M. We have

$$y_0 \operatorname{tg}\theta_0 = HP \quad \text{and} \quad |y| \operatorname{tg}\theta_1 = H_1P$$

where H, H_1 are the source and observation point projections to the interface and P is the interface-ray intersection point (Figure 5.9).

Since $\sin\theta < n$ at the stationary point $\theta = \theta_0$ of the phase function (eqn. 5.38) this point is on the real axis between the branch points A_+ and A_-, the phase function derivative being negative at A_- and positive at A_+. Therefore following the considerations of Section 1.4 the branch points do not contribute into the asymptotic U and, hence, the asymptotic U_1 includes a single summand owing to the stationary point contribution. The phase Φ at this point coincides with

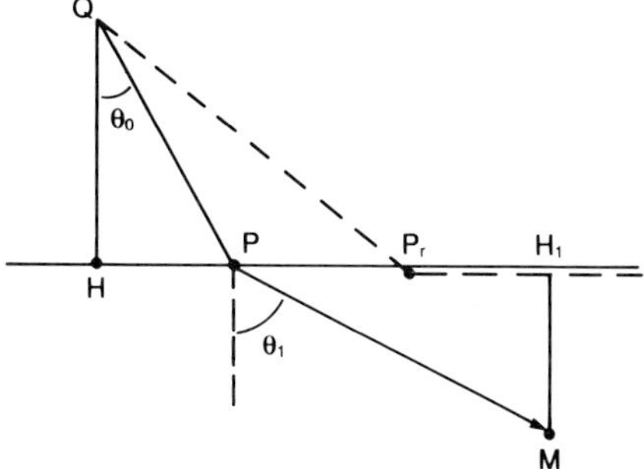

Figure 5.9 Ray of refracted wave

an optical path length along the ray connecting the source Q and the observation point M

$$\Phi = QP + nPM \tag{5.40}$$

Consider how θ_θ varies as M tends to the interface, i.e. as $y \to 0$ and x is fixed. Here, one should analyse two cases. If $x < y_0 \operatorname{tg} \theta_r$ (where $\theta_r = \arcsin n$ is the total internal reflection angle), the limiting value θ_0 is equal to $\operatorname{arctg}(x/y_0)$ and the refraction point P coincides with the limiting position of the point M.

If $x > y_0 \operatorname{tg} \theta_r$, the limiting value of θ_0 is equal to θ_r, the refraction point P tends to a limiting position P_r, while the interval P_rM of the refracted ray approaches the ray sliding along the interface.

In integral eqn. 5.37 for the refracted wave the stationary point of the phase function tends (under these conditions) to a branch point of this function. The foregoing did not consider the asymptotic of such integrals, but this case is reduced to that considered by the integration variable change.

After the integration variable change $\xi = \sqrt{(n^2 - \sin^2\theta)}$ one obtains the integral with the integrand which is the analytical function for small ξ, while the phase function Φ

$$\Phi = x\sqrt{(n^2 - \xi^2)} + y_0\sqrt{(1 - n^2 + \xi^2)} - \xi y$$

has a stationary point close to zero. The contribution of this point into the asymptotic of U_r is evaluated according to Section 1.2. But the asymptotic of U_r evaluated in such a way is not uniform over y as $y \to 0$ since the phase function (eqn. 5.38) as $x \to y_0 \operatorname{tg} \theta_r$ has one more complex stationary point θ_i tending to the real axis as $y \to 0$, $\theta_i = \theta_x + iay + O(y^2)$ where

$$\theta_x = \operatorname{arctg}(x/y_0) > \theta_r$$

$$\alpha = \frac{\sin\theta_x \cos\theta_x}{\sqrt{(x^2 + y_0^2)}\sqrt{(\sin^2\theta_x - n^2)}}$$

The contribution of this point into the asymptotic U_1 is of the order of

$$\exp\left[-k|y|\sqrt{(\sin^2\theta_x - n^2)}\right],$$

it is negligibly small for $|y|$ bounded in magnitude from below, but it appears to be essential as $|y| \to 0$. It corresponds to the wave from the source that propagates through the upper medium at an angle θ_x exceeding the total internal reflection angle θ_r, the wave after the refraction being decayed exponentially when moving in the lower medium from the interface. The situation has been considered in Reference 36 in more detail.

Consider now the high-frequency asymptotic of the reflected field. Here, the phase function is

$$\Phi = x \sin \theta + (y + y_0) \cos \theta$$

and the stationary point $\theta = \theta_\mu$ is defined by

$$\theta_\mu = \text{arctg}\,[x/(y + y_0)]$$

The ray outgoing from Q and specularly reflecting from the interface according to standard reflection law (Figure 5.10) corresponds to this stationary point. Evaluating the contribution of this point by theorem 1.2 obtains

$$U_{GO} = \sqrt{(2)}/\sqrt{(\pi ikr)} \exp(ikr) \frac{m \cos \theta_\mu - \sqrt{(n^2 - \sin^2 \theta_\mu)}}{m \cos \theta_\mu + \sqrt{(n^2 - \sin^2 \theta_\mu)}} \tag{5.41}$$

where

$$r = \sqrt{[x^2 + (y + y_0)^2]} = QP + PM$$

Consider next the contribution of a branch point into the asymptotic. For $\theta_\mu < \theta_r$, i.e. when the angle θ_μ is less than the total internal reflection angle θ_r in a vicinity of the point θ_r, the phase function grows monotonically on the integration contour and therefore this point does not contribute into the integral asymptotic. To find the contribution for $\theta_\mu > \theta_r$ write the integral representation of the reflected field U_r in the form

$$\begin{aligned} U_r &= U_1 + U_2 \\ &= -\frac{1}{\pi} \int_C \frac{m^2 \cos^2 \theta + n^2 - \sin^2 \theta}{m^2 \cos^2 \theta - n^2 + \sin^2 \theta} \\ &\quad \times \exp\{ik[x \sin \theta + (y - y_0) \cos \theta]\}\, d\theta \\ &\quad \times \frac{1}{\pi} \int_C \frac{2m \cos \theta \sqrt{(n^2 - \sin^2 \theta)}}{m^2 \cos^2 \theta - n^2 + \sin^2 \theta} \\ &\quad \times \exp\{ik[x \sin \theta + (y + y_0) \cos \theta]\}\, d\theta \end{aligned} \tag{5.42}$$

The point $\theta = \theta_r$ contributes into the asymptotic of the second integral only. This component of the reflected field is termed the head wave. Evaluating it

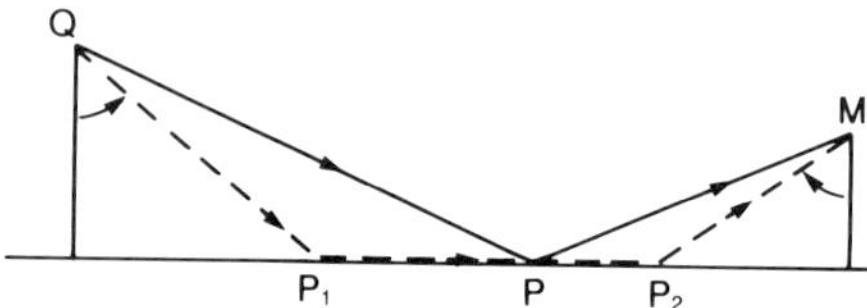

Figure 5.10 Rays of reflected and head waves

according to Section 1.4 obtains

$$U_h \simeq \frac{2\sqrt{(2\,\mathrm{tg}\,\theta_r)}\sqrt{\pi}\,\exp(i\pi/4)\chi(\theta_m - \theta_r)\exp[ikr\cos(\theta_r - \theta_\mu)]}{\pi m k^{3/2} r^{3/2} |\sin(\theta_r - \theta_\mu)|^{3/2}} \tag{5.43}$$

where $\chi = 1$ for $\theta_r > \theta_\mu$ and $\chi = 0$ for $\theta_r < \theta_\mu$, while

$$r = \sqrt{[x^2 + (y + y_0)^2]}, \qquad \theta_\mu = \mathrm{arctg}\,[x/(y + y_0)]$$

Therefore the high-frequency asymptotic of the reflected field is

$$U_r = U_{GO} + U_h$$

Consider the behaviour of the constant-phase lines (wavefronts), $s(x, y) = \tau$, for some components of the total field when τ increases. The function $s(x, y)$ satisfies the eikonal equation in the upper medium

$$|\mathrm{grad}\, s| = 1$$

and

$$|\mathrm{grad}\, s| = n < 1$$

in the lower medium. Therefore the wavefronts propagate at the unit velocity in the upper medium and at a velocity $c = (1/n) > 1$ in the lower medium. The incident wavefront is a circle $[x^2 + (y - y_0)^2] = \tau^2$, while the reflected wavefront is a reflection of this circle from the interface, i.e. it is the circle arc $[x^2 + (y + y_0)^2] = \tau^2$. The refracted wavefront propagates only for $\tau > y_0$. Points M for which the optical path length (eqn. 5.40) is equal to τ correspond to the front. Its parametric equation is

$$x = y_0\,\mathrm{tg}\,\theta + (1/n)(\tau - y_0/\cos\theta)\sin\theta_1$$

$$y = -(1/n)(\tau - y_0/\cos\theta)\cos\theta_1$$

where $\theta_1 = \arcsin(\sin\theta/n)$ is the direction of a reflected ray. The parameter θ (an angle at which the ray is leaving the source Q) changes within the limits $0 < \theta < \arccos(y/\tau)$ when $y_0 < \tau < y_0/\cos\theta_r$. The refraction point M on the interface corresponds to $\theta = \arccos(y_0/\tau)$. The direction of the refracted ray at this point is

$$\theta_1 = \arcsin[\sqrt{(\tau^2 - y_0^2)}/n\tau]$$

When τ increases the angle θ_1 increases and for $\tau = y_0/\cos\theta_r$ one obtains $\theta_1 = \pi/2$, i.e. the refracted ray slides along the interface. For $\tau > y_0/\cos\theta_r$ the refracted wavefront propagates along the boundary at a velocity $c = (1/n)$ and

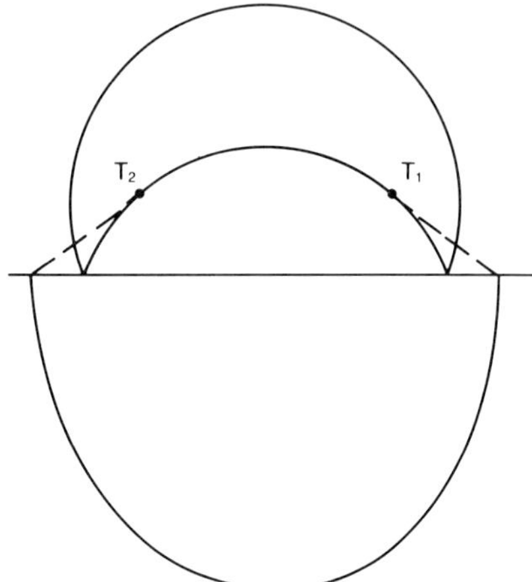

Figure 5.11 Wavefronts of refracted, reflected and head waves

is detached from the incident wavefront. One obtains the situation of Figure 5.11, where the head wavefront

$$\tau = x \sin \theta_r + (y + y_0) \cos \theta_r$$

is shown by the dotted line, beginning at the same point $y=0$; $x = \tau \sin \theta_r - y_0 c \operatorname{tg} \theta_r$ as the refracted wavefront and is tangential to the reflected wavefront at the point

$$T\colon (x = \tau \sin \theta_r; y = \tau \cos \theta_r - y_0)$$

To explain why the head wave arises, the disturbance from the source can reach a point $M(x, y)$ in the upper medium along two ways. First, it can propagate along the specularly reflected ray QPM (Fig. 5.10) outgoing from Q at an angle $\theta = \theta_M = \operatorname{arctg}[x/(y+y_0)]$. Secondly, if the point M is far from Q along the horizontal, i.e. for $x > (y + y_0) \operatorname{tg} \theta_r$, there is a quicker way for a disturbance to propagate, first, along the path at an angle of the total internal reflection from Q to the interface (the point P_1) and then at an angle $\theta_1 = \arcsin(\sin \theta_r / n) = \pi/2$, i.e. along the interface up to the point P_2, and then from P_2 along a ray at an angle θ_r up to the point M. The time gain is a result of the disturbance motion along the interval (P_1P_2) at a velocity $c = (1/n) > 1$ and the motion time is $n|P_1 - P_2|$. Therefore the total time for the disturbance to propagate from Q to M along the path QP_1P_2M appears equal to

$$\begin{aligned} &|QP_1| + n|P_1P_2| + |P_2M| \\ &\quad = [(y + y_0)/\cos \theta_r] + n[x - (y + y_0) \operatorname{tg} \theta_r] \\ &\quad = [(y + y_0)/\cos \theta_r] + \sin \theta_r [x - (y + y_0) \operatorname{tg} \theta_r] \\ &\quad = x \sin \theta_r + (y + y_0) \cos \theta_r \end{aligned}$$

As seen from eqn. 5.43 when approaching the point T (Figure 5.11) where the fronts of the head and reflected waves merge, the amplitude of the head wave grows as $(\theta_r - \theta_\mu)^{-3/2}$, when the observation point M is near the point T, in the second integral U_2 in eqn. 5.42 for U_r the phase function stationary point $\theta=\theta_\mu$ appears to be close to the branch point $\theta=\theta_r$ of the nonexponential factor. Therefore the uniform asymptotic of the function U_2 is determined by theorem 2.4, expressed in terms of the parabolic cylinder functions $D_{1/2}$ and $D_{3/2}$ and of the form

$$U_2 \simeq \exp\{ikr\cos^2[(\theta_\mu-\theta_r)/2]\}k^{-3/4}$$
$$\times\left\{AD_{1/2}\left[2\sqrt{(kr)}\exp(-\pi i/4)\sin\frac{\theta_\mu-\theta_r}{2}\right]\right.$$
$$\left.+k^{-1/2}BD_{3/2}\left[2\sqrt{(kr)}\exp(-\pi i/4)\sin\frac{\theta_\mu-\theta_r}{2}\right]\right\}$$

where

$$A=\frac{4\exp(-5\pi i/8)}{mr^{3/4}}\sqrt{\operatorname{tg}\theta_r}\left[\cos\frac{\theta_\mu-\theta_r}{2}\right]^{-3/2}+O(k^{-1})$$

$$B=\frac{\sqrt{2}\exp(\pi i/8)}{r^{5/4}\sin\dfrac{\theta_\mu-\theta_r}{2}}\left[\frac{m\cos\theta_\mu\sqrt{\{\sin(\theta_r+\theta_\mu)\cos[(\theta_r-\theta_\mu)/2]\}}}{m^2\cos^2\theta_\mu+\sin^2\theta_\mu-\sin^2\theta_r}\right.$$
$$\left.-\frac{\sqrt{2}}{m}\sqrt{\operatorname{tg}\theta_r}\left(\cos\frac{\theta_\mu-\theta_r}{2}\right)^{-3/2}\right]+O(k^{-1})$$

5.6 Reflection from concave surface

Let the field excited by a point source Q hit the concave mirror AB. The resulting scattered field is of a complicated structure. It consists of the field reflected from the illuminated mirror surface, of the primary edge waves generated when an incident wave hits the edges A and B, of re-reflection of these waves from the mirror internal surface, of the secondary edge waves excited due to diffraction of each of the primary waves at the opposite edge, etc. The reflected field dominates in amplitudes, especially in the zones of its focusing. This Section uses the Kirchhoff approximation to obtain the uniform asymptotic of the reflected field.

Consider, first, a ray structure of the field reflected from a concave surface. Generally speaking, the field rays are convergent. If the mirror surface is a part of an ellipse and the source Q is at one of its foci, all the reflected rays converge at the second focus. In all other cases there is no total focusing and the reflected rays have an envelope, a caustic. Typical congruencies of the reflected rays are shown in Figures 5.12*a* and *b*. BB_3 and AA_3 are outer rays outgoing from the edges B, A. The curves B_1DA_1 are caustics, D is a cusp point of the caustics, B_1, A_1 are the endpoints of the caustics.

Inside the mirror (to the left of the outer rays BC and AC in Figure 5.12*a*) a reflected ray passes through each observation point. In the triangles B_1A_2C and A_1B_2C (Figure 5.12*a*) two rays tangent to the caustic arcs B_1D and A_1D

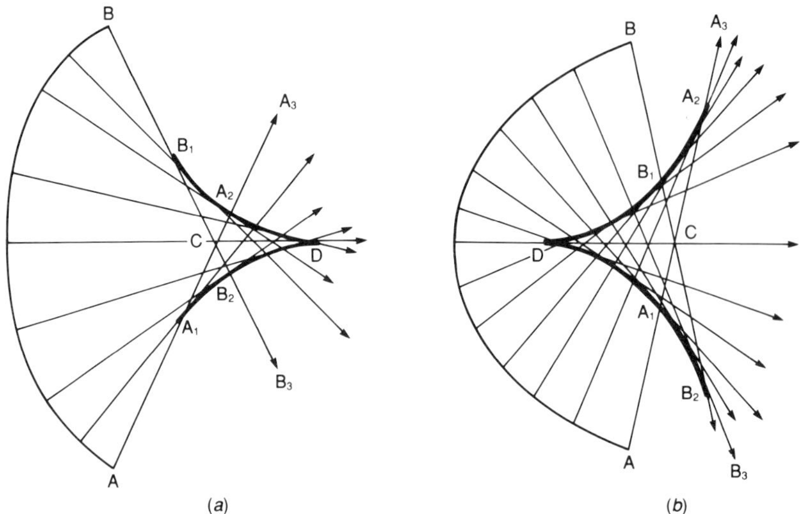

Figure 5.12 Reflection from concave surface

pass through each observation point and inside the region CA_2DB_2C three rays pass. Outside the caustic in sectors $BB_1A_2A_3$ and $AA_1B_2B_3$ there are no reflected rays at all, while to the right of the line $A_3A_2DB_2B_3$ one ray passes through each observation point.

From the ray structure of the reflected field proceed to the construction of its uniform asymptotic. In the Kirchhoff approximation the reflected field is of the form

$$U_r(M)=\int_A^B \exp\{ik[s_i(P)+|P-M|]\}f(P,M)\,dp \tag{5.44}$$

where M is the observation point, S_i is the incident wave eikonal, dp means the integration along the mirror surface. All the amplitude factors are included in $f(P, M)$.

The phase function stationary points are the points where the rays passing through M leave the mirror. If M is near the mirror and only one ray passes through it, then the phase function also has the only stationary point. The contribution of this point into the asymptotic of eqn. 5.44 is of the form of eqn. 5.19 and coincides with the leading term of the exact asymptotic of the reflected wave. When the observation point M approaches the outer rays (e.g. the ray AA_3) then the stationary point $P(M)$ approaches the edge A. Under these conditions the uniform asymptotic of eqn. 5.44 is expressed in terms of the Fresnel integral (see theorem 2.1). As in the case of diffraction by a wedge considered in Section 5.3, this asymptotic describes correctly the field near the light–shadow boundary and differs from the exact asymptotic of the reflected field in a summand with the edge-wave phase which is regular on the light–shadow interface (see, for example, Reference 35 and the references therein).

Since the Kirchhoff approximation expressed by eqn. 5.44 gives correctly the leading term of the reflected wave asymptotic near the mirror (where the system

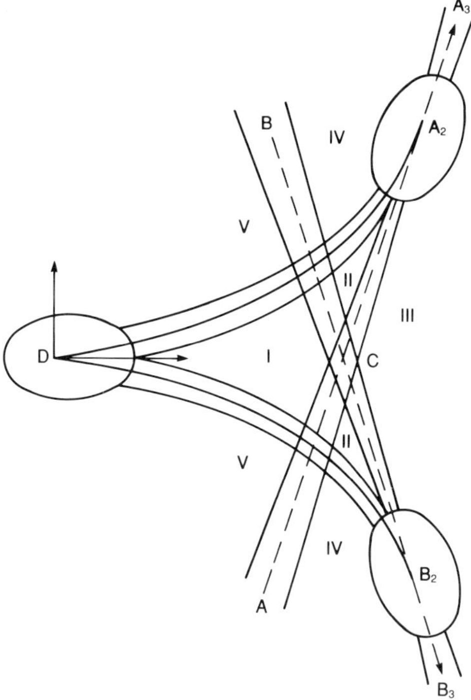

Figure 5.13 Transient zones in vicinity of caustic

of the reflected ray is regular), it also describes correctly the leading term of the reflected field far from the mirror including the transient zones (vicinities of the caustic and outer rays) as well as the regions where these transient zones overlap.

When the observation point M is in transient zones the critical points in eqn. 5.44 appear to be close and the uniform asymptotic of the integral describing the field is expressed in terms of proper special functions. Consider this situation in more detail.

In Figure 5.13 (which reproduces part of Figure 5.12*b* shown in large scale) the line A_2DB_2 is a caustic, AA_3 and BB_3 are outer rays outgoing from the edges A, B. The transient zones which are vicinities of these lines are singled out as well as vicinities of the points A_2, B_2, D where the zones merge.

Outside the transient zones the fixed number of rays hit each observation point. In region V only one ray hits the observation point. When intersecting a caustic (i.e. passing from region V into region I and from IV into region II) the number of the rays passing through the observation point is increased by two. When intersecting each of the outer rays AA_3 and BB_3 the number of these rays is decreased by unity.

Write the nonuniform asymptotic of eqn. 5.44 applicable outside the transient zones and determined by a critical points contributions. In so doing fix the point P from which the reflected ray emanates and move the observation point M along this ray. If M is near the mirror, the second derivative of the summand

$|P-M|$ in the phase function $\Phi(P)=s_i(P)+|P-M|$, at the point P_M as well as the second derivative of the total phase function are large and positive. When removing the point M from P_M the derivative decreases and when M hits the caustic it vanishes. When further moving from the point P_M the second derivative becomes negative, i.e. the point P becomes a maximum point of the phase function. Taking this into account one obtains the following expression for the asymptotic contribution $U_r(P_M, M)$ of the stationary point P_M in eqn. 5.44

$$U_r(P_M, M) = \varepsilon A(P_M)\sqrt{(d/|d_M-d|)}\exp\{ik[s_i(P_M)+d]-i\pi\delta/2\} \tag{5.45a}$$

Here, d is a distance along the ray from P_M to the caustic, $d_M=|P_M-M|$ is a distance from P_M to the observation point M; $\delta=0$, if $d_M<d$ (i.e. when the ray is not tangent to the caustics between P_M and M), and $\delta=1$ for $d_M>d$, the amplitude of the incident field at the point P is $A(P)$. Here, $\varepsilon=1$ for the Neumann boundary condition and $\varepsilon=-1$ for the Dirichlét condition.

The nonuniform asymptotic $U_r(M)$ outside the transient zones can be written as

$$U_r(M)=\sum_e U_r(P_e, M) \tag{5.45b}$$

where summing is taken over all the rays passing through M, i.e. over all the stationary points $P_e=P_e(M)$ of the phase function in eqn. 5.44, each summand being of the form of eqn. 5.45*a*.

Now describe the uniform asymptotic applicable in each of the transient zones shown in Figure 5.13. Outside the vicinities of the points A_2, B_2, D (marked by ovals) this asymptotic on the arcs A_2D and DB_2 of the caustic is expressed in terms of the Airy function, on the intervals AA_2 and BB_2 of the outer rays it is expressed in terms of the Fresnel integral, while on the intervals A_2A_3 and B_2B_3 in terms of the conjugate Fresnel integral. In the vicinities of points A_2 and B_2 the uniform asymptotic is expressed in terms of the incomplete Airy function and in the vicinity of point D in terms of the Piercey integral.

The appropriate expressions are as follows:

(i) If the observation point M is near the caustic arcs A_2D and DB_2 (being outside them), then in sum (eqn. 5.45*b*) there are two summands with close eikonals $s_+(M)=s(P_+)+|P_+-M|$ and $s_-(M)=s_i(P_-)+|P_--M|$, where $s_+(M)>s_-(M)$, the eikonal $s_+(M)$ corresponding to the ray L_+ which is tangential to the caustic between P_+ and M, while the eikonal $s_-(M)$ corresponds to the ray L_-, which is tangential to the caustics behind the point P_-. According to the results of Section 2.3 these two summands should be replaced by

$$2\exp(ik\varphi_0-i\pi/4)[Bk^{1/6}Ai(-k^{2/3}\xi)+Ck^{-1/6}Ai'(-k^{2/3}\xi)] \tag{5.46}$$

where Ai and Ai' are the Airy function and its derivative

$$\varphi_0=\varphi_0(P, M)=\frac{s_++s_-}{2}, \quad \xi=\xi(P, M)=[3(s_+-s_-)/4]^{2/3} \tag{5.47a}$$

Here, B and C are asymptotic series in powers of $1/k$ the leading terms of which are expressed in terms of the amplitude of the GO approximation

(eqn. 5.45*a*)

$$B_0 = \sqrt{\pi}\{A(P_+)\sqrt{[d_+/(d_M^+ - d_+)]} + A(P_-)\sqrt{[d_M^-/(d_M^- - d_-)]}\}\xi^{1/4}$$
$$C_0 = \sqrt{\pi}\{A(P_+)\sqrt{[d_+/(d_M^+ - d_+)]} - A(P_-)\sqrt{[d_-/(d_M^- - d_-)]}\}\xi^{-1/4} \quad (5.47b)$$

where P_+, d_+, d_M^+ are referred to the ray L_+, while P_-, d_-, d_M^- to the ray L_-.

Eqns. 5.47 are defined only outside the caustic, but they are analytically continued into the zones inside the caustic (outside the region A_2DB_2).

(ii) If M is in a vicinity of the interval BB_2 of the outer ray, then the point P_M of the ray origin is near the edge B and the summand $U_r(P_M, M)$ is replaced by the expression

$$U_r(P_M, M)F\{\sqrt{k}[-s_i(P_M) - |P-M| + s_i(B) + |B-M|]\} \quad (5.48a)$$

where $F(\xi)$ is the Fresnel integral. Here the radical sign is taken positive in the illuminated region (to the left of the interval BB_2) and negative in the shade. Similarly, on the interval B_2B_3 the summand U_r in the nonuniform asymptotic is replaced by

$$U_r(P_M, M)F^*\{\sqrt{k}[s_i(P_M) + |P-M| - s_i(B) - |B-M|]\} \quad (5.48b)$$

where again the radical sign is taken positive in the illuminated region (to the right of the interval B_2B_3) and negative in the shade.

Similar expressions obtained by changing indices in eqn. 5.48 are valid in the vicinity of the outer ray AA_3.

If the observation point M is in the vicinity of the caustic endpoint (A_2, B_2 in Figure 5.12*b* or A_1, B_1 in Figure 5.12*a*), then in eqn. 5.44 two close stationary points appear to be near the boundary of the integration interval. Under these conditions the uniform asymptotic of the integral is expressed in terms of the incomplete Airy function and its derivative. The form of the asymptotic depends on the sign of the third derivative of the phase function $\varphi(P, M) = s_i(P) + |P-M|$ in Kirchhoff's integral (eqn. 5.44) which is considered as a function of a length α measured from the edge B along the mirror.

If this derivative is positive (Figure 5.12*a*), then to write the uniform asymptotic of the field in the vicinity of B_1 one should replace the Airy function $Ai(-k^{2/3}\xi)$ and its derivative in eqn. 5.46 by the incomplete Airy function, $V(-k^{2/3}\xi, k^{1/3}q)$ and its derivative $V_p'(-k^{2/3}\xi, k^{1/3}q)$, with respect to $p = -k^{2/3}\xi$, retaining the values of the phase function φ_0 and the amplitudes A and B. If M is inside the caustic, then the argument $q = q(P, M)$ of the function V is defined as the only root of the equation

$$\varphi_0 + (q^3/3) - q\xi = \varphi(B, M) = s_i(B) + |B-M| \quad (5.49)$$

where φ_0 and ξ are determined from eqn. 5.47*a*. If M is outside the caustic where eqn. 5.49 can have three roots, then that root is taken which is an analytical continuation of the function $q(P, M)$ from the region inside the caustic.

When the third derivative of the function $\varphi(P, M)$ is negative (Figure 5.12*b*) the Airy function and its derivative in eqn. 5.46 are replaced by $Ai(-k^{2/3}\xi) - V(-k^{2/3}\xi, -k^{1/3}q)$ and $Ai'(-k^{2/3}\xi) - V_p'(-k^{2/3}\xi, -k^{1/3}q)$, respectively.

In addition to the expression obtained, the uniform asymptotic of Kirchhoff's integral (eqn. 5.44) involves a summand with the edge-wave phase

$$(i/k)\exp\{ik[|B-M|+s_i(B)]\}A(M)$$

where $A(M)$ is regular in the vicinity of the caustic's endpoint. This summand is a small correction to the first two terms of the uniform asymptotic.

Finally, if the observation point is near the caustic cusp D, the uniform asymptotic of the field is expressed in terms of the Piercey integral and its derivatives (Section 3.4). Here, working is restricted to the local asymptotic valid for small x, y where x, y are the co-ordinates in the reference system with D as the origin and the x-axis directed along the ray passing through the point D. In the situation shown in Figure 5.12*b*, i.e. when the caustic cusp is directed to the mirror, the caustic equation is

$$x=\alpha y^{2/3}+O(y^{5/3}) \tag{5.50}$$

for $\alpha>0$ and the field asymptotic is expressed in terms of the Piercey integral $I(x, y)$

$$U\simeq\left[\frac{27kd^2}{4\alpha^3}\right]^{1/4}\frac{A_0}{\sqrt{(i\pi)}}\exp\left[ik(s_0+x)\right]I[-x\sqrt{(27k/4\alpha^3)}, y(27k/\alpha^3)^{1/4}]$$

If the cusp of the caustic is directed from the mirror (Figure 5.12*a*), then $\alpha<0$ in the caustic eqn. 5.50 and the asymptotic is expressed in terms of the function $I^*(x, y)$ complex conjugate to the Piercey integral

$$U\simeq\left[\frac{27ld^2}{4|\alpha|^3}\right]^{1/4}\frac{A}{\sqrt{(i\pi)}}\exp\left[ik(s_0+x)\right]$$
$$\times I^*[x\sqrt{(27k/4|\alpha|^3)}, -y(27k/|\alpha|^3)]^{1/4}$$

5.7 Radiation from aperture (nonuniform asymptotic)

Let a primary field U be incident from the region $x<0$ on a plane screen with an aperture Ω, the screen being located in the plane $x=0$. One needs to find the short-wave asymptotic of the field U passed through the aperture. Assume the Dirichlét or Neumann boundary conditions to be posed on a screen and the aperture edge to be a piecewise analytical curve.

Write the solution of this problem in the Kirchhoff approximation, find its short-wave asymptotic and compare the results with those obtained according to the exact asymptotic of the solution. Estimate the applicability limits for the nonuniform asymptotic and make it clear how the near (Fresnel) and far (Fraunhofer) zones arise when evaluating the Kirchhoff integral asymptotic.

Green's function (eqn. 5.14) is taken as an initial step for a field to the right of the screen, i.e. for $x>0$

$$U(M)=\iint_{x=+0}\left[\frac{\partial U(p)}{\partial x}G(p, M)-\frac{\partial G(p, M)}{\partial x}U(p)\right]dp$$

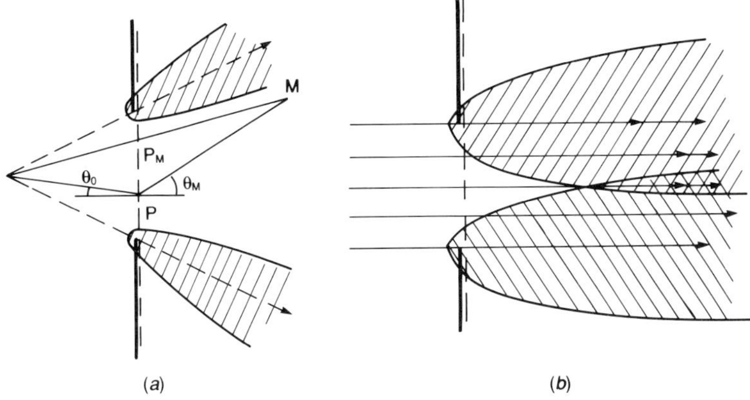

Figure 5.14 Diffraction by aperture

where $U(p)$ and $\partial U(p)/\partial x$ are the values of the total field arising when the field U_0 is incident on the screen. The integration is performed over the halfspace boundary, i.e. over the shadow side of the screen and over the aperture Ω (Figure 5.14). The evaluation of the short-wave asymptotic for the field $U(M)$ in the Kirchhoff approximation is based on the assumption that the field $U(M)$ and its derivative $\partial U/\partial x$ are negligibly small on the shadow side of the screen everywhere, except an immediate vicinity of the edge S, while in the aperture Ω the field coincides everywhere (except a similar vicinity S) with the incident field U_0. Therefore for the field $U(M)$ passed through the screen

$$U(M)=\iint_{\Omega}\left[\frac{\partial U_0(p)}{\partial x}G(p, M)-\frac{\partial G(p, M)}{\partial x}U_0(p)\right]dp \tag{5.51}$$

Here, $G(p, M)$ is the Green's function for the Helmholtz equation in three-dimensional space

$$G(p, M)=-\frac{1}{4\pi}\frac{\exp(ik|p-M|)}{|p-M|}$$

Since only the leading term of the asymptotic for the field $U(M)$ is of interest one can replace the incident field $U_0(M)$ in eqn. 5.51 by its GO approximation

$$U_0(p)\simeq A(p)\exp[iks_0(p)]$$

(where s_0, A are the eikonal and the incident-wave amplitude) and only the exponential term is differentiated with respect to x. This results in the following expression

$$U(M)\simeq-\frac{ik}{4\pi}\iint_{\Omega}\frac{A(p)}{|p-M|}(\cos\theta_0+\cos\theta_M)$$
$$\times\exp[ik(s_0(p)+|p-M|)]\,dp \tag{5.52}$$

where θ_0 and θ_M are the angles between the x-axis and the rays at the point M, namely, the ray of the incident wave and the ray going from M to P, respectively.

Denote the corner points of the boundary S by T_m while the parts of S between T_m and T_{m+1} are denoted by S_m. Each S_m is an interval of an analytical curve. Assume that in the region $x>0$ the ray system of the field U_M is divergent, i.e. in this region the primary field does not form caustics and only one ray passes through each point M.

Consider the short-wave asymptotic of eqn. 5.52. It is the sum of the contributions of the phase-function stationary point P_0 in the region Ω, the stationary points P_q of the intervals S_m and the corner points T_m

$$\begin{aligned} U(M) = {} & \chi_0 A_0(M) \exp\{ik[s_0(p_0) + |p_0 - M|]\} \\ & + \sqrt{(i/k)} \sum_q A_q(M) \exp\{ik[s_0(p_q) + |p_q - M|]\} \\ & + (i/k) \sum_m B_m(M) \exp\{ik[s_0(T_m) + |T_m - M|]\} \end{aligned} \tag{5.53}$$

Here, $\chi_0 = 1$, if P_0 is inside Ω, otherwise $\chi_0 = 0$. The first summation is performed over all stationary points P_q on the intervals S_m. The number of these points depends, generally speaking, on the observation point position.

Compare the written asymptotic of the field $U(M)$ in the Kirchhoff approximation with the known exact asymptotic (see, for instance, Reference 35).

$$\begin{aligned} U(M) = {} & \chi_0 U_0(M) + \sqrt{(i/k)} \sum_q \tilde{A}_q(M) \exp[iks_e^0(M)] \\ & + (i/k) \sum_m \tilde{B}_m(M) \exp\{ik[s_0(T_m) + |T_m - M|]\} + \cdots \end{aligned} \tag{5.54}$$

The first summand is a primary field multiplied by the cutting function χ_0 which describes the shadow effect. It is equal to unity in a region illuminated by the incident field rays passed through Ω and to zero in a region shaded by the screen. The sum over q describes the edge waves that arise in the case of primary field diffraction by smooth parts S_m of the screen boundary. From each point P on such an interval the cone of diffraction rays emanates, the rays being formed at the same angle θ with the tangent to the edge as the primary field ray incident at P. The phase of the edge-wave ray at the point M on the ray is

$$s_e(M) = s_0(P) + |P - M|$$

where P is the point of the outgoing ray. The first summation in eqn. 5.54 is over all edge-wave rays passing through M. Each summand in the second sum describes the primary spherical diffraction wave arising when the incident primary field reaches the boundary corner point T_m. The primary edge and spherical diffraction waves again reaching the edge S_m generate the secondary diffraction waves. Similarly, the subsequent diffraction waves arise. The appropriate summands in eqn. 5.54 are denoted by dots.

Indicating the correspondence between the field (eqn. 5.53) in the Kirchhoff approximation and exact asymptotic (eqn. 5.54), show that the first summands in these equations coincide, i.e. the contribution of the phase-function stationary point in eqn. 5.52 coincides with $U_0(M)$ (or more precisely, with the leading term of the asymptotic of eqn. 5.51). In so doing it is sufficient to write Green's function for $U_0(M)$

$$U_0(M) = \iint_{x=0} \left[\frac{\partial U_0(P)}{\partial x} G(p, M) - \frac{\partial G(p, M)}{\partial x} U_0(p) \right] dp \tag{5.55}$$

where, as previously, the point M is in the halfspace $x > 0$.

Replacing $U_0(p)$, $G(p, M)$ and their derivatives by the leading terms of their asymptotics obtains an expression which differs from eqn. 5.52 only in replacing the integration domain Ω by the entire plane $x = 0$. Therefore the contributions of the phase-function stationary point (coinciding with the point of the plane $x = 0$ from which the primary ray passing through the point M emanates) coincide in these integrals and are equal due to eqn. 5.55 to the values at the point M for the short-wave asymptotic of the incident wave.

Now it is shown that in the sum over q in eqns. 5.53 and 5.54 there are summands with the same phases. Indeed, stationary points in eqn. 5.53 are determined from the condition

$$\frac{\partial}{\partial I}[s_0(p) + |p - M|] = \cos\theta_0 - \cos\theta_\mu = 0$$

where $\partial/\partial I$ means differentiating along the edge, while θ_0 and θ_μ are the angles between the tangent to the edge and the incident wave ray and PM, respectively. At the stationary point $p = p_q$, obviously, $\theta_0 = \theta_\mu$, i.e. P_qM is an edge-wave ray emanating from p_q. Therefore the phase function $s_0(p) + |p - M|$ coincides with $s^{(q)}(M)$.

Thus, the second and third sums in eqn. 5.53 include the summands with the phase of the primary edge and spherical diffraction waves passing through the observation point M. In the Kirchhoff approximation the amplitude factors A_q and B_m are determined incorrectly.

The asymptotics of eqns. 5.53 and 5.54 are nonuniform. They are inapplicable in vicinities of the light–shadow boundaries for incident and edge waves and near-edge-wave caustics. In these transient zones two or more critical points in eqn. 5.53 appear to be close to each other and the uniform asymptotic of the field $U(M)$ in the Kirchhoff approximation as well as the exact uniform asymptotic are expressed in terms of the appropriate special functions according to the algorithms of Chapter 4.

Consider the region Σ where the nonuniform asymptotic of eqn. 5.53 is applicable, i.e. the region where the field $U(M)$ is a sum of the primary field $U_0(M)$ and the correction which consists of the edge and spherical diffraction waves excited on the edge S and is essentially less (in amplitude) than the primary field (excluding the case of edge-wave focusing). This region is called the region of geometrical optics.

It follows that observation point M is in the geometrical optics zone if, first, the phase-function stationary point P_M is in the integration domain Ω and, secondly, if the influence domain of this point does not intersect the integration domain boundary S, the influence domain size being of the essential role here. For instance, if it is greater than the diameter of Ω, the influence domain does intersect S and one cannot single out the incident wave from the field $U(M)$.

According to Section 4.6 the influence domain can be specified by the ellipse with the stationary point as a centre, the axes (directed along eigenvectors of the matrix whose elements are the phase function second derivatives) and the halfaxis lengths, $d_{1,2} = \sqrt{(\lambda/\xi_{1,2})}/2$, where $\lambda = 2\pi/k$ is the wavelength and $\xi_{1,2}$ are the eigenvalues of the matrix.

Consider how the size of the influence domain changes when observation point M moves from Ω along a primary field ray passing from the fixed point P_M of the region Ω. The phase function is

$$\varphi(P) = s_i(P) + |P - M|$$

Its stationary point coincides with P_M. Since when M moves away along a ray outgoing from P_M the second derivatives of the function $|P - M|$ tend to zero as $|P - M|^{-1}$, then the eigenvalues of the matrix (composed of the second derivatives of the function $\varphi(P)$) decrease at the point P_M and the influence domain of the point P_M increases. Therefore the point M initially located in the geometrical optics zone can leave the zone moving along a primary field ray. Two qualitatively different situations should be distinguished.

If the ray congruence of the incident field is divergent, then the function $s_i(P)$ is convex, the matrix of its second derivatives is nongenerate and when M is removed from Ω, then the size of the influence domain of the point P_μ tends to a finite limit which is determined by the eigenvalues of this matrix. If the limiting influence region of the point P_μ does not intersect S, all the points of a ray outgoing from P_μ are in the geometrical optics zone, i.e. this zone extends to infinity. For a given convex function $s_i(P)$ such a situation occurs always if the wavelength λ is sufficiently small (see Figure 5.14*a* where outer rays, the light–shadow boundaries, are marked by dotted lines, while transient zones surrounding them are shadowed).

If there is a parallel ray beam incident on Ω, i.e. $s_i(P)$ is a linear function of the co-ordinates ξ, η of the point P, the matrix of the phase-function second derivatives is determined by the derivatives of $|P - M|$ and decreases when M moves away and the size of the influence region tends to infinity. In this case, the point M when moving along any ray leaves the GO zone, i.e. the zone does not go to infinity but stops at some distance from the aperture (Figure 5.14*b*).

To estimate this distance, assume for simplicity that the incident field rays are normal to the screen, i.e. $S_0 = 0$. Then the phase function $\varphi(P) = \sqrt{[x^2 + (y - \xi)^2 + (z - \eta)^2]}$ where x, y, z are the co-ordinates of the point M, while ξ, η are the co-ordinates of P. The stationary point P has co-ordinates $\xi = y$, $\eta = z$ and the matrix of the second derivatives at this point is diagonal, $\Phi''_{\xi\xi} = \Phi''_{\eta\eta} = (1/x)$, $\Phi''_{\eta\xi} = 0$. Therefore the influence region is a circle of the radius r

$$r = (1/2)\sqrt{(\lambda x)}$$

The geometrical optics zone is terminated at a distance $x = l$ for which $r \simeq D/2$, where D is the diameter of the domain Ω, i.e. at a distance

$$l \simeq D^2/\lambda \tag{5.56}$$

Here, l is a parameter well known in optics. The condition $R > l$ where R is the distance from the aperture determines the Fraunhofer zone, i.e. the region where the field passed through the aperture can be considered as a directed spherical wave passed through the aperture

$$U \simeq \frac{\exp(ikr)}{r} f(k, \theta, \varphi) + O(r^{-2}) \tag{5.57}$$

where r, θ, φ are the spherical co-ordinates of observation point M and $f(k, \theta, \varphi)$ is the radiation pattern for a field passed through the aperture. The region $R < l$ determines the Fresnel zone where everywhere except relatively narrow transient zones the primary field passed through the aperture is the leading term of the asymptotic for the field passed through the aperture.

These considerations are valid for the analysis of the radiation structure for the parabolic antenna as well. The plane of the antenna aperture is taken as a plane $x=0$ after which the radiated field $U(M)$ in Green's function is taken to be equal to zero outside the antenna aperture and to the GO reflected field in the aperture. As a result, the radiated field is represented in the form of eqn. 5.53 where $A(P)$ and $s_0(P)$ are the amplitude and phase of the radiation source field reflected from the antenna surface and the integral is taken over the antenna aperture. The definition of the Fraunhofer zones $R > l$ is the same, l and D in eqn. 5.56 mean the antenna diameter.

5.8 Radiation from aperture (uniform asymptotic)

This Section describes the uniform asymptotic expressed in terms of the special functions introduced in Sections 4.2–4.5. It is convenient to write the field passed through the aperture (in the Kirchhoff approximation) as

$$U(M) = \iint_{\Omega} \exp\{ik[s_0(P) + |P - M|]\} A(P, M)\, dp \tag{5.58}$$

where the integration is performed over the smooth surface Ω with a piecewise smooth boundary. In terms of such integrals the more general problems (as compared with those considered in Section 5.7) are described, for example, the problem of field reflection from a mirror antenna, Ω being the antenna surface, $s_0(P)$ being the value of the incident wave eikonal on this surface. As in Section 5.7, the corner points of the boundary S (enclosing the domain Ω) are denoted by T_m and a smooth part of the boundary between T_m and T_{m+1} by S_m.

The nonuniform asymptotic of eqn. 5.58 is a sum with summands corresponding to the phase-function stationary points P_p inside Ω, to the stationary points Q on the smooth intervals S_m of the boundary of Ω, and to the corner points T_m

$$\begin{aligned} U(M) = {} & \sum_p A_p \exp\{ik[S(P_p) + |P_p - M|]\} \\ & + \sqrt{(i/k)} \sum_q B_q \exp\{ik[S(Q_q) + |Q_q - M|]\} \\ & + (i/k) \sum_m C_m \exp\{ik[S(T_m) + |T_m - M|]\} \end{aligned} \tag{5.59}$$

P_p are points from which GO rays (passing through M) emanate; these are rays of primary (in the case of radiation from an aperture in the screen) or reflected (in the case of a reflector antenna) fields. The points Q_q are edge points from which the edge waves passing through M emanate, and the corner points T_m are centres of spherical diffraction waves. At these critical points the values of the phase function for eqn. 5.58 are equal to the appropriate wave eikonals at M.

The number of summands in the first and second sums in eqn. 5.59 depends on the observation point position. The asymptotic fails near singular surfaces and lines; when intersected the critical points P_p, Q_q and T_m merge and the number of summands changes. On singular surfaces two critical points merge, at their intersections (singular lines) three or four critical points merge. The physical classification of these surfaces and lines is given in the simplest and most frequent cases, and special functions describing the uniform asymptotic in the vicinity of these singular lines and surfaces are indicated.

Four types of singular surface arise:

(i) Caustic of GO field: merging of two phase-function stationary points inside Ω.
(ii) Light–shadow boundary of GO field: merging of stationary point P inside Ω with point Q on a smooth part of boundary S of aperture Ω. This singular surface consists of all rays of the GO field emanating from points of S.
(iii) Edge-wave caustic: merging of two phase-function stationary points at a smooth part of the boundary.
(iv) Light–shadow boundaries of the edge wave: merging of stationary point on a smooth part of boundary S and a corner point. The boundary consists of all edge-wave rays emanating from the corner points.

In the vicinity of the caustic of the GO fields and of the edge waves the asymptotic of the field $U(M)$ is expressed in terms of the Airy function and its derivative. In the vicinity of the light–shadow boundaries of both GO waves and edge waves the uniform asymptotic of the field $U(M)$ is expressed in terms of the Fresnel integral or conjugate Fresnel integral.

Consider now singular lines that are intersections of the surfaces of types (i)–(iv) where three or four critical points merge:

(i) + (i): *Caustic cuspidal edge for the GO rays.* Three summands corresponding to close stationary points in Ω are replaced by the Piercey integral. Under these conditions one can choose the integration variables ξ, η in vicinities of these points in such a way that the phase-function second derivative with respect to ξ should be bounded from below and the function should have stationary point for any sufficiently small η. After asymptotic integration with respect to ξ the one-dimensional integral considered in Section 3.1 is obtained.

(i) + (ii): *Intersection of the light–shadow boundary of the GO field and the GO field caustic.* Here, three critical points appear to be close, P_1 and P_2 in the domain Ω, and Q on the boundary S. In the vicinities of these points one can introduce such local co-ordinates ξ, η that the phase-function second derivative with respect to ξ is bounded in magnitude from below, while the boundary S has the equation $\eta = 0$. After asymptotic integration with respect to ξ one obtains one-dimensional integral with two phase-function stationary points close to the integration interval boundary. The uniform asymptotic is expressed in terms of the incomplete Airy function and its derivative.

For the combinations (i) + (iii) and (i) + (iv) there is no critical point merging.

(ii) + (iii): *Intersection of the edge-wave caustic with the light–shadow boundary of the geometrical optics field.* Three critical points appear to be close, two stationary points M_1 and M_2 on a smooth part of the boundary S and a stationary point O inside the integration domain Ω. This stationary point appears to be a saddle point (otherwise only one stationary point could be located near O on S), while the direction S near O appears to be close to the phase-function level line starting from the point O. According to the results of Section 4.3 the uniform asymptotics of such integrals are expressed in terms of the Airy-Fresnel integral, Airy function and their derivatives.

(ii) + (iv): *Intersection of the light–shadow boundary of the GO field with the light–shadow boundary of the edge wave.* This intersection is a primary field ray starting off the corner point of the integration domain boundary. In its vicinity four critical points appear to be close, the phase-function stationary point O in the domain Ω, one point on each of the boundary segments terminating at the corner point and the corner point itself.

If O is a minimum point of the phase function (i.e. there are no GO ray caustics between O and M), then the uniform asymptotic of $U(M)$ is expressed in terms of the generalised Fresnel integral. If O is a maximum point of the phase function (i.e. there are two GO ray caustics between O and M), then the asymptotic $U(M)$ is expressed in terms of the complex conjugate generalised Fresnel integral.

If O is a saddle point (i.e. there is one caustic between O and M), then $U(M)$ is expressed in terms of the generalised Fresnel integral of imaginary arguments, i.e. in terms of the *Ff*-integral. It is assumed that the directions of the boundary segments emanating from the corner point are not close to the directions of the level lines for the phase function, otherwise three singular surfaces would overlap (the light–shadow boundaries of the GO field and edge waves) as well as the edge-wave caustic and therefore more complicated form of the uniform asymptotic.

(iii) + (iii): *Cuspidal edge of the edge-wave caustic.* The uniform asymptotic of the edge wave is expressed in terms of the Piercey integral.

(iii) + (iv): *Intersection of the caustic and the light–shadow interface of the edge wave.* The uniform asymptotic of the edge wave is expressed in terms of the incomplete Airy function.

5.9 Diffraction by two wedges

Consider successive diffractions of the isotropic source field $U_0(P) = G(P - Q)$ by the wedges S_1 and S_2 located between the source Q and observation point P (Figure 5.15). The form of the short-wave asymptotic for the field depends on the interlocation of the source Q, observation point P and wedges S_1 and S_2. If P and Q are above the straight line passing through the apices A_1, A_2 of the wedges S_1, S_2 (called the horizon) and the wedges do not shade observation point P, the leading term of the field asymptotic at P coincides at this point with the incident field $U_0(P) = G(P - Q)$. If P is below the horizon while Q is above the line, two situations are possible: P is not shaded by the wedge S_2, then at P the leading term of the field asymptotic is equal to $G(P - Q)$, or the point P is shaded by the wedge, then the leading term of the asymptotic is an

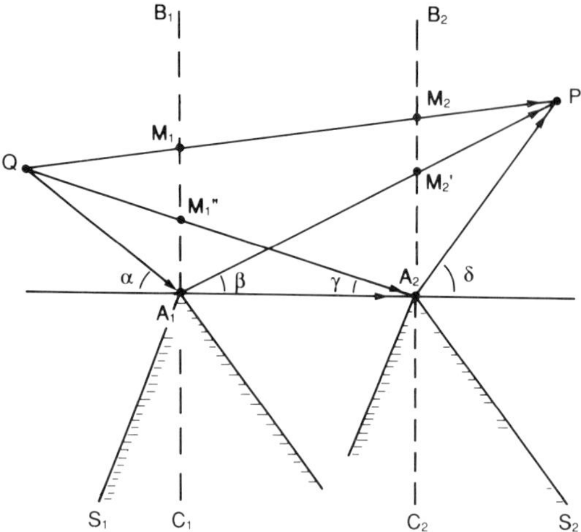

Figure 5.15 Diffraction by two wedges

edge wave arising when the primary field is diffracted by the wedge S_2. Similarly, if Q is lower than the horizon while P is above, the field at P coincides (in its leading term) with the edge wave starting from A_1 or with $G(P, Q)$ depending on whether P is in the shadow of the wedge or out of it. Finally, if P and Q are lower than the horizon, the secondary edge wave arising due to the diffraction of the primary edge wave from A_1 at A_2. In what follows the expression for the uniform field asymptotic at P is found assuming P and Q to be close to the horizon. The evaluation is performed in the Kirchhoff approximation and how the written equations change when going from the Kirchhoff approximation to the exact uniform asymptotic is indicated.

Denote the field arising when the primary wave is diffracted at the wedge S_1 by $U_1(M)$. After two diffractions the sought-for field $U(P)$ at the observation point can be written in the Kirchhoff approximation as

$$U(P) = \int_0^\infty \left[\frac{\partial U_1(M_2)}{\partial x} G(M_2, P) - U_1(M_2) \frac{\partial G(M_2, P)}{\partial x} \right] d\xi_2$$

The integration is performed over the aperture A_2, B_2, and ξ_2 is a distance measured along the aperture from the edge A_2. In turn, the field $U_1(M_2)$, i.e. the field excited due to the incident wave diffraction by the wedge S_1, can be evaluated in the Kirchhoff approximation

$$U_1(M_2) = \int_0^\infty \left[\frac{\partial G(Q, M_1)}{\partial x} G(M_1, M_2) - G(Q, M_1) \frac{\partial G(M_1, M_2)}{\partial x} \right] d\xi_1$$

If in the written integrals one replaces the functions G and their derivatives with respect to x by the leading terms of the short-wave asymptotics one obtains the expression for $U_1(M_2)$ in the form of the one-dimensional integral

$$U_1(M_2) = \frac{-1}{8\pi} \int_0^\infty \exp\left[ik\left(|QM_1| + |M_1M_2|\right)\right] \frac{\cos\theta_q + \cos\theta}{\sqrt{|QM_1|}\sqrt{|M_1M_2|}} \, d\xi_1$$

while $U(P)$ is expressed as two-fold integral

$$U(P) = \frac{\sqrt{k}\exp(3\pi i/4)}{2^{9/2}\pi^{3/2}} \int d\xi_1 \int d\xi_2 \frac{(\cos\theta_q + \cos\theta)(\cos\theta + \cos\theta_p)}{\sqrt{|QM_1|}\sqrt{|M_1M_2|}\sqrt{|M_2P|}} \times \exp\left[ik\left(|QM_1| + |M_1M_2| + |M_2P|\right)\right] \quad (5.60)$$

integrating over the aperture B_1A_1 (variable ξ_1) and over B_2A_2 (variable ξ_2) and where θ_q, θ, θ_p are the angles between the horizon and segments QM_1, M_1M_2, M_2P, respectively.

Write the nonuniform asymptotic of the integral; its integration domain Ω is the quadrant $\xi_1 > 0$, $\xi_2 > 0$. The phase function has a minimum at $\xi_1 = \xi_1^0$ and $\xi_2 = \xi_2^0$, where ξ_1^0 and ξ_2^0 correspond to the points M_1, M_2 that are intersections of the segment QP with the apertures B_1A_1, B_2A_2 (or their continuations A_1C_1, A_2C_2). The stationary point contribution in the integral asymptotic coincides with the primary field value at P.

On the part of the integration domain boundary $\xi_1 = 0$, $\xi_2 > 0$, the phase function is

$$\varphi = |QA_1| + |A_1M_2| + |M_2P|$$

The stationary point of the phase function considered as a function of ξ_2 for $\xi_1 = 0$ corresponds to the point $M_2 = M_2'$ which is the intersection of the straight line B_2C_2 and the segment A_1P. If this point is above the edge A_2, i.e. P is illuminated by the edge wave from A_1, the stationary point contributes to the asymptotic of eqn. 5.60 when integrating along the part of the integration domain boundary $\xi_1 = 0$, $\xi_2 > 0$.

Similarly, the intersection of the segment QA_2 and the aperture A_1B_1 is a stationary point M_1'' of the phase function

$$\varphi = |QM_1| + |M_1A_2| + |A_2P|$$

on the part of the integration domain boundary, $\xi_2 = 0$, $\xi_1 > 0$. It exists provided the edge A_2 is illuminated by the primary field, i.e. under the condition that the primary edge wave (of the edge A_2) exists.

So the equation for the nonuniform asymptotic of eqn. 5.60 follows:

$$\begin{aligned} U(P) = {} & \sqrt{(i/k)} A \exp(ik|QP|)\chi(\alpha + \beta)\chi(\gamma + \delta) \\ & + (i/k) B \exp\left[ik\left(|QA_1| + |A_1P|\right)\right]\chi(\beta) \\ & + (i/k) C \exp\left[ik\left(|QA_2| + |A_2P|\right)\right]\chi(\theta) \\ & + (i/k)^{3/2} D \exp\left[ik\left(|QA_1| + |A_1A_2| + |A_2P|\right)\right] \end{aligned} \quad (5.61)$$

Angles α, β, γ, δ are shown in Figure 5.15. Each of the functions χ is equal to unity if its argument is positive, and to zero if it is negative. These functions describe the shadowing effect. The first summand in eqn. 5.61 (the primary field at observation point P) is a contribution of the phase-function stationary point inside the integration domain. The second summand (the primary edge wave of the edge A_1 at observation point P) is a contribution of the stationary point on the integration domain boundary $\xi_1 = 0$, $\xi_2 > 0$. The third summand is a primary edge wave of the edge A_2. The last summand (the contribution of the integration domain corner point) is a secondary edge wave excited when there is diffraction of the primary edge wave of the edge A_1 by the edge A_2.

Eqn. 5.61 denoted the amplitudes of the relevant waves by A, B, C, D; the exact values are of no importance now, but note that

$$A = \frac{1}{2\sqrt{(2\pi|QP|)}}$$

According to Section 4.4 the uniform asymptotic

$$\begin{aligned} U(P) = [i/k]^{1/2} A\, e^{ik|QP|}[&\chi(\alpha+\beta)\chi(\gamma+\delta) \\ &-G\{-[k(|A_1A_2| + |A_2P| - |A_1P|)]^{1/2}, \\ &\quad [k(|QA_1| + |A_1P| - |QP|)]^{1/2}\} \\ &-G\{-[k(|QA_1| + |A_1A_2| - |QA_2|)]^{1/2}, \\ &\quad [k(|QA_2| + |A_2P| - |QP|)]^{1/2}\}] \\ &+[i/k]B^*\, e^{ik(|QA_1| + |A_1P|)} \\ &\times F\{-[k(|A_1A_2| + |A_2P| - |A_1P|)]^{1/2}\} \\ &+[i/k]C^*\, e^{ik(|QA_2| + |A_2P|)} \\ &\times F\{-[k(|QA_1| + |A_1A_2| - |QA_2|)]^{1/2}\} \\ &+[i/k]^{3/2} D^*\, e^{ik(|QA_1| + |A_1A_2| + |A_2P|)} \end{aligned} \qquad (5.62)$$

corresponds to the nonuniform asymptotic of eqn. 5.61. Here, $G(x, y)$ is the generalised Fresnel integral, the radical signs being chosen as

$$\begin{aligned} &\sqrt{[k(|A_1A_2| + |A_2P| - |A_1P|)]} && \text{the same sign as angle } \beta \\ &\sqrt{[k(|QA_1| + |A_1A_2| - |QA_2|)]} && \text{the same sign as angle } \gamma \\ &\sqrt{[k(|QA_1| + |A_1P| - |QP|)]} && \text{the same sign as angle } \alpha+\beta \\ &\sqrt{[k(|QA_2| + |A_2P| - |QP|)]} && \text{the same sign as angle } \delta+\gamma \end{aligned}$$

So defined arguments of special functions in eqn. 5.62 are analytical functions of the angles a, β, γ, δ, i.e. of positions of the source Q and observation point P. The amplitude factors B^*, C^*, D^* are evaluated by asymptotic matching using the known functions A, B, C, D in eqn. 5.6.

The uniform asymptotic eqn. 5.62 was obtained in the Kirchhoff approximation, but the same equation is also valid for the uniform asymptotic. The problem of sequential diffractions by two halfplanes has been solved in Reference 37 (see also Reference 38); the equations for the uniform asymptotic in the problem of wave diffraction by two wedges are given in Reference 35.

The exact asymptotic differs from the Kirchhoff approximation in the amplitudes B, C, D in eqn. 5.61 (they have, in a similar way to the problem of the diffraction by a wedge Sections 5.3 and 5.4, the same poles at the light–shadow boundary). They are evaluated, for instance, by the geometrical theory of diffraction, i.e. in terms of the Keller diffraction coefficients. These amplitude changes lead to the corresponding change of amplitudes B^*, C^*, D^* in eqn. 5.62.

Chapter 6

Special functions

This Chapter treats those special functions in terms of which the uniform asymptotics of the integrals considered in Chapters 2–4 are expressed. Mainly, only that information is given which can be useful when numerically evaluating these functions. As before, there are no references to the literature when the equations are well known (see, for example, References 19 and 26).

6.1 Fresnel integral

The Fresnel integral $F(\xi)$ is defined by the expression

$$F(\xi)=\frac{\exp(-\pi i/4)}{\sqrt{\pi}}\int_{-\infty}^{\xi}\exp(is^2)\,ds \tag{6.1a}$$

The equality

$$F(\xi)=\frac{i\exp(i\xi^2)}{2\pi}\int_{-\infty+iO}^{\infty+iO}\exp(-ix^2)\frac{dx}{x-\xi} \tag{6.1b}$$

also holds (Section 2.2) where the integration contour bypasses the pole $x=\xi$ in the upper halfplane. The function $F(\xi)$ satisfies the relations

$$F(\xi)+F(-\xi)=1$$

$$\begin{aligned}F(i\xi)=F^*(\xi)&=\frac{\exp(\pi i/4)}{\sqrt{\pi}}\int_{-\infty}^{\xi}\exp(-is^2)\,ds\\&=\frac{-i}{2\pi}\exp(-i\xi^2)\int_{-\infty-iO}^{\infty-iO}\frac{\exp(ix^2)\,dx}{x-\xi}\end{aligned} \tag{6.2}$$

where F^* is the complex conjugate function to the Fresnel integral and the pole $x=\xi$ is bypassed in the lower halfplane. Hence

$$F^*(-i\xi)=F(\xi)$$

Using eqn. 6.2 for $\xi>0$ to evaluate $F^*(i\xi)$ one obtains

$$\begin{aligned}F(-\xi)&=\frac{-i\exp(i\xi^2)}{2\pi}\int_{-\infty}^{\infty}\exp(ix^2)\frac{dx}{x-i\xi}\\&=\frac{\xi}{2\pi}\exp(i\xi^2)\int_{-\infty}^{\infty}\exp(ix^2)\frac{dx}{x^2+\xi^2}\\&=\frac{\exp(i\xi^2)}{2\pi}\int_{-\infty}^{\infty}\exp(i\xi^2x^2)\frac{dx}{x^2+1}\end{aligned} \tag{6.3}$$

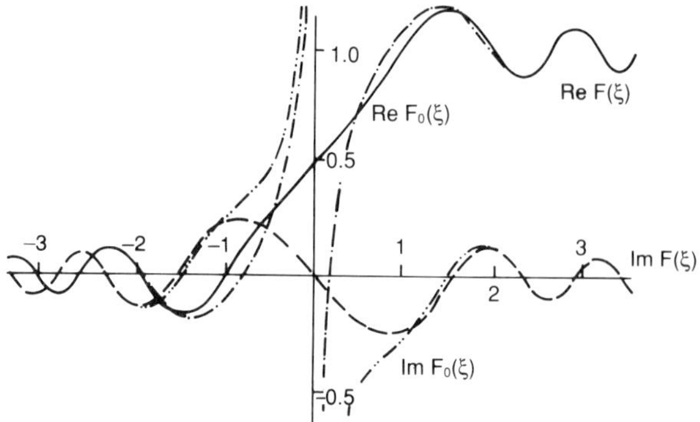

Figure 6.1 Fresnel integral and its asymptotic

The function $F(\xi)$ is expanded into the series in powers of ξ:

$$F(\xi)=\frac{1}{2}+\frac{\exp(-\pi i/4)}{\sqrt{\pi}}\sum_{n=0}^{\infty}\frac{i^n\xi^{2n+1}}{(2n+1)n!} \tag{6.4}$$

For $|\xi|\to\infty$ the integral $F(\xi)$ has the asymptotic

$$F(\xi)\simeq\chi(\xi)-\exp(i\xi^2)T(\xi) \tag{6.5}$$

where $T(\xi)$ is the asymptotic series

$$T(\xi)=\frac{\exp(\pi i/4)}{2\pi\xi}\sum_{n=0}^{\infty}\frac{\Gamma(n+1/2)}{(i\xi^2)^n} \tag{6.6}$$

The functions Re $F(\xi)$, Im $F(\xi)$, as well as Re $F_0(\xi)$, Im $F_0(\xi)$, are plotted in Figure 6.1, where $F_0(\xi)$ is the leading part of the asymptotic eqn. 6.5:

$$F_0(\xi)=\chi(\xi)-\frac{\exp(\pi i/4+i\xi^2)}{2\sqrt{\pi}\xi}$$

The interrelated integrals introduced by Fresnel

$$C_1(x)=\sqrt{2/\pi}\int_0^x\cos t^2\,dt; \qquad S_1(x)=\sqrt{2/\pi}\int_0^x\sin t^2\,dt$$

$$C_2(x)=\frac{1}{\sqrt{2\pi}}\int_0^x\frac{\cos t\,dt}{\sqrt{t}}; \qquad S_2(x)=\frac{1}{\sqrt{2\pi}}\int_0^x\frac{\sin t\,dt}{\sqrt{t}}$$

are also used. The integral $F(\xi)$ is related to $C_2(\xi)$ and $S_2(\xi)$ as

$$F(\xi)=\frac{1}{2}+\frac{1-i}{2}\,\text{sign}\,\xi[C_2(\sqrt{|\xi|})+iS_2(\sqrt{|\xi|})] \tag{6.7}$$

The coefficients of the expansions for the functions

$$C_3(x)=\int_0^x t^{-1/2}\cos t\,dt; \qquad S_3(x)=\int_0^x t^{-1/2}\sin t\,dt$$

into the series in Chebyshev's polynomials $T_{2n}(x/8)$ and $T_{2n+1}(x/8)$ in the interval $0<x<8$ as well as for expansions of

$$C_4(x)+iS_4(x)=\int_x^\infty t^{-1/2}\exp it\,dt=\sqrt{\pi}\exp(\pi i/4)-C_3(x)-iS_3(x)$$

in Chebyshev's polynomials $T_n(10/x-1)$ in the interval $5<x<\infty$ are given in the Appendix.

6.2 Airy functions

The Airy function Ai (ξ) is specified by the integral

$$\mathrm{Ai}(\xi)=\frac{1}{2\pi}\int_{-\infty}^{\infty}\exp i(t^3/3+\xi t)\,dt=\frac{1}{\pi}\int_0^\infty \cos(t^3/3+\xi t)\,dt$$

The function $v(\xi)$ introduced by Fock is also used:

$$v(\xi)=\sqrt{\pi}\mathrm{Ai}(\xi)=\frac{1}{2\sqrt{\pi}}\int_{-\infty}^{\infty}\exp i(t^3/3+\xi t)\,dt$$

The Airy function is expanded into the power series in ξ

$$\mathrm{Ai}(\xi)=c_1 f(\xi)-c_2 g(\xi)$$

where

$$\left.\begin{aligned} f(\xi)&=1+\frac{1}{3!}\xi^3+\frac{1\cdot 4}{6!}\xi^6+\frac{1\cdot 4\cdot 7}{9!}\xi^9+\cdots \\ g(\xi)&=\xi+\frac{2}{4!}\xi^4+\frac{2\cdot 5}{7!}\xi^7+\frac{2\cdot 5\cdot 8}{10!}\xi^{10}+\cdots \\ c_1&=\mathrm{Ai}(0)=0.35502; \qquad c_2=\mathrm{Ai}'(0)=0.25881 \end{aligned}\right\} \tag{6.8}$$

As $\xi\to\infty$ the Airy function has the asymptotics

$$\begin{aligned} \mathrm{Ai}(\xi)&\simeq\frac{1}{2\pi}\xi^{-1/4}\exp\left(-\tfrac{2}{3}\xi^{3/2}\right)\sum_{n=0}^{\infty}\frac{\Gamma(3n+1/2)(-1)^n}{(2n)!\,(9\xi^{3/2})^n} \\ \mathrm{Ai}(-\xi)&\simeq\mathrm{Im}\left(\frac{1}{\pi}\xi^{-1/4}\exp\left(\tfrac{2}{3}i\xi^{3/2}+\frac{\pi i}{4}\right)\cdot\sum_{n=0}^{\infty}\frac{\Gamma(3n+1/2)(-i)^n}{(2n)!\,(9\xi^{3/2})^n}\right) \end{aligned} \tag{6.9}$$

The Airy function satisfies the Airy equation

$$y''(\xi) - \xi y(\xi) = 0 \tag{6.10}$$

and can be defined as a solution of this equation bounded on the whole real axis and equal to c_1 at $\xi = 1$. The second solution Bi (ξ) of the equation

$$\mathrm{Bi}\,(\xi) = \sqrt{3}[c_1 f(\xi) + c_2 g(\xi)]$$

(where the functions $f(\xi)$, $g(\xi)$ and constants c_1, c_2 are defined in eqn. 6.8) is also used. As $\xi \to \infty$ this function has the asymptotics

$$\mathrm{Bi}\,(\xi) \simeq \frac{1}{\pi}\,\xi^{-1/4} \exp\left(\tfrac{2}{3}\xi^{3/2}\right) \sum_{n=0}^{\infty} \frac{\Gamma(3n+1/2)(-1)^n}{(2n)!\,(9\xi^{3/2})^n}$$

$$\mathrm{Bi}\,(-\xi) \simeq \mathrm{Re}\left[\frac{1}{\pi}\,\xi^{-1/4} \exp\left(\tfrac{2}{3} i \xi^{3/2} + \frac{\pi i}{4}\right) \cdot \sum_{n=0}^{\infty} \frac{\Gamma(3n+1/2)(-i)^n}{(2n)!\,(9\xi^{3/2})^n}\right]$$

Since the Airy eqn. 6.10 is reduced to the Bessel equation by the variable change $\xi = \pm(3z/2)^{2/3}$ the functions Ai (ξ), Bi (ξ) and their derivatives are expressed in terms of the Bessel functions $J_{\pm 1/3}(z)$, $J_{\pm 2/3}(z)$ for $\xi < 0$, and in terms of the modified Bessel functions $I_{\pm 1/3}(z)$, $I_{\pm 2/3}(z)$ for $\xi > 0$

$$\mathrm{Ai}\,(-\xi) = \tfrac{1}{3}\sqrt{\xi}[J_{1/3}(\tfrac{2}{3}\xi^{3/2}) + J_{-1/3}(\tfrac{2}{3}\xi^{3/2})]$$
$$\mathrm{Ai}\,(\xi) = \tfrac{1}{3}\sqrt{\xi}[I_{-1/3}(\tfrac{2}{3}\xi^{3/2}) - I_{1/3}(\tfrac{2}{3}\xi^{3/2})]$$
$$\mathrm{Bi}\,(-\xi) = \sqrt{\xi/3}\,[J_{-1/3}(\tfrac{2}{3}\xi^{3/2}) - J_{1/3}(\tfrac{2}{3}\xi^{3/2})]$$
$$\mathrm{Bi}\,(\xi) = \sqrt{\xi/3}[I_{1/3}(\tfrac{2}{3}\xi^{3/2}) + I_{-1/3}(\tfrac{2}{3}\xi^{3/2})]$$
$$\mathrm{Ai}'\,(-\xi) = -\tfrac{1}{3}\xi[J_{-2/3}(\tfrac{1}{3}\xi^{3/2}) - J_{2/3}(\tfrac{2}{3}\xi^{3/2})]$$
$$\mathrm{Ai}'\,(\xi) = -\tfrac{1}{3}\xi[I_{-2/3}(\tfrac{2}{3}\xi^{3/2}) - I_{2/3}(\tfrac{2}{3}\xi^{3/2})]$$
$$\mathrm{Bi}'\,(-\xi) = (\xi/\sqrt{3})[J_{-2/3}(\tfrac{2}{3}\xi^{3/2}) + J_{2/3}(\tfrac{2}{3}\xi^{3/2})]$$
$$\mathrm{Bi}'\,(\xi) = (\xi/\sqrt{3})[I_{2/3}(\tfrac{2}{3}\xi^{3/2}) + I_{-2/3}(\tfrac{2}{3}\xi^{3/2})]$$

The function Ai (ξ) and the leading terms of its asymptotics for $\xi > 0$ and $\xi < 0$ are given in Figure 6.2 taken from Reference 39. The coefficients of the expansions of $J_{\pm 1/3}(x)$, $I_{\pm 1/3}(x)$ and $J_{\pm 2/3}(x)$, $I_{\pm 2/3}(x)$ into the series in Chebyshev's polynomials are given in the Appendix.

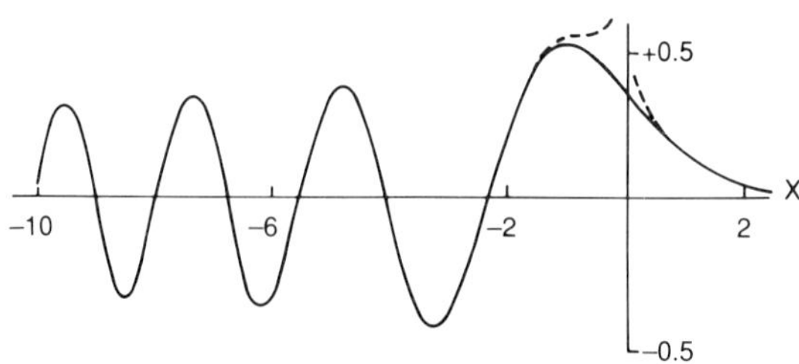

Figure 6.2 Airy function and its asymptotics

6.3 Parabolic cylinder functions

The parabolic cylinder function $D_p(z)$ is defined by the integral

$$D_p(z) = \pi^{-1/2} 2^{p+1/2} \exp(-\pi i p/2 + z^2/4) \times \int_{-\infty+iO}^{\infty+iO} x^p \exp(-2x^2 + 2ixz)\, dx$$

where the integration contour bypasses the branch point $x=0$ in the upper halfplane. This results in

$$\begin{aligned} D_p[(1+i)z] &= \pi^{-1/2} 2^{p/2} \exp(-3\pi i p/4 - \pi i/4) \exp(iz^2/2) \times \int_{-\infty+iO}^{\infty+iO} x^p \exp i(x^2 + 2xz)\, dx \\ &= 2^{1+p/2} \pi^{-1/2} \exp[-\pi i(p+1)/4) \exp(iz^2/2) \times \int_0^{\infty} x^p e^{ix^2} \cos(2xz - \pi p/2)\, dx \end{aligned} \tag{6.11a}$$

$$\begin{aligned} D_p[(1-i)z] &= \pi^{-1/2} 2^{p/2} \exp(-\pi i p/4 + \pi i/4) \exp(-iz^2/2) \times \int_{-\infty+iO}^{\infty+iO} x^p \exp(-ix^2 + 2ixz)\, dx \\ &= 2^{1+p/2} \pi^{-1/2} \exp[\pi i(p+1)/4] \exp(-iz^2/2) \times \int_0^{\infty} x^p e^{-ix^2} \cos(2xz - \pi p/2)\, dx \end{aligned} \tag{6.11b}$$

To evaluate $D_p[(1 \pm i)z]$ for sufficiently small $|z|$ use the expansion

$$D_p[(1 \pm i)z] = 2^{p/2} \exp(\mp iz^2/2) \left\{ \frac{\sqrt{\pi}}{\Gamma[(1-p)/2]}\, {}_1F_1(-p/2, 1/2, \pm iz^2) - \frac{\sqrt{2\pi}(1 \pm i)z}{\Gamma(-p/2)}\, {}_1F_1[(1-p)/2, 3/2, \pm iz^2] \right\} \tag{6.12}$$

where ${}_1F_1$ is the degenerate hypergeometrical function

$${}_1F_1(\alpha, \beta, z) = 1 + \frac{\alpha}{\beta} \cdot z + \frac{\alpha(\alpha+1)}{\beta(\beta+1)} \cdot \frac{z^2}{2!} + \frac{\alpha(\alpha+1)(\alpha+2)}{\beta(\beta+1)(\beta+2)} \cdot \frac{z^3}{3!} + \cdots$$

As $z \to \pm\infty$ the function $D_p[(1 \pm i)z]$ has the asymptotics

$$D_p[(1 \pm i)z] \simeq \exp i(\mp z^2/2 \pm \pi p/4) \cdot 2^{p/2} z^p A_p(\mp 4iz^2)$$

as $z \to \infty$ and

$$\begin{aligned} D_p[(1 \pm i)z] \simeq{} & \exp i(\mp z^2/2 \mp 3\pi p/4) \cdot 2^{p/2} |z|^P A_p(\mp 4iz^2) \\ & + \frac{\sqrt{\pi}}{\Gamma(-p)} \exp i(\pm z^2/2 \mp \pi(p+1)/4) \\ & \times 2^{-P/2} |z|^{-p-1} B_p(\mp 4iz^2) \end{aligned} \tag{6.13a}$$

as $z \to -\infty$.

Here, A_p and B_p are asymptotic series in reciprocal powers of $t = \mp 4iz^2$

$$A_p(t) \simeq 1 + \frac{p(p-1)}{t} + \frac{p(p-1)(p-2)(p-3)}{2!\,t^2} + \frac{p(p-1)\cdots(p-5)}{3!\,t^3} + \cdots$$

$$B_p(t) \simeq 1 - \frac{(p+1)(p+2)}{t} + \frac{(p+1)(p+2)(p+3)(p+4)}{2!\,t^2} - \frac{(p+1)\cdots(p+6)}{3!\,t^3} + \cdots \qquad (6.13b)$$

The function $u(z) = D_p[(1 \pm i)z]$ satisfies the equation

$$u''_{zz} + [\pm i(2p+1) + z^2]u = 0$$

The derivative of the function $D_p[(1 \pm i)z]$ with respect to z is expressed in terms of D_p, D_{p+1}, D_{p-1} as

$$\frac{d}{dz} D_p[(1 \pm i)z] = \pm izD_p[(1 \pm i)z] - (1 \pm i)D_{p+1}[(1 \pm i)z]$$

$$\frac{d}{dz} D_p[(1 \pm i)z] = \mp izD_p[(1 \pm i)z] + p(1 \pm i)D_{p-1}[(1 \pm i)z]$$

For positive integers $p = n$ the function D_p is expressed in terms of the Hermitte polynomials

$$D_n[(1 \pm i)z] = 2^{-n/2} \exp(\mp iz^2/2) H_n[\exp(\pm \pi i/4)z]$$

where

$$H_n(x) = (-1)^n \exp(x^2) \frac{d^n}{dx^n}[\exp(-x^2)]$$

For negative integers p the functions $D_p[(1-i)z]$ and $D_p[(1+i)z]$ are expressed in terms of the Fresnel integral and its complex conjugate, respectively,

$$\begin{aligned} D_{-1}[(1-i)z] &= \exp(-iz^2/2)\sqrt{2\pi}F(-z) \\ D_{-1}[(1+i)z] &= \exp(iz^2/2)\sqrt{2\pi}F^*(-z) \end{aligned} \qquad (6.14a)$$

$$\begin{aligned} D_{-2}[(1-i)z] &= \exp(-iz^2/2)\sqrt{2\pi}(1-i)zF(-z) \\ &\quad -\sqrt{(2/\pi)}\exp(iz^2/2) \\ D_{-2}[(1+i)z] &= \exp(iz^2/2)\sqrt{2\pi}(1+i)zF^*(-z) \\ &\quad -\sqrt{(2/\pi)}\exp(-iz^2/2) \end{aligned} \qquad (6.14b)$$

and the recurrence equation

$$D_{p+1}[(1 \pm i)z] - (1 \pm i)zD_p[(1 \pm i)z] + pD_{p-1}[(1 \pm i)z] = 0$$

is used for $p < -2$. The coefficients of the expansions of $D_{\pm 1/2}[(1 \pm i)z]$ and $D_{\pm 3/2}[(1 \pm i)z]$ in Chebyshev's polynomials are given in the Appendix.

6.4 Incomplete gamma function

The incomplete γ-function $\gamma^*(\alpha, z)$ is defined by

$$\gamma^*(\alpha, z)=\frac{z^{-\alpha}}{\Gamma(\alpha)}\int_0^z \exp(-t)t^{\alpha-1}\,dt$$

where $\Gamma(\alpha)$ is the standard Γ-function. The following forms of the incomplete γ-function are also used

$$\gamma(\alpha, z)=\int_0^z \exp(-t)t^{\alpha-1}\,dt$$

$$\Gamma(\alpha, z)=\int_z^\infty \exp(-t)t^{\alpha-1}\,dt=\Gamma(\alpha)-\gamma(\alpha, z)$$

The function $\gamma^*(\alpha, z)$ is an integer analytical function of α and z and it is expanded into the power series in z

$$\gamma^*(\alpha, z)=\frac{1}{\Gamma(\alpha)}\sum_{n=0}^{\infty}\frac{(-z)^n}{n!\,(\alpha+n)} \tag{6.15a}$$

The expansion

$$\gamma^*(\alpha, z)=\exp(-z)\sum_{n=0}^{\infty}\frac{z^n}{\Gamma(\alpha+n+1)} \tag{6.15b}$$

also holds. The function $\Gamma(\alpha, ix)$ is expressed in terms of the generalised Fresnel functions $C(x, \alpha)$ and $S(x, \alpha)$ (see Reference 19, eqns. 6.5.7 and 6.5.8)

$$C(x, \alpha)=\int_x^\infty t^{\alpha-1}\cos t\,dt; \qquad S(x, \alpha)=\int_x^\infty t^{\alpha-1}\sin t\,dt$$

using the relation

$$\Gamma(\alpha, ix)=\exp(\pi i\alpha/2)[C(x, \alpha)-iS(x, \alpha)]$$

The functions $\gamma(\alpha, z)$ and $\Gamma(\alpha, z)$ are many-valued functions of z having a branch point $z=0$; $\Gamma(\alpha, z)$ is an integer function of α.

It is convenient to write the asymptotic expansion of the incomplete γ-function as $|z|\to\infty$ for $\Gamma(\alpha, z)$

$$\Gamma(\alpha, z)\simeq z^{\alpha-1}\exp(-z)\left[1+\frac{(\alpha-1)}{z}+\frac{(\alpha-1)(\alpha-2)}{z^2}+\cdots\right] \tag{6.16}$$

where $|\arg z|<3\pi/2$. The function $\Gamma(\alpha, z)$ is expanded into the continued fraction

$$\Gamma(\alpha, z)=\exp(-z)z^{\alpha}\left[\frac{1}{z+}\;\frac{1-\alpha}{1+}\;\frac{1}{z+}\;\frac{2-\alpha}{1+}\;\frac{2}{z+}\;\frac{3-\alpha}{1+}\cdots\right] \tag{6.17}$$

The functions γ, γ^*, Γ satisfy the recurrence relations

$$\gamma(\alpha+1, z)=\alpha\gamma(\alpha, z)-z^{\alpha}\exp(-z) \tag{6.18a}$$

$$\gamma^*(\alpha-1, z)=z\gamma^*(\alpha, z)+\exp(-z)/\Gamma(\alpha) \tag{6.18b}$$

$$\Gamma(\alpha+1, z)=\alpha\Gamma(\alpha, z)+z^{\alpha}\exp(-z) \tag{6.18c}$$

For integers $\alpha=-n$ $(n=0, 1, \ldots)$

$$\gamma^*(-n, z)=z^n$$

$$\Gamma(-n, z)=\frac{(-1)^n}{n!}\left[\Gamma(0, z)-\exp(-z)\sum_{m=0}^{n-1}(-1)^m\frac{m!}{z^{m+1}}\right]$$

The function $\Gamma(0, z)$ for $z>0$ is expressed in terms of the integral exponential function

$$\Gamma(0, z)=\int_z^{\infty}\frac{\exp(-t)\,dt}{t}=E_1(z)$$

For imaginary z the function $\Gamma(0, z)$ is expressed in terms of the integral sine and cosine (Section 6.5), while the function $\Gamma(1/2, z)$ in terms of the Fresnel integral for $\xi>0$

$$\Gamma(0, \mp i\xi)=-\mathrm{Ci}(z)\pm i[\pi/2-\mathrm{Si}(z)]$$

$$\Gamma(1/2, -i\xi)=2\sqrt{\pi}F(-\sqrt{\xi})$$

$$\Gamma(1/2, i\xi)=2\sqrt{\pi}F^*(-\sqrt{\xi})$$

The problems of evaluating the incomplete γ-function are considered in Reference 40, Chapter 4.

6.5 Integral sine and cosine

The integral sine $\mathrm{Si}(\xi)$ and cosine $\mathrm{Ci}(\xi)$ are defined by the equations

$$\mathrm{Si}(\xi)=\int_0^{\xi}\frac{\sin t\,dt}{t}=\frac{\pi}{2}-\int_{\xi}^{\infty}\frac{\sin t\,dt}{t}$$

$$\mathrm{Ci}(\xi)=C+\ln\xi+\int_0^{\xi}\frac{\cos t-1}{t}\,dt=-\int_{\xi}^{\infty}\frac{\cos t\,dt}{t}$$

where $C=0.5772157$ is the Euler constant. The functions $\mathrm{Si}(z)$ and $\mathrm{Ci}(z)$ are expanded into converging series

$$\mathrm{Si}(\xi)=\sum_{n=0}^{\infty}\frac{(-1)^n\xi^{2n+1}}{(2n+1)(2n+1)!} \tag{6.19a}$$

$$\mathrm{Ci}(\zeta)=C+\ln\xi+\sum_{n=0}^{\infty}\frac{(-1)^n\xi^{2n}}{2n(2n)!} \tag{6.19b}$$

For $\xi > 0$ the functions Si (z) and Ci (z) are expressed in terms of integral exponential $E_1(z)$

$$\text{Ci}\,(\xi) + i\,\text{Si}\,(\xi) = \frac{\pi i}{2} - E_1(-i\xi)$$

and in terms of auxiliary functions $f(\xi)$ and $g(\xi)$

$$\begin{aligned} \text{Si}\,(\xi) &= \pi/2 - f(\xi)\cos\xi - g(\xi)\sin\xi \\ \text{Ci}\,(\xi) &= f(\xi)\sin\xi - g(\xi)\cos\xi \end{aligned} \tag{6.20a}$$

where

$$f(\xi) = \int_0^\infty \frac{\exp(-t\xi)\,dt}{t^2+1}; \qquad g(\xi) = \int_0^\infty \frac{t\exp(-t\xi)\,dt}{t^2+1}$$

As $\xi \to \infty$ the functions $f(\xi)$ and $g(\xi)$ are expanded into the asymptotic series

$$\begin{aligned} f(\xi) &\simeq \frac{1}{\xi}\left[1 - \frac{2!}{\xi^2} + \frac{4!}{\xi^4} - \frac{6!}{\xi^6} + \cdots\right] \\ g(\xi) &\simeq \frac{1}{\xi^2}\left[1 - \frac{3!}{\xi^2} + \frac{5!}{\xi^4} - \frac{7!}{\xi^6} + \cdots\right] \end{aligned} \tag{6.20b}$$

The coefficients of the expansion in Chebyshev's polynomials for the integrals in terms of which Si (ξ) and Ci (ξ) are expressed are given in the Appendix.

6.6 Hankel functions

The Hankel function $H_\mu^{(1)}(z)$ is a cylindrical function of the index μ defined for $z > 0$ by the integral

$$H_\mu^{(1)}(z) = -\frac{1}{\pi}\int_{-i\infty}^{-\pi+i\infty} \exp(-iz\sin\theta + i\theta\mu)\,d\theta$$

where the integration contour is shown in Figure 5.1 (contour C_1). The function $H_\mu^{(1)}(z)$ satisfies the Bessel equation

$$\frac{\partial^2}{\partial z^2} H_\mu^{(1)}(z) + \frac{1}{z}\frac{\partial}{\partial z} H_\mu^{(1)}(z) + \left(1 - \frac{\mu^2}{z^2}\right) H_\mu^{(1)}(z) = 0$$

and is expressed in terms of the Bessel functions $J_\mu(z)$ and $N_\mu(z)$ by the relation

$$H_\mu^{(1)}(z) = J_\mu(z) + iN_\mu(z) \tag{6.21}$$

The function $J_\mu(z)$ is expanded into the converging series

$$J_\mu(z) = \frac{z^\mu}{2^\mu}\sum_{k=0}^{\infty} (-1)^k \frac{z^{2k}}{2^{2k}k!\,\Gamma(\mu+k+1)}; \qquad (|\arg z| < \pi) \tag{6.22a}$$

If μ is not an integer

$$N_\mu(z) = \frac{1}{\sin\pi\mu}\left[\cos\pi\mu J_\mu(z) - J_{-\mu}(z)\right] \tag{6.22b}$$

If μ is an integer, $\mu = n$

$$\pi N_n(z) = 2J_n(z)\left(\ln\frac{z}{2} + C\right) - \sum_{k=0}^{n-1}\frac{(n-k-1)!}{k!}\left(\frac{z}{2}\right)^{2k-n} - \frac{1}{n!}\left(\frac{z}{2}\right)^n \sum_{k=1}^{n}\frac{1}{k}$$

$$-\sum_{k=1}^{\infty}\frac{(-1)^k(z/2)^{n+2k}}{k!\,(k+n)!}\left[\sum_{m=1}^{n+k}\frac{1}{m} + \sum_{m=1}^{k}\frac{1}{m}\right] \qquad (6.22c)$$

where C is the Euler constant. As $z\to\infty$ the Hankel function has the asymptotic

$$H_\mu^{(1)}(z) \simeq \left(\frac{2}{\pi z}\right)^{1/2} \exp i\left(z - \frac{\pi\mu}{2} - \frac{\pi}{4}\right)$$

$$\times \sum_{k=0}^{\infty}\frac{(-1)^k\Gamma(\mu+k+1/2)}{(2iz)^k k!\,\Gamma(\mu-k+1/2)} \qquad (6.23)$$

If $\mu = n - 1/2$ $(n \geq 1)$ is a halfinteger, the series has a finite number of summands and the written equation is an exact one

$$H_{n-1/2}^{(1)}(z) = \left(\frac{2}{\pi z}\right)^{1/2} \exp i\left(z - \frac{\pi n}{2}\right)\sum_{k=0}^{n-1}\frac{(-1)^k(n+k-1)!}{(2iz)^k k!\,(n-k-1)!}$$

In particular,

$$H_{1/2}^{(1)}(z) = -i\left(\frac{2}{\pi z}\right)^{1/2} \exp\,(iz)$$

$$H_{3/2}^{(1)}(z) = -\left(\frac{2}{\pi z}\right)^{1/2} \exp\,(iz)(1 + i/z)$$

For $H_0^{(1)}(z)$, $H_{\pm 1/4}^{(1)}(z)$, $H_{\pm 1/3}^{(1)}(z)$, $H_{\pm 2/3}^{(1)}(z)$, $H_{\pm 3/4}^{(1)}(z)$ and $H_1^{(1)}(z)$ the tables of expansions in Chebyshev's polynomials are given in the Appendix. The relation

$$z[Z_{\mu+1}(z) + Z_{\mu-1}(z)] = 2\mu Z_\mu(z)$$

valid for any cylindrical function of the index μ makes it possible to find (using these tables) the values of $H_\mu^{(1)}(z)$ for any μ of the form $\mu = \pm p$, $\mu = \pm p/3$, $\mu = \pm p \pm 1/4$, $\mu = \pm p \pm 3/4$ where p is a positive integer. The derivative of any cylindrical function Z_μ is expressed in terms of Z_μ and $Z_{\mu\pm 1}$ by the relation

$$z\frac{d}{dz}Z_\mu(z) \pm \mu Z_\mu(z) = \pm z Z_{\mu\mp 1}(z)$$

6.7 Piercey integral

The Piercey integral is defined by

$$I(p, q) = \int_{-\infty}^{\infty} \exp i(pt + qt^2 + t^4)\,dt$$

It is an integer function of p, q and is expanded into the power series in p and q:

$$I(p, q) = \tfrac{1}{2} \sum_{m,n=0}^{\infty} \frac{p^{2m} q^n}{(2m)!\, n!} \Gamma\left(\frac{m+n}{2} + \frac{1}{4}\right) \exp \frac{\pi i}{2}\left(\frac{5}{2} m + \frac{3}{2} n + \frac{1}{4}\right)$$

The Piercey integral is an even function of p. For large p and q the asymptotic of $I(p, q)$ is found by the stationary phase methods. The appropriate equations are extremely cumbersome since the determination of stationary points of the phase function $\varphi(t) = pt + qt^2 + t^4$ is reduced to solving the cubic equation

$$\Phi_t = p + 2qt + 4t^3 = 0 \tag{6.24}$$

For $p^2 + 8q^3/27 > 0$ the equation has one real root, for $p^2 + 8q^3/27 < 0$ there are three roots. For $|p|, |q| \gg 1$ the asymptotic of $I(p, q)$ includes three summands when $p^2 + 8q^3/27 < 0$ and one summand when $p^2 + 8q^3/27 > 0$. For $p^2 + 8q^3/27 = 0$ the phase function $\Phi(t)$ has a simple stationary point (a local minimum) at $t = -p^{1/3}$ and a degenerate stationary point (with the second derivative vanishing) at $t = p^{1/3}/2$. In a vicinity of the semicube parabola $q = -3p^{2/3}/2$ for $|p| \gg 1$ and ξ bounded in magnitude the asymptotic

$$\begin{aligned} &I(p, -(3/2)p^{2/3} - \xi) \\ &\simeq \frac{\sqrt{2\pi}}{3p^{1/3}} \exp i[-(3/2)p^{4/3} - p^{2/3}\xi - (2/9)\xi^2 + \pi/4] \\ &\cdot \left[1 - \frac{5}{27}\xi p^{-2/3} + O(\xi^2 p^{-4/3})\right] \\ &+ \exp i[(3/16)p^{4/3} - (\xi/4)p^{2/3} - 5\xi^2/36] \\ &\cdot [P \cdot \mathrm{Ai}(\eta) + iQ \cdot \mathrm{Ai}'(\eta)] \end{aligned} \tag{6.25}$$

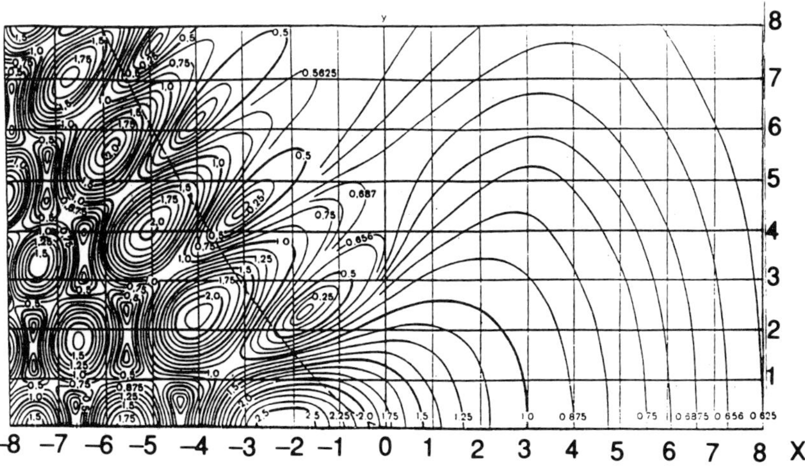

Figure 6.3 Level lines of Piercey integral modulus

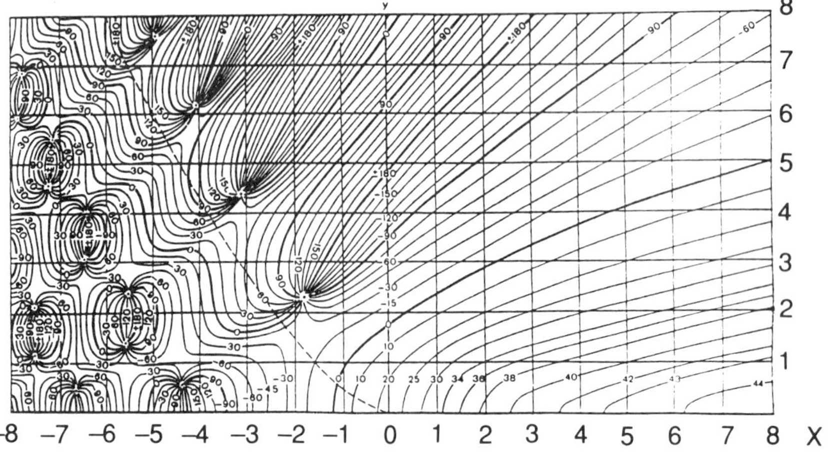

Figure 6.4 Level lines of Piercey integral phase

holds where

$$P = 2^{5/3}3^{-1/3}p^{-1/9}\pi[1 - (5/108)\xi p^{-2/3} + O(\xi^2 p^{-4/3})]$$
$$Q = -2^{4/3}3^{-5/3}p^{-5/9}\pi[1 - (35/108)\xi p^{-2/3} + O(\xi^2 p^{-4/3})]$$
$$\eta = -6^{-1/3}\xi p^{2/9}[1 + (2/27)\xi p^{-2/3} + O(\xi^2 p^{-4/3})]$$

The level lines of the modulus and phase of the function $I(p, q)$ are given in Figures 6.3 and 6.4, taken from Reference 31.

6.8 Incomplete Airy function

The incomplete Airy function is defined by

$$V(p, q) = \frac{1}{2\pi}\int_q^\infty \exp i(t^3/3 + tp)\, dt$$

and satisfies the equation

$$V''_{pp} - pV = \frac{-i}{2\pi}\exp i(pq + q^3/3)$$

and the relation

$$V(p, q) + V^*(p, -q) = \mathrm{Ai}(p)$$

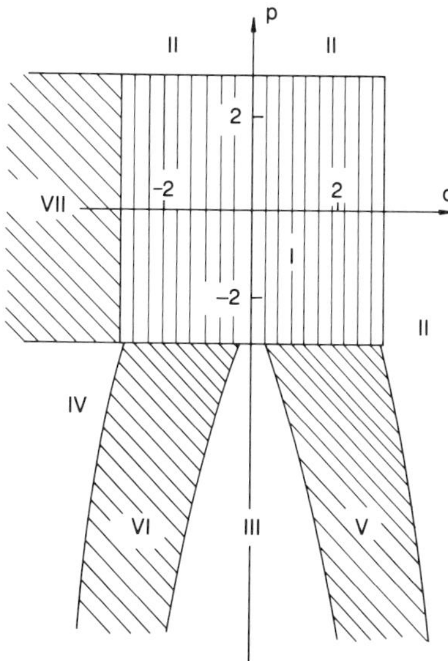

Figure 6.5 Evaluation of incomplete Airy function

This is an integer function of p, q and it is expanded into the series in powers of p, q

$$V(p, q) = \frac{1}{2\pi} \sum_{n=0}^{\infty} \frac{p^n i^n}{n!} \left[\exp\,[\pi i(n+1)/6]\,\Gamma\left(\frac{n+1}{3}\right) 3^{(n-2)/3} - \sum_{m=0}^{\infty} \frac{i^m q^{3m+n+1}}{m!\, 3^m (3m+n+1)} \right]$$

For large p, q the asymptotic of $I(p, q)$ is determined by the mutual location of the phase-function stationary points $t = \pm\sqrt{-p}$ (for $p \ll -1$) and the integration domain boundary q. Region I (Figure 6.5) is a rectangle including the origin. If rectangle sizes are sufficiently large (see later), outside of it one can use the following asymptotic equation. For $p \gg 0$, for small $p > 0$ and $q \gg 0$, as well as for $p < 0$ and $q \gg \sqrt{-p}$ (region II in Figure 6.5) the phase function has no stationary points and the asymptotic $V(p, q)$ coincides with the contribution of the integration domain boundary $t = q$

$$V(p, q) \simeq \mathcal{J}_0 = \frac{i \exp i(q^3/3 + pq)}{2\pi(q^2 + p)} \tag{6.26a}$$

(here and subsequently only leading terms of the relevant asymptotic expansions are written). If $p \ll 0$ and $-\sqrt{-p} \ll q \ll \sqrt{-p}$ (region III in Figure 6.5), there is a stationary point $t = \sqrt{-p}$ in the integration interval and the asymptotic $V(p, q)$ is of the form

$$V(p, q) \simeq \mathcal{J}_0 + \frac{1}{2\sqrt{\pi}} |p|^{-1/4} \exp i[-(2/3)|p|^{3/2} + \pi/4] \tag{6.26b}$$

where the first summand is a contribution of the integration domain boundary. For $p \ll 0$ and $q \ll -\sqrt{-p}$ (region IV) there are two stationary points in the integration domain, $t = \pm\sqrt{-p}$ and

$$V(p, q) \simeq \mathcal{J}_0 + \frac{1}{2\sqrt{\pi}} |p|^{-1/4} \{\exp i[-(2/3)|p|^{3/2} + \pi/4] + \exp i[(2/3)|p|^{3/2} - \pi/4]\} \tag{6.26c}$$

For q close to the stationary point $t = \sqrt{-p}$ (region V) the asymptotic V is expressed in terms of the Fresnel integral

$$V(p, q) \simeq \frac{F(\tau_1)}{2\sqrt{\pi}|p|^{1/4}} \exp i[-(2/3)|p|^{3/2} + \pi/4] + \frac{i \exp i(q^3/3 + pq)}{2\pi} \left(\frac{1}{q^2 + p} + \frac{1}{2|p|^{1/4}\tau_1} \right) \tag{6.26d}$$

where

$$\tau_1 = (\sqrt{|p|} - q)(q/3 + 2\sqrt{|p|}/3)^{1/2} \tag{6.27a}$$

Here the second summand is found by asymptotic matching, i.e. it is chosen from the condition that the equation is transformed into eqn. 6.26*a* for $q > \sqrt{|p|}$ and into eqn. 6.26*b* for $q < \sqrt{|p|}$ when replacing the Fresnel integral in eqn. 6.26*d* by its asymptotic. It is easy to verify that this summand remains regular as $q \to \sqrt{|p|}$.

Similarly, in region VI for q close to $-\sqrt{|p|}$ the asymptotic V is of the form

$$V(p, q) \simeq \frac{F^*(\tau_2) \exp i[(2/3)|p|^{3/2} - \pi/4]}{2\sqrt{\pi}|p|^{1/4}} + \frac{\exp i[-(2/3)|p|^{3/2} + \pi/4]}{2\sqrt{\pi}|p|^{1/4}} + \frac{i \exp i(q^3/3 + pq)}{2\pi} \left(\frac{1}{q^2 + p} - \frac{1}{2\tau_2|p|^{1/4}} \right) \tag{6.26e}$$

where

$$\tau_2 = (\sqrt{|p|} + q)(2\sqrt{|p|}/3 - q/3)^{1/2} \tag{6.27b}$$

Finally, for $q \ll 0$ and small $|p|$ the asymptotic V in region VII is expressed in terms of the Airy function

$$V(p, q) \simeq \mathcal{J}_0 + Ai(p) \tag{6.26f}$$

The boundaries of regions I, V, VI and VII are determined by the accuracy required. If engineering accuracy (up to 0.01) is sufficient, region I can be specified by the inequalities $-3 < q < 3$, $-3 < p < 3$ and region VII by the inequalities $|p| < 3$, $q < -3$. The boundaries of regions V and VI, i.e. transient zones between regions II and III and III and IV, respectively, are determined by those p, q for which the arguments τ_1, τ_2 of the Fresnel integrals in eqns. 6.26*d* and 6.26*e* are so large that one can replace integrals by their asymptotics. For the accuracy accepted it is sufficient to require these arguments to be equal in modulo $\sqrt{\pi}$. As a result, the condition for the boundary of region V is obtained

$$\tau_1^2 = \frac{q^3}{3} + pq + \tfrac{2}{3}|p|^{3/2} = \Phi_Q - \Phi^- = \pi \qquad (6.28a)$$

Here Φ_Q is a value of the phase function at the integration domain boundary $t=q$ and Φ^- is a value of the phase function at the local minimum point $t=\sqrt{|p|}$. The physical meaning of eqn 6.27a is the requirement for the phase difference at the stationary point $t=\sqrt{|p|}$ and on the integration domain boundary to be equal to π.

The boundary of region VI is determined by the similar condition

$$\tau_2^2 = -\frac{q^3}{3} - pq + \tfrac{2}{3}|p|^{3/2} = \Phi^+ - \Phi_Q = \pi \qquad (6.28b)$$

but there is a phase difference between the integration domain boundary and the local maximum point $t=-\sqrt{|p|}$.

6.9 Airy–Fresnel integral

The Airy–Fresnel integral is defined by

$$Af(p, q) = \frac{i}{2\pi} \int_{-\infty + iO}^{\infty + iO} \exp i\left(\frac{t^3}{3} + tp\right) \frac{dt}{t-q} \qquad (6.29)$$

where the pole $t=q$ is passed around in the upper halfplane. The integral is an integer function of variables p, q and is expanded into the power series in p, q:

$$Af(p, q) = \sum_{m,n=0}^{\infty} \frac{3^{(n-m)/3-1}(-1)^n i^{-m} p^n q^m}{\Gamma\left(1 - \frac{n-m}{3}\right) n!} \qquad (6.30)$$

It satisfies the differential equations

$$\frac{\partial Af(p, q)}{\partial p} = iqAf(p, q) - Ai(p) \qquad (6.31a)$$

$$\frac{\partial Af(p, q)}{\partial q} = i(q^2 + p)Af(p, q) + iAi'(p) - qAi(p) \qquad (6.31b)$$

and the relation

$$Af(p, q) = Af^*(p, -q) \qquad (6.32)$$

where $Af^*(p, q)$ is the complex conjugate function to $Af(p, q)$.

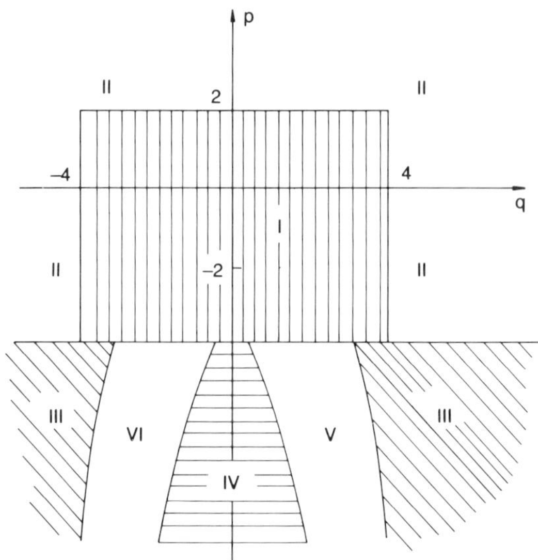

Figure 6.6 Evaluation of Airy–Fresnel integral

The asymptotic of $Af(p, q)$ for large p, q is determined by the mutual location of stationary points $t=\pm\sqrt{-p}$ of the phase function (for $p\ll -1$) and the pole $t=q$. In Figure 6.6 region I is a rectangle including the origin. If the size of the rectangle is sufficiently large (see later), outside of it one can use the following equations:

As $p\to\infty$ eqn. 6.29 falls exponentially (since its phase function grows monotonically) and has the asymptotic

$$Af(p, q)\simeq\frac{iqAi(p)}{p+q^2}+\frac{Ai'(p)}{p+q^2} \tag{6.33a}$$

(only leading terms of the asymptotics are written). The asymptotic is valid for small $|p|$ and $q^2+p\gg 0$, i.e. it is applicable in whole region II shown in Figure 6.6.

For $p\ll 0$ the phase function has the stationary points $t=\pm\sqrt{-p}$. For $|q|>\sqrt{-p}$ (region III) the asymptotic Af consists of the sum of contributions of these stationary points. For $|q|<\sqrt{-p}$ (region IV), i.e. when the pole $t=q$ is between stationary points and the phase function falls off at these t, the pole contribution is added

$$\begin{aligned} Af(p, q)\simeq{}&\chi(-p-q^2)\exp i(q^3/3+pq)\\ &+\frac{i}{2\sqrt{\pi}|p|^{1/4}}\left[\frac{\exp i(-\frac{2}{3}|p|^{3/2}+\pi/4)}{\sqrt{-p}-q}\right.\\ &\left.+\frac{\exp i(\frac{2}{3}|p|^{3/2}-\pi/4)}{-\sqrt{-p}-q}\right] \end{aligned} \tag{6.33b}$$

This asymptotic is invalid for $|p|\gg 1$ and small q^2+p when the pole $t=q$ is close to the phase-function stationary point. Then the asymptotic of Af is expressed in terms of the Fresnel integral. For q close to $\sqrt{|p|}$ (region V) one has

$$Af(p,q) \simeq \exp i\left(\frac{q^3}{3}+pq\right)F^*(\tau_1)+\frac{i}{2\sqrt{\pi}}\left[\frac{\exp i\,(\frac{2}{3}|p|^{3/2}-\pi/4)}{|p|^{1/4}(-\sqrt{|p|}-q)}\right.$$
$$\left.+\left(-\frac{1}{\tau_1}+\frac{|p|^{-1/4}}{\sqrt{|p|}-q}\right)\exp i\,(-\tfrac{2}{3}|p|^{3/2}+\pi/4)\right] \qquad (6.33c)$$

where τ_1 is determined by

$$\tau_1=(\sqrt{|p|}-q)\sqrt{[2\sqrt{|p|}/3+q/3]}$$

In eqn. 6.33c the second summand is calculated by asymptotic matching, i.e. it is chosen from the condition that this formula should be transformed into eqn. 6.33b when replacing the Fresnel integral by its asymptotic.

Similarly, for q close to $-\sqrt{|p|}$ (region VI)

$$Af(p,q) \simeq \exp i(q^3/3+pq)F(\tau_2)+\frac{i}{2\sqrt{\pi}}\left[\frac{\exp i\,(-\frac{2}{3}|p|^{3/2}+\pi/4)}{|p|^{1/4}(\sqrt{|p|}-q)}\right.$$
$$\left.+\left(\frac{1}{\tau_2}-\frac{|p|^{-1/4}}{\sqrt{|p|}+q}\right)\exp i\,(\tfrac{2}{3}|p|^{3/2}-\pi/4)\right] \qquad (6.33d)$$

where

$$\tau_2=(\sqrt{|p|}+q)\,[\tfrac{2}{3}\sqrt{|p|}-q/3]^{1/2}$$

The boundaries of regions I, V, VI are determined by the accuracy required. If one restricts oneself to an error of a few per cent, region I can be given by the inequalities $-4<q<4$, $-4<p<2$. The boundaries of regions V and VI, i.e. the transient zones between regions II, III and III, IV, respectively, are defined by those p, q for which the arguments τ_1, τ_2 of the Fresnel integrals in eqns. 6.33c and 6.33d are so large that the integrals could be replaced by their asymptotics. For the accepted accuracy it is sufficient to require the arguments to be equal in modulo $\sqrt{\pi}$ for the boundary of region V. One obtains the condition

$$\tau_1^2=\frac{q^3}{3}+pq+\tfrac{2}{3}|p|^{3/2}=\Phi_Q-\Phi^-=\pi \qquad (6.34a)$$

Here Φ_Q is a value of the phase function at the pole $t=q$, Φ^- is a phase function value at the local minimum point $t=\sqrt{|p|}$. The physical meaning of eqn. 6.34a is to require the phase difference between the stationary point $t=\sqrt{|p|}$ and the pole $t=q$ to be equal to π.

A similar condition gives the boundary of region VI

$$\tau_2^2=-\frac{q^3}{3}-pq+\tfrac{2}{3}|p|^{3/2}=\Phi^+-\Phi_Q=\pi \qquad (6.34b)$$

but here the phase difference between the pole and the local maximum point $t=-\sqrt{|p|}$ is taken.

The coefficients of the $Af(p, q)$ expansions in the double series in Chebyshev's polynomials for $-3<x<4$ and $0<y<4$ are given in the appendix. Since $Af\,1(x,y)=Af^*(x,-y)$, the data are sufficient to evaluate $Af(x,y)$ for any $-3<x<4$, $|y|<4$.

6.10 Generalised Fresnel integral

The generalised Fresnel integral $G(x, y)$ is defined by

$$G(x,y)=\frac{y}{2\pi}\int_x^{\infty}\frac{\exp i(\xi^2+y^2)}{(\xi^2+y^2)}\,d\xi \tag{6.35}$$

The function G is an odd function of y. The functions $G(x, y)$ and $G(-x, y)$ are related by

$$G(x,y)+G(-x,y)=\frac{y}{2\pi}\int_{\infty}^{\infty}\frac{\exp i(\xi^2+y^2)}{(\xi^2+y^2)}\,d\xi$$

$$=\operatorname{sign} y\cdot F(-|y|) \tag{6.36}$$

where $F(\xi)$ is the Fresnel integral. It follows that

$$G(0,y)=\tfrac{1}{2}\operatorname{sign} y\cdot F(-|y|) \tag{6.37}$$

Writing $G(x, y)$ as

$$G(x,y)=G(0,y)-\frac{y}{2\pi}\int_0^x\frac{d\xi}{(\xi^2+y^2)}-\frac{y}{2\pi}\int_0^x\frac{\exp i(\xi^2+y^2)-1}{(\xi^2+y^2)}\,d\xi$$

one obtains

$$G(x,y)=\tfrac{1}{2}\operatorname{sign} y\cdot F(-|y|)-\frac{1}{2\pi}\operatorname{arctg}\frac{x}{y}+\tilde{G}(x,y) \tag{6.38a}$$

where arctg varies is taken from the interval $(-\pi/2, \pi/2)$ and $\tilde{G}(x,y)$ is the integer analytical function of x, y.

$$\tilde{G}(x,y)=-\frac{y}{2\pi}\int_0^x\frac{\exp[i(\xi^2+y^2)]-1}{(\xi^2+y^2)}\,d\xi$$

$$=\sum_{m,n=0}^{\infty}x^{2m+1}y^{2n+1}$$

$$\times\frac{i^{m+n-1}}{2\pi\Gamma(m+1)\Gamma(n+1)(2m+1)(n+m+1)} \tag{6.38b}$$

Eqns. 6.38 are convenient to evaluate $G(x, y)$ for small x, y. For $x, y>0$ the equality

$$G(x,y)+G(y,x)=F(-x)F(-y) \tag{6.39}$$

holds. To prove the equality note that both the right-hand side and the left-hand one of the equality are regular for $x, y>0$ and tend to zero as $y\to\infty$.

Therefore it is sufficient to verify the equality obtained from eqn. 6.39 when differentiating it with respect to y:

$$\frac{\partial G(x,y)}{dy}=\frac{-x\exp i(x^2+y^2)}{2\pi(x^2+y^2)}-F(-x)\cdot\frac{\exp iy^2}{\sqrt{\pi i}} \tag{6.40a}$$

It is easy to verify that

$$\frac{\partial G(x,y)}{\partial y}+F(-x)\cdot\frac{\exp iy^2}{\sqrt{\pi i}}=-\frac{1}{2\pi}\int_x^\infty d\left[\frac{\xi\exp i(\xi^2+y^2)}{\xi^2+y^2}\right]$$

from which eqn. 6.40a and hence eqn. 6.39 follow. The next expression is evident:

$$\frac{\partial G(x,y)}{\partial x}=-\frac{y\exp i(x^2+y^2)}{2\pi(x^2+y^2)} \tag{6.40b}$$

The asymptotic of $G(x,y)$ for large $|x|$, $|y|$ has the form

$$G(x,y)\simeq\operatorname{sign}y\cdot\chi(-x)F(-|y|)+M(x,y)\exp i(x^2+y^2) \tag{6.41a}$$

as $|x|\to\infty$, where $\chi(x)=1$ for $x>0$ and $\chi(x)=0$ for $x<0$, and $M(x,y)$ is the double asymptotic series

$$M(x,y)=\frac{iy}{4\pi^{3/2}}\sum_{s,q=0}^{\infty}\frac{(-i)^{s+q}\Gamma(s+1/2)\Gamma(s+q+1)}{\Gamma(s+1)x^{2s+1}(x^2+y^2)^{q+1}} \tag{6.42}$$

$$G(x,y)\simeq F(-x)T(-y)\exp iy^2-M(y,x)\exp i(x^2+y^2) \tag{6.41b}$$

as $|y|\to\infty$, where the asymptotic series $T(\xi)$ is defined by eqn. 6.6. If $|x|\to\infty$ and $|y|\to\infty$ simultaneously,

$$G(x,y)\simeq\chi(-x)\exp iy^2T(y)+M(x,y)\exp i(x^2+y^2) \tag{6.41c}$$

The coefficients of the $G(x,y)$ expansion in the double series in Chebyshev's polynomials in rectangles $-2<x,y<2$; $2<x<5$, $0<y<2$ and $2<x,y<5$, respectively, are given in the Appendix. Since $G(x,y)$ is an odd function of y, $G(x,y)$ and $G(-x,y)$ are related by eqn. 6.36, while $G(x,y)$ and $G(y,x)$ by eqn. 6.39, the data given are sufficient to evaluate $G(x,y)$ in the square $|x|$, $|y|>5$. If $|x|>5$ or $|y|>5$, then the asymptotic eqns. 6.41a and b are applicable within an accuracy up to 10^{-8}.

6.11 Ff-integral

The function $Ff(x,y)$ is defined by

$$Ff(x,y)=\frac{1}{2\pi}\int_{-\infty}^{x}\frac{\exp it^2}{t-y+iO}\,dt \tag{6.43}$$

where the pole $t=y$ is passed around in the upper halfplane. Evaluating $\partial Ff/\partial y$ and integrating the integral obtained by parts one gets

$$\frac{\partial Ff(x,y)}{\partial y}=\frac{-\exp ix^2}{2\pi(x-y+iO)}+\frac{i^{3/2}F(x)}{\sqrt{\pi}}+2iyFf(x,y) \tag{6.44}$$

Taking into account that the function Ff tends to $-i\exp(iy^2)$ as $y\to-\infty$ and treating eqn. 6.44 as a differential equation with respect to $Ff(x,y)$ one obtains

$$Ff(x,y)=\exp i(x^2+y^2)Ff^*(y,x)+i\exp iy^2[F(x)F^*(y)-1] \tag{6.45}$$

In this expression Ff^* is a complex conjugate function to the Ff-integral, F and F^* are the Fresnel integral and the complex conjugate function to it, respectively. The following relation also holds

$$Ff(x,y)-Ff(-x,-y)=i\exp(iy^2)[F^*(y)-\chi(x-y)] \tag{6.46}$$

As $x\to y$ the function Ff has a logarithmic singularity. Set $r=x-y$. Then changing the variable in eqn. 6.43, $\tau=x-t$, one obtains

$$Ff(x,y)=-\frac{\exp ix^2}{2\pi}\int_0^\infty\frac{\exp i(\tau^2-2\tau x)\,d\tau}{\tau-r-iO}=\mathcal{J}_1+\mathcal{J}_2$$

where

$$\mathcal{J}_1=-\frac{\exp i(2xy-x^2)}{2\pi}\int_0^\infty\frac{\exp i\tau^2\,d\tau}{\tau-r-iO}$$

$$\mathcal{J}_2=-\frac{\exp ix^2}{2\pi}\int_0^\infty\frac{\exp i\tau^2[\exp(-2i\tau x)-\exp(-2irx)]\,d\tau}{\tau-r} \tag{6.47}$$

Shifting the integration in $\mathcal{J}_1$ to the positive imaginary axis one gets

$$Ff(x,y)=\exp iy^2\left[\frac{-1}{4\pi}E_1(ir^2)-\frac{i}{2}F^*(r)-\frac{i}{2}\chi(r)\right]+\mathcal{J}_2(x,y)$$

$$=\exp(iy^2)\left[\frac{1}{4\pi}(\mathrm{Ci}\,(r^2)-i\cdot\mathrm{Si}\,(r^2))+\frac{i}{8}\right.$$

$$\left.-\frac{i}{2}F^*(r)-\frac{i}{2}\chi(r)\right]+\mathcal{J}_2(x,y) \tag{6.48a}$$

Here, $E_1(\xi)$ is the integral exponential, Ci (ξ) and Si (ξ) are the integral sine and cosine (Section 6.5); F^* is the complex conjugate Fresnel integral. The function $\mathcal{J}_2$ is specified by eqn. 6.47 and is an integer function of x, y. It is expanded into the double series in powers of x, $r=x-y$

$$\mathcal{J}_2(x,y)=-\frac{\exp ix^2}{2\pi}\sum_{m,n=0}^{\infty}r^m x^{m+n+1}2^{m+n+1}$$

$$\times\exp\left[-\frac{\pi i}{2}\left(m+\frac{n+1}{2}\right)\right]\frac{\Gamma\left(\frac{n+1}{2}\right)}{\Gamma(m+n+2)} \tag{6.48b}$$

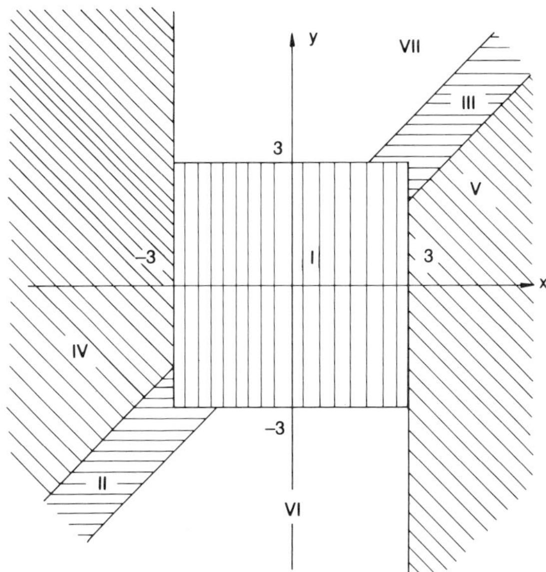

Figure 6.7 Evaluation of Ff-integral

Eqns. 6.48 are convenient to evaluate $Ff(x, y)$ for small x, y (region I in Figure 6.7). As $|x|$, $|y| \to \infty$ the following asymptotic expressions are valid:

for $x \ll -1$ *and bounded* $|r| = |x - y|$ (region II in Figure 6.7):

$$Ff(x, y) \simeq -\frac{\exp iy^2 E_1(-2ixr)}{2\pi} - i\, e^{iy^2}\chi(r) + \mathcal{J}_4(r, x)$$

$$\simeq \exp iy^2 \left(\frac{\mathrm{Ci}\,(2|xr|) + i \cdot \mathrm{Si}\,(2xr)}{2\pi} - \frac{i}{4}\,\mathrm{sign}\,(r) - \frac{i}{2}\right) + \mathcal{J}_4(r, x) \qquad (6.49a)$$

where $\mathcal{J}_4$ is expanded into the asymptotic series in powers of x^{-1}

$$\mathcal{J}_4(r, x) \simeq -\frac{\exp ix^2}{2\pi} \sum_{n=0}^{\infty} (2ix)^{-n-1} c_n$$

and

$$c_n = \frac{d^n}{d\xi^n} \frac{\exp i\xi^2 - \exp ir^2}{\xi - r}\bigg|_{\xi=0}$$

$$= \Gamma(n+1) r^{-n-1} \begin{cases} \exp(ir^2) - 1 - \cdots - \dfrac{(ir^2)^{n/2}}{\Gamma(n/2 + 1))} & \text{for even } n \\ \exp(ir^2) - 1 - \cdots - \dfrac{(ir^2)^{(n-1)/2}}{\Gamma((n+1)/2))} & \text{for odd } n \end{cases} \qquad (6.49b)$$

To prove eqn. 6.49*a* one should write $Ff(x, y)$ (after the variable change $\tau = x - t$), in the form

$$Ff(x, y) = -\frac{\exp ix^2}{2\pi}\int_0^\infty \frac{\exp i(\tau^2 - 2\tau x)\, d\tau}{\tau - r - i0} = \mathcal{J}_3 + \mathcal{J}_4$$

where

$$\mathcal{J}_3 = -\frac{\exp i(x^2 + r^2)}{2\pi}\int_0^\infty \frac{\exp(-2ix\tau)\, d\tau}{\tau - r - i0}$$

$$\mathcal{J}_4 = -\frac{\exp ix^2}{2\pi}\int_0^\infty \exp(-2ix\tau)\cdot\left(\frac{\exp i\tau^2 - \exp ir^2}{\tau - r}\right) d\tau$$

Reducing $\mathcal{J}_3$ to the integral sine and cosine and evaluating the asymptotics of $\mathcal{J}_4$ as $x \to \infty$ obtains eqn. 6.49*a*.

Using eqns. 6.46 and 6.46*a*, receive:

for $x \gg 1$ *and bounded* $|r| = |x - y|$ (region III in Figure 6.7):

$$Ff(x, y) \simeq -\frac{\exp iy^2}{2\pi} E_1(2ixr) + \mathcal{J}_4(r, x)$$

$$\simeq \exp iy^2\left(\frac{\text{Ci}\,(2|xr|) + i\,\text{Si}\,(2xr)}{2\pi} - \frac{i}{4}\,\text{sign}\, r\right) + \mathcal{J}_4(r, x) \qquad (6.49c)$$

for $x \ll -1$ *and* $y > x + c$, where c is a fixed constant (region IV):

$$Ff(x, y) \simeq \exp ix^2 \cdot A(x, y) \qquad (6.50a)$$

where $A(x, y)$ is the asymptotic series

$$A(x, y) \simeq \frac{x}{4\pi}\sum_{n=0}^{\infty} (ix^{-2})^{n+1}\frac{d^n}{d\xi^n}\frac{1}{-x\xi + y\sqrt{\xi}}\bigg|_{\xi=1} \qquad (6.50b)$$

To prove eqn. 6.50*a* it is sufficient to change the variable $t = x\sqrt{\xi}$ in eqn. 6.43 after which the asymptotic of the obtained integral is found as $x \to -\infty$. Then using eqns. 6.45 and 6.46 one gets

for $x \gg 1$ *and* $y < x - c$ (region V):

$$Ff(x, y) \simeq \exp ix^2 \cdot A(-x, -y) - i\exp(iy^2)F^*(-y) \qquad (6.50c)$$

for $y \ll -1$ *and* $x > y + c$ (region VI):

$$Ff(x, y) \simeq \exp ix^2 \cdot A^*(y, x) + i\exp(iy^2)[F(x)F^*(y) - 1] \qquad (6.50d)$$

for $y \gg 1$ *and* $x < y - c$ (region VII):

$$Ff(x, y) \simeq \exp ix^2 \cdot A^*(-y, -x) - i\exp(iy^2)F^*(-y)F(x) \qquad (6.50e)$$

For $x \ll -1$, $y \gg 1$ both asymptotics of eqns. 6.50*a* and 6.50*e* are applicable and these expressions turn out to be asymptotically equivalent. Similarly, for $x \gg 1$, $y \ll -1$ eqns. 6.50*c* and 6.50*d* are asymptotically equivalent. In the zone where regions IV and VII overlap it is convenient to use the simpler eqn. 6.50*a*. Similarly, in the zone where regions V and VI overlap it is convenient to use eqn. 6.50*b*. In other words, regions IV, VII and V, VI are defined as shown in Figure 6.7.

The boundaries of regions I–VII are determined by the accuracy required. If the error 0.01 is admissible, and in eqns. 6.49 and 6.50 only the first terms of the asymptotic series of $J_4(r, x)$ and $A(x, y)$ are retained, these regions can be defined by

I: $-3 < x < 3$, $-3 < y < 3$

II: $|x - y| < 1$, $x < -3$ or $y < -3$

III: $|x - y| < 1$, $x > 3$ or $y > 3$

IV: $x < -3$, $y > x + 1$

V: $x > 3$, $y < x - 1$

VI: $y < -3$, $y + 1 < x < 3$

VII: $y > 3$, $-3 < x < y - 1$

The coefficients of the $Ff(x, y)$ expansion in the double series in Chebyshev's polynomials for $-5 < x < 0$ and all $r = x - y$ are given in the appendix. Since $Ff(x, y)$ and $Ff(-x, -y)$ are related by eqn. 6.46, the data are sufficient to evaluate Ff for $|x| < 5$. If $|x| > 5$, asymptotics for $|r| < 3.5$ (eqn. 6.49) and for $|r| > 3.5$ (eqn. 6.50) are applicable for an accuracy up to 10^{-8}.

References

1 ERDELYI, A.: 'Asymptotic expansions' (Dover, New York, 1956)

2 COPSON, E.T.: 'Asymptotic expansions' (Cambridge University Press, 1965)

3 FEDORYUK, M.V.: 'Saddle-point method' (Nauka, Moscow, 1977)

4 KRYUKOVSKY, A.S., LUKIN, D.S. and PALKIN, E.A.: 'Wave catastrophe theory'. To be published in 1996 by Amer. Math. Society

5 BORN, M. and WOLF, E.: 'Principles of optics' (Pergamon Press, 1964)

6 FELSEN, L.B. and MARCUVITZ, N.: 'Radiation and scattering of waves' (Prentice-Hall, Englewood Cliffs, New Jersey, 1973)

7 SOLIMANO, S., CROSINGNANI, B. and DIPORTO, P.: 'Guiding, diffraction and confinement of optical radiation' (Academic Press, 1986)

8 VAN CAMPEN, N. C.: 'The method of the stationary phase and the method of Fresnel zones', *Physica*, 1958, **24,** (6) p. 437

9 FRESNEL, O.: 'Oeuvres completes d'Augustin Fresnel'. (M.M. Henri de Senarmont, Emile Verdét et Leons Fresnel. Paris, 1864), vol. 1 no. xi; 'Annales de Chimie et de Physique', 1819, **11,** pp. 246–296

10 AIRY, G.: 'On the intensity of light in the neighbourhood of caustic', *Trans. Cambr. Phil. Soc.* 1838, **6,** pp. 379–402

11 LEVEY, L. and FELSEN, L.B.: 'On incomplete Airy functions and their applications to diffraction problems', *Radio Sci.*, 1969, **4,** (10) pp. 959–969

12 ORLOV, YU. I.: 'Modification of GTD near the edge-wave caustic', *Radiotekh and Electron.*, 1976, **21,** pp. 62–71

13 POINCARÉ, H.: 'Sur les intégrales irrégulières des equations lineaires', *Acta Math.*, 1886, **10,** pp. 295–344

14 STIELTJES, T.: 'Recherches sur quelques series semi convergentes', *Ann. de l'Ec. Norm. Sup (3)*, 1886, **3,** pp. 201–258

15 EULER, L.: *Novi commentarii Ac. Sci. Petropolitanae*, 1754, **5,** pp. 205–237; *Opera omnia, ser. 1*, Lpz., 1911–1932, **14,** pp. 585–617

16 STIRLING: 'Methodus differentialis' (London, 1730)

17 MACLAURIN, C.: 'A complete treatise of fluxions' (Edinburgh, 1742)

18 LAPLACE, P.S.: *Oeuvres* **9,** pp. 422–446; **10,** pp. 209–295, Paris, Gauthier-Villars

19 ABRAMOVITZ, M. and STEGAN, I.A. (Eds.): 'Handbook of mathematical functions with formulas, graphs and mathematical tables'. National Bureau of Standards applied mathematics series, vol. 55, June 1964

20 TIKHONOV, A.N.: 'On asymptotic behaviour of integrals with Bessel functions', *Dokl. Akad. Nauk SSSR*, 1959, **125,** (5) pp. 982–985

21 LARICHEV, V.A.: 'Asymptotic behavior of integrals with large parameters in the Bessel function arguments', *J. Comput. Math. Phys.*, 1973, **13,** (4) pp. 1029–1035

22 CHESTER, C., FRIEDMAN, B. and URSELL, F.: 'An extension of the method of the steepest descent', *Proc. Cambr. Phyl. Soc.*, 1957, **53,** pp. 599–611

23 BOROVIKOV, V.A., VLADIMIROV, JU.V. and KEL'BERT, M. JA.: 'The fields of internal gravity waves excited by localized sources', *Izvestija Akademii Nauk SSSR*, 1984, **20,** (6) pp. 526–532

24 BLESTEIN, N.: 'Uniform asymptotic expansion of integrals with stationary points near algebraic singularity', *Comm. Pure Appl. Math.* 1966, **19,** pp. 353–370

25 BOROVIKOV, V.A.: 'Critical layer formation in stratified media with mean shear flows', *Izvestija Akademii Nauk SSSR, MZhG*, 1990, (3), pp. 82–93

26 GRADSTEIN, I.S. and RYZHIK, I.M.: 'Tables of integrals, series and products' (Academic Press, New York, 1980)

27 HANDELSMAN, R.A. and BLEISTEIN, N.: 'Uniform asymptotic expansions of integrals that arise in the analysis of precursors', *Arch. Rat. Mech. Anal.*, 1969, **35,** (4) pp. 284–298
28 ARNOLD, V.I., VARCHENKO, A.N. and GUSSEIN-ZADE, S.M.: 'Singularities of differentiable mappings' (Birkhäuser, Boston, USA, 1985) vol. 1
29 POSTON, T. and STEWART, I.: 'Catastrophe theory and its applications' (Pitman Press, London, San Francisco, Melbourne, 1978)
30 BOROVIKOV, V.A.: 'Stationary phase method for a saddle point near the boundary of integration region', *Mat. Zametki*, 1989, **45,** (2) pp. 3–14
31 KARATYGIN, V.A. and ROSOV, V.A.: 'Method of stationary phase for double integral with arbitrary located stationary point', *J. Comput. Math. Phys.*, 1972, **12,** (6) pp. 1391–1405
32 BOROVIKOV, V.A.: 'Stationary phase method for two-dimensional regions with corner points', *Mat. Zametki*, 1984, **36,** (5) pp. 777–789
33 YARYGIN, A.P.: 'Stationary phase method for 'saddle' stationary point in problems of diffraction in inhomogeneous media', *Radiotech and Electron.*, 1970, **15,** (6) pp. 1826–1832
34 BABICH, V.M. and BULDYREV, V.S.: 'Asymptotic methods in short wave diffraction theory' (Springer, Berlin & Heidelberg, 1989)
35 BOROVIKOV, V.A. and KINBER, B.YE.: 'Geometrical theory of diffraction' (Peter Peregrinus, 1994)
36 BREKHOVSKIKH, L.M.: 'Waves in layered media' (Academic Press, New York, 1980) 2nd edn.
37 JONES, D.S.: 'Double knife-edge diffraction and ray theory', *Quart. J. Mech. Appl. Math.* 1973, **26,** pp. 1–13
38 TISCHENKO, V.A. and KHESTANOV, R.KH.: 'Diffraction of a field with shadow boundary by a half-plane', *Dokl. Akad. Nauk SSSR*, 1973, **212,** (4) pp. 842–845
39 LIGHTHILL, J.: 'Waves in fluids' (Cambridge University Press, 1978)
40 LUKE, Y.L.: 'Mathematical functions and their approximations' (Acad. Press, New York, San Francisco, London, 1975)

Appendix

Coefficients of special function expansions in Chebyshev's polynomials

Chebyshev's polynomials $T_n(x)$ are defined (for $|x|<1$) by

$$T_n(x) = \cos\,(n \arccos x)$$

and satisfy the recurrence relations

$$T_0(x)=1; \qquad T_1(x)=x; \qquad T_n(x)=2xT_{n-1}(x)-T_{n-2}(x)$$

The polynomials T_n with even and odd n are even and odd functions of x, respectively

$$T_n(x)=\sum_{k=0}^{m} a_{k,n}x^{n-2k}$$

where $m=n/2$ for even n and $m=(n-1)/2$ for odd n. In particular, the coefficients $a_{k,n}$ for $0<n<15$ are of the form

n	$a_{0,n},\ a_{1,n}\ .\ .\ .\ a_{m,n}$
0	1
1	1
2	2, −1
3	4, −3
4	8, −8, 1
5	16, −20, 5
6	32, −48, 18, −1
7	64, −112, 56, −7
8	128, −256, 160, −32, 1
9	256, −576, 432, −120, 9
10	512, −1280, 1120, −400, 50, −1
11	1024, −2816, 2816, −1232, 220, −11
12	2048, −6144, 6912, −3584, 840, −72, 1
13	4096, −13312, 16640, −9984, 2912, −364, 13
14	8192, −28672, 39424, −26880, 9408, −1568, −98, −1
15	16384, −61440, 92160, −70400, 28800, −6048, 560, −15

The polynomials $Q_n(x)=2^{-n}T_n(x)$ have minimum deviation from zero in the interval $-1<x<1$ among those of nth degree with the coefficient of x^n equal to unity

$$\max_{|x|<1} Q_n(x)=2^{-n}$$

The function $f(x)$, summable over the interval $|x|<1$ with weight $(1-x^2)^{-1/2}$, can be expanded into the series in Chebyshev's polynomials

$$f(x)=\sum_{n=0}^{\infty} a_n T_n(x)$$

The approximate representation of a function as a finite sum of the series

$$f(x)\simeq S_N(x)=\sum_{n=0}^{N} a_n T_n(x) \tag{A1}$$

appears to be convenient to obtain computer-oriented tables of special functions. As experience shows, the sum $S_N(x)$ approximates the function $f(x)$ (if the function is sufficiently smooth) with high accuracy even for small N. Also, it is easy to write programs to evaluate a function according to eqn. A1.

In the case of decreasing coefficients a_n (n increasing) it is convenient to evaluate $S_N(x)$ with the Clenshaw algorithm according to which a partial sum $S_N(x)$ and its derivative

$$S'_N(x)=\sum_{n=0}^{N} a_n T'_n(x)$$

are evaluated by the recurrence relations

$$\begin{aligned} &B_{N+1}=B_{N+2}=0; \qquad C_N=C_{N+1}=0\\ &\quad B_n=2xB_{n+1}-B_{n+2}+a_n; \qquad n=N, N-1, \ldots, 0\\ &\quad C_n=2xC_{n+1}-C_{n+2}-2B_{n+1}; \qquad n=N-1, N-2, \ldots, 0\\ &S_N(x)=B_0-xB_1; \qquad S'_N(x)=-B_1-C_0+xC_1 \end{aligned} \tag{A2}$$

If in eqn. A1 there is a summation over the even Chebyshev polynomials only

$$S_N(x)=\sum_{n=0}^{N} b_n T_{2n}(x)$$

or over the odd polynomials

$$S_N(x)=\sum_{n=0}^{N} c_n T_{2n+1}(x)$$

eqn. A2 relations are slightly changed

$$\begin{aligned} &B_{N+1}=B_{N+2}=0; \qquad C_N=C_{N+1}=0\\ &\quad B_n=2(2x^2-1)B_{n+1}-B_{n+2}+b_n; \qquad n=N, N-1, \ldots, 0\\ &\quad C_n=2(2x^2-1)C_{n+1}-C_{n+2}-8xB_{n+1};\\ &\quad n=N-1, N-2, \ldots, 0\\ &S_N(x)=B_0-(2x^2-1)B_1; \qquad S'_N(x)=-4xB_1-C_0+(2x^2-1)C_1 \end{aligned}$$

and

$$\begin{aligned} &B_{N+1}=B_{N+2}=0; \qquad C_N=C_{N+1}=0\\ &\quad B_n=2(2x^2-1)B_{n+1}-B_{n+2}+c_n; \qquad n=N, N-1, \ldots, 0\\ &\quad C_n=2(2x^2-1)C_{n+1}-C_{n+2}-8xB_{n+1}; \qquad n=N-1, N-2, \ldots, 0\\ &S_N(x)=x(B_0-B_1); \qquad S'_N(x)=B_0-B_1-x(C_0-C_1) \end{aligned}$$

If one needs to evaluate the $f(x)$ in the interval different from $-1<x<1$, this interval is reduced to -1 $1<x<1$ by means of a linear or linear-fractional transformation. In this case it turns to be useful to single out increasing summands and rapidly oscillating factors from the functions $f(x)$, i.e. to represent the function in the form

$$f(x)=A(x)+B(x)\sum_{n} a_n T_n(x)$$

where $A(x)$ and $B(x)$ are elementary functions.

This method to evaluate special functions is considered in detail in Reference 40 where, in particular, the tables for the coefficients a_n for many special functions are included. The appendix includes tables for parabolic cylinder functions $D_p[(1 \pm i)x]$, Airy–Fresnel integral $Af(x, y)$, generalised Fresnel integral $G(x, y)$, and Ff-integral which were evaluated by V.V. Zalipayev and A.V. Kostin for this book; the rest of the tables in the Appendix are taken from Reference 40.

A.1 Fresnel integrals

$$\int_0^x t^{-1/2} \cos t \, dt = x^{1/2} \sum_{n=0}^{\infty} a_n T_{2n}(x/8)$$

$$\int_0^x t^{-1/2} \sin t \, dt = x^{1/2} \sum_{n=0}^{\infty} b_n T_{2n+1}(x/8)$$

	$\lvert x \rvert < 8$	
n	a_n	b_n
0	0.76435 13866 41860	0.63041 40431 45705
1	−0.43135 54754 76602	−0.42344 51140 57053
2	0.43288 19997 97267	0.37617 17264 33437
3	−0.26973 31033 83871	−0.16249 48915 45096
4	0.08416 04532 08769	0.03822 25577 86330
5	−0.01546 52448 44614	−0.00564 56347 71322
6	0.00187 85542 34398	0.00057 45495 19769
7	−0.00016 26497 76189	−0.00004 28707 15321
8	0.00001 05739 76564	0.00000 24512 07499
9	−0.00000 05360 93399	−0.00000 01109 88418
10	0.00000 00218 16585	0.00000 00040 82497
11	−0.00000 00007 29016	−0.00000 00001 24498
12	0.00000 00000 20373	0.00000 00000 03200
13	−0.00000 00000 00483	−0.00000 00000 00070
14	0.00000 00000 00009	0.00000 00000 00001

$$\int_x^{\infty} t^{-1/2} \exp(it)\,dt = ix^{-1/2} \exp(ix) \sum_{n=0}^{\infty} (a_n + ib_n) T_n\left(\frac{10}{x} - 1\right)$$

$x>5$		
n	a_n	b_n
0	0.99056 04793 73498	−0.04655 77987 37516
1	−0.01218 35098 31479	−0.04499 21302 01239
2	−0.00248 27428 23113	0.00175 42871 39651
3	0.00026 60949 52647	0.00014 65340 02581
4	−0.00000 10790 68987	−0.00003 91330 40863
5	−0.00000 48836 81754	0.00000 34932 28660
6	0.00000 09990 55266	0.00000 03153 53003
7	−0.00000 00750 92717	−0.00000 01876 58201
8	−0.00000 00190 79488	0.00000 00377 55280
9	0.00000 00090 90797	−0.00000 00026 65517
10	−0.00000 00019 66236	−0.00000 00010 88145
11	0.00000 00001 64773	0.00000 00005 35501
12	0.00000 00000 63080	−0.00000 00001 31577
13	−0.00000 00000 36423	0.00000 00000 15286
14	0.00000 00000 10537	0.00000 00000 03395
15	−0.00000 00000 01716	−0.00000 00000 02702
16	−0.00000 00000 00107	0.00000 00000 00946
17	0.00000 00000 00204	−0.00000 00000 00207
18	−0.00000 00000 00090	0.00000 00000 00013
19	0.00000 00000 00026	0.00000 00000 00014
20	−0.00000 00000 00004	−0.00000 00000 00009
21	−0.00000 00000 00001	0.00000 00000 00003
22	0.00000 00000 00001	−0.00000 00000 00001

A.2 Parabolic cylinder functions

$x > 0$:

$$D_{1/2}[(1 \pm i)x] = \exp(\mp ix^2/2) \sum_{n=0}^{\infty} (a_n \pm ib_n) T_n\left(\frac{2x}{3} - 1\right)$$

n	a_n ($0 < x < 3$)	b_n
0	1.30486 12050 96373	0.46105 18916 77552
1	0.65863 13921 60126	0.36809 11640 82738
2	−0.06061 32703 95309	−0.06854 79402 67986
3	0.00600 00265 53771	0.01866 11825 17274
4	0.00069 19934 97722	−0.00469 64894 86542
5	−0.00070 24023 65408	0.00095 54840 72570
6	0.00026 95744 03355	−0.00012 25719 29681
7	−0.00007 25940 57301	−0.00000 73654 28617
8	0.00001 41103 98293	0.00001 05819 58151
9	−0.00000 15194 36115	−0.00000 38850 12879
10	−0.00000 01841 76816	0.00000 09467 01073
11	0.00000 01511 06090	−0.00000 01570 23792
12	−0.00000 00469 48852	0.00000 00106 23488
13	0.00000 00097 07815	0.00000 00037 28396
14	−0.00000 00012 56736	−0.00000 00018 04052
15	0.00000 00000 03174	0.00000 00004 55205
16	0.00000 00000 52137	−0.00000 00000 76238
17	−0.00000 00000 17202	0.00000 00000 06462
18	0.00000 00000 03470	0.00000 00000 00911
19	−0.00000 00000 00440	−0.00000 00000 00529
20	0.00000 00000 00009	0.00000 00000 00129
21	0.00000 00000 00013	−0.00000 00000 00020
22	−0.00000 00000 00004	0.00000 00000 00002
23	0.00000 00000 00001	

$$D_{1/2}[(1 \pm i)x] = \exp(\mp ix^2/2)x^{1/2} \sum_{n=0}^{\infty} (a_n \pm ib_n)T_n\left(1 - \frac{6}{x}\right)$$

$x > 3$		
n	a_n	b_n
0	1.09997 02701 15473	0.45228 49834 30685
1	−0.00174 08309 72456	0.00372 36718 51544
2	0.00047 34923 22363	−0.00090 59973 56980
3	−0.00002 08646 83898	−0.00001 52261 87492
4	0.00000 17998 16380	0.00000 25940 76648
5	0.00000 02767 13767	−0.00000 01574 07777
6	−0.00000 00293 89552	−0.00000 00131 53476
7	−0.00000 00011 54218	0.00000 00054 26603
8	0.00000 00007 58916	−0.00000 00000 04340
9	−0.00000 00000 40281	−0.00000 00001 26675
10	−0.00000 00000 21816	0.00000 00000 10586
11	0.00000 00000 02640	0.00000 00000 03624
12	0.00000 00000 00652	−0.00000 00000 00638
13	−0.00000 00000 00146	−0.00000 00000 00121
14	−0.00000 00000 00023	0.00000 00000 00034
15	0.00000 00000 00008	0.00000 00000 00005
16	0.00000 00000 00001	−0.00000 00000 00002

$$D_{3/2}[(1 \pm i)x] = \exp(\mp ix^2/2) \sum_{n=0}^{\infty} (a_n \pm ib_n) T_n\left(\frac{2x}{3} - 1\right)$$

$0<x<3$		
n	a_n	b_n
0	1.11649 80239 19605	3.50019 00549 32329
1	1.88964 44515 85587	4.11861 08297 37539
2	0.17055 07785 23001	0.56365 95077 84149
3	0.00095 48200 83155	−0.04693 37680 02929
4	−0.00309 46695 54054	0.00686 47858 21092
5	0.00112 41757 46526	−0.00093 40871 54061
6	−0.00029 86233 39242	0.00006 24452 23702
7	0.00006 24564 47072	0.00001 96569 83126
8	−0.00000 95054 35141	−0.00001 04843 02067
9	0.00000 05824 14754	0.00000 29912 29023
10	0.00000 02314 62774	−0.00000 06073 34770
11	−0.00000 01097 69886	0.00000 00817 00530
12	0.00000 00283 26606	−0.00000 00018 14429
13	−0.00000 00050 29586	−0.00000 00028 78665
14	0.00000 00005 16813	0.00000 00010 42861
15	0.00000 00000 31421	−0.00000 00002 27752
16	−0.00000 00000 30120	0.00000 00000 32985
17	0.00000 00000 08326	−0.00000 00000 01885
18	−0.00000 00000 01485	−0.00000 00000 00611
19	0.00000 00000 00159	0.00000 00000 00251
20	0.00000 00000 00003	−0.00000 00000 00054
21	−0.00000 00000 00006	0.00000 00000 00008
22	0.00000 00000 00002	

$$D_{3/2}[(1 \pm i)x] = \exp(\mp ix^2/2)x^{3/2} \sum_{n=0}^{\infty} (a_n \pm ib_n) T_n\left(1 - \frac{6}{x}\right)$$

$x > 3$		
n	a_n	b_n
0	0.63142 35234 74568	1.55870 82415 58663
1	0.01623 54141 57864	−0.00655 38810 87552
2	−0.00407 00647 98834	0.00160 06983 74423
3	0.00000 57209 72219	0.00002 16567 39600
4	−0.00000 01941 23033	−0.00000 27382 46935
5	−0.00000 01577 71641	−0.00000 00116 64432
6	0.00000 00099 50328	0.00000 00099 09191
7	0.00000 00012 41309	−0.00000 00015 07601
8	−0.00000 00001 99910	−0.00000 00000 79091
9	−0.00000 00000 02375	0.00000 00000 34985
10	0.00000 00000 05493	−0.00000 00000 00425
11	−0.00000 00000 00271	−0.00000 00000 00885
12	−0.00000 00000 00157	0.00000 00000 00075
13	0.00000 00000 00018	0.00000 00000 00028
14	0.00000 00000 00005	−0.00000 00000 00004
15	−0.00000 00000 00001	−0.00000 00000 00001

$x<0$:

$$D_{1/2}[(1\pm i)x] = \mp i D_{1/2}[(1\pm i)|x|] + \exp(\pm i x^2/2)\sum_{n=0}^{\infty}(a_n \pm i b_n)T_n\left(\frac{-2x}{3}-1\right)$$

$-3<x<0$

n	a_n	b_n
0	0.10969 38039 84208	0.27655 35832 53580
1	−0.23227 09168 50843	−0.26289 07760 98319
2	0.14864 55860 75542	0.06732 28013 42454
3	−0.06620 95824 17035	0.00249 87192 15964
4	0.02105 59990 38809	−0.01231 07137 43177
5	−0.00419 15741 04148	0.00717 19163 28918
6	−0.00000 28065 81926	−0.00265 11592 83417
7	0.00041 48283 39729	0.00068 40209 49714
8	−0.00020 48749 17644	−0.00010 39677 40759
9	0.00006 28258 98141	−0.00000 60951 80112
10	−0.00001 30367 25358	0.00001 03711 89210
11	0.00000 13151 28370	−0.00000 40285 03402
12	0.00000 02827 22528	0.00000 10029 15610
13	−0.00000 01899 83160	−0.00000 01598 03889
14	0.00000 00571 59246	0.00000 00055 53771
15	−0.00000 00112 72273	0.00000 00063 18366
16	0.00000 00012 25397	−0.00000 00025 54159
17	0.00000 00001 02337	0.00000 00005 97217
18	−0.00000 00000 90011	−0.00000 00000 89031
19	0.00000 00000 25561	0.00000 00000 03992
20	−0.00000 00000 04606	0.00000 00000 02361
21	0.00000 00000 00455	−0.00000 00000 00907
22	0.00000 00000 00033	0.00000 00000 00192
23	−0.00000 00000 00027	−0.00000 00000 00025
24	0.00000 00000 00007	0.00000 00000 00001
25	−0.00000 00000 00001	0.00000 00000 00001

$$D_{1/2}[(1 \pm i)x] = \mp i D_{1/2}[(1 \pm i)|x|] + \exp(\pm i x^2/2)|x|^{-3/2} \sum_{n=0}^{\infty} (a_n \pm i b_n) T_n\left(1 + \frac{6}{x}\right)$$

$x < -3$

n	a_n	b_n
0	−0.14537 47107 88522	0.39227 61107 48801
1	−0.02072 15914 77913	−0.00446 59711 93323
2	0.00517 32953 85137	0.00016 05171 16800
3	0.00006 56692 69171	0.00052 74333 57036
4	−0.00005 20001 43318	−0.00004 70247 03046
5	0.00001 03303 46193	−0.00000 93939 37492
6	0.00000 08527 71065	0.00000 18503 55670
7	−0.00000 04255 28284	0.00000 00484 51217
8	0.00000 00024 61455	−0.00000 00839 86064
9	0.00000 00165 97435	0.00000 00043 13729
10	−0.00000 00015 21989	0.00000 00035 38895
11	−0.00000 00007 44188	−0.00000 00004 32028
12	0.00000 00001 20801	−0.00000 00001 60861
13	0.00000 00000 36424	0.00000 00000 32314
14	−0.00000 00000 08511	0.00000 00000 08447
15	−0.00000 00000 02034	−0.00000 00000 02256
16	0.00000 00000 00598	−0.00000 00000 00511
17	0.00000 00000 00133	0.00000 00000 00159
18	−0.00000 00000 00043	0.00000 00000 00036
19	−0.00000 00000 00010	−0.00000 00000 00011
20	0.00000 00000 00003	−0.00000 00000 00003
21	0.00000 00000 00001	0.00000 00000 00001

$$D_{3/2}[(1\pm i)x] = \pm iD_{3/2}[(1\pm i)|x|] + \exp(\pm ix^2/2)\sum_{n=0}^{\infty}(a_n+ib_n)T_n\left(\frac{-2x}{3}-1\right)$$

$-3<x<0$		
n	a_n	b_n
0	−0.22107 48353 87510	0.07784 28135 32566
1	0.29900 86702 09966	−0.20730 22271 43027
2	−0.11204 18754 75578	0.17609 88665 63449
3	0.01105 08203 19306	−0.09887 18474 98910
4	0.01537 98509 26563	0.03868 22632 97450
5	−0.01226 99404 69046	−0.00989 39467 46945
6	0.00544 53768 43995	0.00080 39618 53896
7	−0.00165 40770 07675	0.00069 94640 59017
8	0.00031 74134 93262	−0.00045 22703 26035
9	−0.00000 69161 62861	0.00016 12924 48964
10	−0.00002 29708 14910	−0.00003 87438 56515
11	0.00001 08540 78130	0.00000 52396 85179
12	−0.00000 30727 31347	0.00000 04427 76479
13	0.00000 05689 73032	−0.00000 05218 59478
14	−0.00000 00412 43585	0.00000 01812 22793
15	−0.00000 00163 48464	−0.00000 00402 08375
16	0.00000 00082 95487	0.00000 00053 16411
17	−0.00000 00021 75006	0.00000 00001 06883
18	0.00000 00003 67212	−0.00000 00002 92227
19	−0.00000 00000 26512	0.00000 00000 95124
20	−0.00000 00000 07127	−0.00000 00000 19021
21	0.00000 00000 03423	0.00000 00000 02236
22	−0.00000 00000 00805	0.00000 00000 00047
23	0.00000 00000 00120	−0.00000 00000 00100
24	−0.00000 00000 00007	0.00000 00000 00029
25	−0.00000 00000 00002	−0.00000 00000 00005
26	0.00000 00000 00001	

$$D_{3/2}[(1 \pm i)x] = \pm i D_{3/2}[(1 \pm i)|x|] + \exp(\pm ix^2/2)|x|^{-5/2} \sum_{n=0}^{\infty} (a_n \pm ib_n) T_n\left(1 + \frac{6}{x}\right)$$

$x < -3$		
n	a_n	b_n
0	−0.20202 26240 01870	−0.38946 52085 48711
1	0.03999 00387 31131	−0.03160 97202 79710
2	−0.00717 21322 66602	0.01003 70003 56888
3	−0.00171 72126 42132	−0.00091 83153 21400
4	0.00025 89506 75212	−0.00011 44345 75958
5	0.00001 50063 67515	0.00006 62346 18762
6	−0.00001 19676 60643	0.00000 01138 68575
7	0.00000 07939 73452	−0.00000 25513 67215
8	0.00000 05436 49107	0.00000 02559 99901
9	−0.00000 00779 90355	0.00000 01085 61916
10	−0.00000 00232 21405	−0.00000 00225 63703
11	0.00000 00060 37176	−0.00000 00050 53621
12	0.00000 00011 03859	0.00000 00016 23900
13	−0.00000 00004 38211	0.00000 00002 53684
14	−0.00000 00000 60543	−0.00000 00001 17897
15	0.00000 00000 32080	−0.00000 00000 15028
16	0.00000 00000 03947	0.00000 00000 08819
17	−0.00000 00000 02444	0.00000 00000 01091
18	−0.00000 00000 00316	−0.00000 00000 00684
19	0.00000 00000 00193	−0.00000 00000 00096
20	0.00000 00000 00030	0.00000 00000 00055
21	−0.00000 00000 00016	0.00000 00000 00010
22	−0.00000 00000 00003	−0.00000 00000 00004
23	0.00000 00000 00001	−0.00000 00000 00001

A.3 Integral sine and cosine

$$\text{Ci}(x) = C + \ln x - \int_0^x t^{-1}(1 - \cos t)\,dt$$

$$= C + \ln x - \sum_{n=0}^{\infty} a_n T_{2n}\left(\frac{x}{8}\right)$$

$$\text{Si}(x) = \int_0^x t^{-1} \sin t\,dt = \sum_{n=0}^{\infty} b_n T_{2n+1}\left(\frac{x}{8}\right)$$

$0<x<8$		
n	a_n	b_n
0	1.94054 91464 83555	1.95222 09759 53071
1	0.94134 09132 86521	−0.68840 42321 25715
2	−0.57984 50342 92993	0.45518 55132 25585
3	0.30915 72011 15927	−0.18045 71236 83878
4	−0.09161 01792 20771	0.04104 22133 75859
5	0.01644 37407 51546	−0.00595 86169 55589
6	−0.00197 13091 95216	0.00060 01427 41414
7	0.00016 92538 85083	−0.00004 44708 32911
8	−0.00001 09393 29573	0.00000 25300 78231
9	0.00000 05522 38575	−0.00000 01141 30759
10	−0.00000 00223 99493	0.00000 00041 85784
11	0.00000 00007 46533	−0.00000 00001 27347
12	−0.00000 00000 20818	0.00000 00000 03267
13	0.00000 00000 00493	−0.00000 00000 00072
14	−0.00000 00000 00010	0.00000 00000 00001

$$E_1(ix) = \int_x^\infty t^{-1} \exp(-it)\, dt = -ix \exp(-ix) \times \sum_{n=0}^{\infty} (a_n + ib_n) T_n\left(\frac{10}{x} - 1\right)$$

$x > 5$

n	a_n	b_n
0	0.97615 52711 28712	0.08968 45854 91642
1	−0.03046 56658 03070	0.08508 92472 92295
2	−0.00578 07368 31484	−0.00507 18267 77757
3	0.00083 86432 56651	−0.00033 42234 15982
4	−0.00002 15746 20728	0.00012 85606 50086
5	−0.00001 56456 41351	−0.00001 52025 51360
6	0.00000 40400 10138	−0.00000 05958 96123
7	−0.00000 04349 85306	0.00000 07134 72534
8	−0.00000 00534 30219	−0.00000 01760 03581
9	0.00000 00385 02886	0.00000 00192 57654
10	−0.00000 00100 73536	0.00000 00033 63592
11	0.00000 00012 80496	−0.00000 00024 25469
12	0.00000 00001 86917	0.00000 00007 13431
13	−0.00000 00001 70673	−0.00000 00001 14604
14	0.00000 00000 58800	−0.00000 00000 06784
15	−0.00000 00000 12157	0.00000 00000 12656
16	0.00000 00000 00475	−0.00000 00000 05323
17	0.00000 00000 00906	0.00000 00000 01400
18	−0.00000 00000 00501	−0.00000 00000 00180
19	0.00000 00000 00166	−0.00000 00000 00050
20	−0.00000 00000 00035	0.00000 00000 00046
21	0.00000 00000 00001	−0.00000 00000 00020
22	0.00000 00000 00004	0.00000 00000 00006
23	−0.00000 00000 00002	−0.00000 00000 00001
24	0.00000 00000 00001	

A.4 Bessel functions

$$\mathcal{J}_0(x) = \sum_{n=0}^{\infty} a_n T_{2n}(x/8)$$

$$\mathcal{N}_0(x) = \frac{2}{\pi}\left(C + \ln\frac{x}{2}\right)\mathcal{J}_0(x) + \sum_{n=0}^{\infty} b_n T_{2n}\left(\frac{x}{8}\right);$$

$$C = 0.57721\ 56649\ 01533$$

$0<x<8$		
n	a_n	b_n
0	0.15772 79714 74890	−0.02150 51114 49658
1	−0.00872 34423 52852	−0.27511 81330 43519
2	0.26517 86132 03337	0.19860 56347 02554
3	−0.37009 49938 72650	0.23425 27461 09022
4	0.15806 71023 32097	−0.16563 59817 13650
5	−0.03489 37694 11409	0.04462 13795 40669
6	0.00481 91800 69468	−0.00693 22862 91523
7	−0.00046 06261 66206	0.00071 91174 03752
8	0.00003 24603 28821	−0.00005 39250 79723
9	−0.00000 17619 46908	0.00000 30764 93288
10	0.00000 00760 81636	−0.00000 01384 57181
11	−0.00000 00026 79254	0.00000 00050 51054
12	0.00000 00000 78487	−0.00000 00001 52583
13	−0.00000 00000 01944	0.00000 00000 03883
14	0.00000 00000 00041	−0.00000 00000 00084
15	−0.00000 00000 00001	0.00000 00000 00002

$$H_0^{(1)}(x) = \mathcal{J}_0(x) + iN_0(x)$$
$$= \left(\frac{2}{\pi x}\right)^{1/2} e^{i(x-\pi/4)} \sum_{n=0}^{\infty} (a_n + ib_n) T_n\left(\frac{10}{x} - 1\right)$$

	$x > 5$	
n	a_n	b_n
0	0.99898 80898 58965	−0.01233 15205 78544
1	−0.00133 84285 49972	−0.01224 94962 81259
2	−0.00031 87898 78062	0.00009 64941 84993
3	0.00000 85112 32211	0.00001 36555 70490
4	0.00000 06915 42349	−0.00000 08518 06644
5	−0.00000 00907 70102	−0.00000 00272 44053
6	0.00000 00014 54928	0.00000 00096 46421
7	0.00000 00009 26762	−0.00000 00006 83348
8	−0.00000 00001 39166	−0.00000 00000 60627
9	0.00000 00000 03238	0.00000 00000 21696
10	0.00000 00000 02535	−0.00000 00000 02305
11	−0.00000 00000 00559	−0.00000 00000 00123
12	0.00000 00000 00042	0.00000 00000 00092
13	0.00000 00000 00009	−0.00000 00000 00017
14	−0.00000 00000 00004	0.00000 00000 00001
15	0.00000 00000 00001	

$$\mathcal{J}_1(x) = \sum_{n=0}^{\infty} a_n T_{2n+1}(x/8)$$

$$\mathcal{N}_1(x) = \frac{2}{\pi}\left(C + \ln\frac{x}{2}\right)\mathcal{J}_1(x) - \frac{2}{\pi x} + \sum_{n=0}^{\infty} b_n T_{2n+1}\left(\frac{x}{8}\right)$$

$0 < x < 8$		
n	a_n	b_n
0	0.05245 81903 34656	−0.04017 29465 44414
1	0.04809 64691 58230	−0.44444 71476 30558
2	0.31327 50823 61567	−0.02271 92444 28418
3	−0.24186 74084 47407	0.20664 45410 17491
4	0.07426 67962 16787	−0.08667 16970 56949
5	−0.01296 76273 11735	0.01763 67030 03163
6	0.00148 99128 96668	−0.00223 56192 94485
7	−0.00012 22786 85054	0.00019 70623 02702
8	0.00000 75626 30230	−0.00001 28858 53299
9	−0.00000 03661 30855	0.00000 06528 47952
10	0.00000 00142 77324	−0.00000 00264 50737
11	−0.00000 00004 58570	0.00000 00008 78030
12	0.00000 00000 12352	−0.00000 00000 24343
13	−0.00000 00000 00283	0.00000 00000 00573
14	0.00000 00000 00006	−0.00000 00000 00012

$$H_1^{(1)}(x) = \mathcal{J}_1(x) + iN_1(x)$$
$$= \left(\frac{2}{\pi x}\right)^{1/2} e^{i(x-3\pi/4)} \sum_{n=0}^{\infty} (a_n + ib_n) T_n\left(\frac{10}{x} - 1\right)$$

	$x > 5$	
n	a_n	b_n
0	1.00170 22348 53821	0.03726 17150 00538
1	0.00225 55728 46561	0.03714 53224 79808
2	0.00054 32164 87508	−0.00013 72632 38202
3	−0.00001 11794 61895	−0.00001 98512 94688
4	−0.00000 09469 01382	0.00000 10700 14057
5	0.00000 01110 32677	0.00000 00383 05262
6	−0.00000 00012 94399	−0.00000 00116 28723
7	−0.00000 00011 14906	0.00000 00007 59733
8	0.00000 00001 57637	0.00000 00000 75476
9	−0.00000 00000 02830	−0.00000 00000 24753
10	−0.00000 00000 02932	0.00000 00000 02494
11	0.00000 00000 00618	0.00000 00000 00156
12	−0.00000 00000 00043	−0.00000 00000 00103
13	−0.00000 00000 00010	0.00000 00000 00018
14	0.00000 00000 00004	−0.00000 00000 00001
15	−0.00000 00000 00001	−0.00000 00000 00001

$$J_\nu(x) = x^\nu \sum_{n=0}^{\infty} a_n(\nu) T_{2n}(x/8), \qquad \nu = \pm 1/4$$

$$N_{1/4}(x) = J_{1/4}(x) - 2^{1/2} J_{-1/4}(x);$$

$$N_{-1/4}(x) = -J_{-1/4}(x) + 2^{1/2} J_{1/4}(x)$$

	$0<x<8$	
n	$a_n(-1/4)$	$a_n(1/4)$
0	0.16292 58079 76064	0.14095 04424 34506
1	0.10511 50518 33156	−0.09889 25831 80257
2	0.13242 24755 27353	0.29237 16573 91931
3	−0.47013 66584 22100	−0.27196 83738 64697
4	0.24169 85159 18687	0.10031 55706 12325
5	−0.05877 81532 27299	−0.02032 93309 73457
6	0.00866 64845 97493	0.00264 33080 03302
7	−0.00087 05421 69721	−0.00024 10962 16786
8	0.00006 38666 80735	0.00001 63496 60758
9	−0.00000 35864 44045	−0.00000 08589 20905
10	0.00000 01594 93492	0.00000 00360 46120
11	−0.00000 00057 64899	−0.00000 00012 37640
12	0.00000 00001 72875	0.00000 00000 35438
13	−0.00000 00000 04373	−0.00000 00000 00860
14	0.00000 00000 00095	0.00000 00000 00018
15	−0.00000 00000 00002	

$$H^{(1)}_{1/4}(x) = J_{1/4}(x) + iN_{1/4}(x)$$
$$= \left(\frac{2}{\pi x}\right)^{1/2} e^{i(x-3\pi/8)} \sum_{n=0}^{\infty} (a_n + ib_n) T_n\left(\frac{10}{x} - 1\right)$$

$x > 5$		
n	a_n	b_n
0	0.99926 16997 83419	−0.00925 32994 85605
1	−0.00097 66473 08282	−0.00919 40369 61714
2	−0.00023 27900 19042	0.00006 97280 65100
3	0.00000 61210 11437	0.00000 98816 25563
4	0.00000 04986 81090	−0.00000 06109 30823
5	−0.00000 00649 96102	−0.00000 00196 87938
6	0.00000 00010 23765	0.00000 00069 01194
7	0.00000 00006 62955	−0.00000 00004 86456
8	−0.00000 00000 99178	−0.00000 00000 43470
9	0.00000 00000 02275	0.00000 00000 15468
10	0.00000 00000 01809	−0.00000 00000 01638
11	−0.00000 00000 00398	−0.00000 00000 00088
12	0.00000 00000 00030	0.00000 00000 00066
13	0.00000 00000 00006	−0.00000 00000 00012
14	−0.00000 00000 00003	0.00000 00000 00001

$$J_\nu(x) = x^\nu \sum_{n=0}^{\infty} a_n(\nu) T_{2n}(x/8) \qquad \nu = \pm 1/3$$

$$N_{1/3}(x) = 3^{-1/2}[J_{1/3}(x) - 2J_{-1/3}(x)]$$

$$N_{-1/3}(x) = -3^{-1/2}[J_{-1/3}(x) + 2J_{2/3}(x)]$$

$0 < x < 8$

n	$a_n(-1/3)$	$a_n(1/3)$
0	0.15921 18677 83190	0.13437 97942 27756
1	0.13935 64007 14896	−0.11984 86974 10733
2	0.05406 07319 77581	0.28710 68687 28687
3	−0.49889 25940 08177	−0.24251 07291 97801
4	0.27634 28099 41437	0.08569 82591 68484
5	−0.06962 17491 14125	−0.01691 33949 03177
6	0.01050 51479 21580	0.00215 74934 76006
7	−0.00107 36049 94092	−0.00019 38490 54789
8	0.00007 98632 63462	0.00001 29832 45927
9	−0.00000 45371 68415	−0.00000 06748 56479
10	0.00000 02038 09351	0.00000 00280 59240
11	−0.00000 00074 32216	−0.00000 00009 55461
12	0.00000 00002 24649	0.00000 00000 27155
13	−0.00000 00000 05724	−0.00000 00000 00654
14	0.00000 00000 00125	0.00000 00000 00014
15	−0.00000 00000 00002	

$$H^{(1)}_{1/3}(x) = J_{1/3}(x) + iN_{1/3}(x)$$
$$= \left(\frac{2}{\pi x}\right)^{1/2} e^{i(x-5\pi/12)} \sum_{n=0}^{\infty} (a_n + ib_n) T_n\left(\frac{10}{x} - 1\right)$$

$x > 5$		
n	a_n	b_n
1	0.99946 50123 35294	−0.00685 69474 27213
2	−0.00070 77638 36250	−0.00681 43334 21756
3	−0.00016 87952 26920	0.00005 01456 42020
4	0.00000 43855 75455	0.00000 71142 29549
5	0.00000 03580 42553	−0.00000 04367 92367
6	−0.00000 00464 10844	−0.00000 00141 58283
7	0.00000 00007 21026	0.00000 00049 24418
8	0.00000 00004 73021	−0.00000 00003 45775
9	−0.00000 00000 70557	−0.00000 00000 31071
10	0.00000 00000 01600	0.00000 00000 11008
11	0.00000 00000 01288	−0.00000 00000 01163
12	−0.00000 00000 00283	−0.00000 00000 00063
13	0.00000 00000 00021	0.00000 00000 00047
14	0.00000 00000 00004	−0.00000 00000 00008
15	−0.00000 00000 00002	

$$\mathcal{J}_\nu(x) = x^\nu \sum_{n=0}^{\infty} a_n(\nu) T_{2n}(x/8) \qquad \nu = \pm 2/3$$

$$\mathcal{N}_{2/3}(x) = -3^{-1/2}[\mathcal{J}_{2/3}(x) + 2\mathcal{J}_{-2/3}(x)]$$

$$\mathcal{N}_{-2/3}(x) = 3^{-1/2}[\mathcal{J}_{-2/3}(x) + 2\mathcal{J}_{2/3}(x)]$$

$0<x<8$

n	$a_n(-2/3)$	$a_n(2/3)$
0	0.08307 65189 70714	0.10706 16423 86466
1	0.14677 82421 85397	−0.15964 55672 68641
2	−0.49064 81271 48274	0.23154 28348 62295
3	−0.53708 96254 93876	−0.14613 03842 74246
4	0.45119 46685 89249	0.04445 81121 09229
5	−0.13366 52251 07965	−0.00795 89793 03528
6	0.02229 27277 97850	0.00094 45189 12901
7	−0.00245 08688 34376	−0.00008 01183 62342
8	0.00019 31741 98309	0.00000 51145 43070
9	−0.00001 15157 21500	−0.00000 02550 98880
10	0.00000 05391 20815	0.00000 00102 28603
11	−0.00000 00203 86971	−0.00000 00003 37192
12	0.00000 00006 36538	0.00000 00000 09306
13	−0.00000 00000 16702	−0.00000 00000 00218
14	0.00000 00000 00374	0.00000 00000 00004
15	−0.00000 00000 00007	

$$H^{(1)}_{2/3}(x)=J_{2/3}(x)+iN_{2/3}(x)$$
$$=\left(\frac{2}{\pi x}\right)^{1/2} e^{i(x-7\pi/12)} \sum_{n=0}^{\infty} (a_n+ib_n)T_n\left(\frac{10}{x}-1\right)$$

$x>5$		
n	a_n	b_n
0	1.00063 42623 22219	0.00962 40948 68137
1	0.00083 96170 22157	0.00957 62502 42783
2	0.00020 10007 10698	−0.00005 63476 40295
3	−0.00000 48012 97393	−0.00000 80532 21334
4	−0.00000 03975 54268	0.00000 04712 25027
5	0.00000 00496 32219	0.00000 00158 78186
6	−0.00000 00006 98220	−0.00000 00052 40952
7	−0.00000 00005 03120	0.00000 00003 58294
8	0.00000 00000 73554	0.00000 00000 33441
9	−0.00000 00000 01538	−0.00000 00000 11503
10	−0.00000 00000 01353	0.00000 00000 01194
11	0.00000 00000 00292	0.00000 00000 00068
12	−0.00000 00000 00021	−0.00000 00000 00049
13	−0.00000 00000 00005	0.00000 00000 00009
14	0.00000 00000 00002	

$$\mathcal{J}_\nu(x) = x^\nu \sum_{n=0}^{\infty} a_n(\nu) T_{2n}(x/8) \qquad \nu = \pm 3/4$$

$$\mathcal{N}_{3/4}(x) = -\mathcal{J}_{3/4}(x) - 2^{1/2}\mathcal{J}_{-3/4}(x)$$

$$\mathcal{N}_{-3/4}(x) = \mathcal{J}_{-3/4}(x) + 2^{1/2}\mathcal{J}_{3/4}(x)$$

$0<x<8$

n	$a_n(-3/4)$	$a_n(3/4)$
0	0.03818 88739 23737	0.10033 19891 39054
1	0.08376 61801 83354	−0.16051 88065 96380
2	−0.69230 31627 37762	0.21380 33597 57252
3	−0.51214 95647 45616	−0.12741 31877 14541
4	0.50327 80857 82480	0.03750 21128 48965
5	−0.15627 56201 74437	−0.00656 39297 55198
6	0.02678 43130 08895	0.00076 57670 81903
7	−0.00300 22381 05764	−0.00006 40663 07078
8	0.00024 02372 95881	0.00000 40426 83397
9	−0.00001 45009 90789	−0.00000 01996 26344
10	0.00000 06861 46612	0.00000 00079 33847
11	−0.00000 00261 89587	−0.00000 00002 59479
12	0.00000 00008 24520	0.00000 00000 07110
13	−0.00000 00000 21797	−0.00000 00000 00166
14	0.00000 00000 00491	0.00000 00000 00003
15	−0.00000 00000 00010	

$$H^{(1)}_{3/4}(x) = J_{3/4}(x) + iN_{3/4}(x)$$
$$= \left(\frac{2}{\pi x}\right)^{1/2} e^{i(x-5\pi/8)} \sum_{n=0}^{\infty} (a_n + ib_n) T_n\left(\frac{10}{x} - 1\right)$$

	$x>5$	
n	a_n	b_n
0	1.00095 37439 52710	0.01548 04303 74290
1	0.00126 28060 52686	0.01540 99141 20987
2	0.00030 27053 90738	−0.00008 30726 67788
3	−0.00000 70122 06892	−0.00001 19032 14814
4	−0.00000 05834 75836	0.00000 06845 99587
5	0.00000 00718 80236	0.00000 00233 76238
6	−0.00000 00009 74118	−0.00000 00075 77172
7	−0.00000 00007 27215	0.00000 00005 13083
8	0.00000 00001 05563	0.00000 00000 48531
9	−0.00000 00000 02141	−0.00000 00000 16523
10	−0.00000 00000 01946	0.00000 00000 01704
11	0.00000 00000 00418	0.00000 00000 00100
12	−0.00000 00000 00030	−0.00000 00000 00070
13	−0.00000 00000 00007	0.00000 00000 00012
14	0.00000 00000 00003	−0.00000 00000 00001

A.5 Bessel functions of imaginary argument

$$I_\nu(x) = x^\nu \sum_{n=0}^{\infty} a_n(\nu) T_{2n}(x/8) \qquad \nu = \pm 1/3$$

$$K_{1/3}(x) = K_{-1/3}(x) = 3^{-1/2}\pi [I_{-1/3}(x) - I_{1/3}(x)]$$

	$0<x<8$	
n	$a_n(-1/3)$	$a_n(1/3)$
0	246.76446 77610 21950	65.17366 54884 16527
1	374.90419 63977 61772	95.23190 27040 92164
2	168.94690 54665 51934	39.59331 03435 21063
3	47.81292 47970 09021	10.20622 18597 97179
4	9.02594 18911 35279	1.75552 58978 34804
5	1.19852 10167 30412	0.21351 04236 00374
6	0.11697 76116 93127	0.01921 26868 62831
7	0.00869 83095 97421	0.00132 57975 71063
8	0.00050 74091 61720	0.00007 22075 54088
9	0.00002 37873 65337	0.00000 31776 13643
10	0.00000 09143 62383	0.00000 01152 07185
11	0.00000 00293 09948	0.00000 00034 97936
12	0.00000 00007 94846	0.00000 00000 90184
13	0.00000 00000 18463	0.00000 00000 01998
14	0.00000 00000 00371	0.00000 00000 00038
15	0.00000 00000 00007	0.00000 00000 00001

$$I_{1/3}(x) = (2\pi x)^{-1/2} e^x \sum_{n=0}^{\infty} a_n T_n\left(\frac{16}{x} - 1\right) \qquad x > 8$$

$$K_{1/3}(x) = \left(\frac{\pi}{2x}\right)^{1/2} e^{-x} \sum_{n=0}^{\infty} b_n T_n\left(\frac{10}{x} - 1\right) \qquad x > 5$$

	$x > 8$	$x > 5$
n	a_n	b_n
0	1.00458 61710 93207	0.99353 64122 76093
1	0.00467 34791 99874	−0.00631 44392 60799
2	0.00009 08034 04815	0.00014 30095 80961
3	0.00000 37262 16111	−0.00000 57870 60592
4	0.00000 02520 73238	0.00000 03265 50333
5	0.00000 00227 82111	−0.00000 00231 23232
6	0.00000 00012 91332	0.00000 00019 39555
7	−0.00000 00006 11915	−0.00000 00001 85898
8	−0.00000 00003 75617	0.00000 00000 19868
9	−0.00000 00001 16415	−0.00000 00000 02327
10	−0.00000 00000 14443	0.00000 00000 00295
11	0.00000 00000 05374	−0.00000 00000 00040
12	0.00000 00000 03074	0.00000 00000 00006
13	0.00000 00000 00298	−0.00000 00000 00001
14	−0.00000 00000 00265	
15	−0.00000 00000 00091	
16	0.00000 00000 00016	
17	0.00000 00000 00014	

$$I_\nu(x) = x^\nu \sum_{n=0}^{\infty} a_n(\nu) T_{2n}(x/8) \qquad \nu = \pm 2/3$$

$$K_{2/3}(x) = K_{-2/3}(x) = 3^{-1/2}\pi[I_{-2/3}(x) - I_{2/3}(x)]$$

	$0<x<8$	
n	$a_n(-2/3)$	$a_n(2/3)$
0	469.95312 79949 20422	32.77479 24030 35026
1	726.03216 50664 95924	46.83793 90742 83087
2	340.10272 33771 79942	18.68656 02563 49831
3	100.90075 85218 30026	4.60121 27076 45103
4	19.98542 28723 93609	0.75671 34574 71448
5	2.77810 11375 55156	0.08824 17620 77464
6	0.28291 28347 05747	0.00763 78220 14940
7	0.02187 48334 95706	0.00050 85533 07618
8	0.00132 26199 57421	0.00002 68011 17197
9	0.00006 40820 86137	0.00000 11441 45898
10	0.00000 25392 92566	0.00000 00403 31192
11	0.00000 00837 21455	0.00000 00011 92919
12	0.00000 00023 30625	0.00000 00000 30014
13	0.00000 00000 55474	0.00000 00000 00650
14	0.00000 00000 01141	0.00000 00000 00012
15	0.00000 00000 00020	

$$I_{2/3}(x) = (2\pi x)^{-1/2}\, e^{x} \sum_{n=0}^{\infty} a_n T_n\left(\frac{16}{x} - 1\right) \qquad x > 8$$

$$K_{2/3}(x) = \left(\frac{\pi}{2x}\right)^{1/2} e^{-x} \sum_{n=0} b_n T_n\left(\frac{10}{x} - 1\right) \qquad x > 5$$

	$x>8$	$x>5$
n	a_n	b_n
0	0.99363 49867 16925	1.00914 95380 72789
1	−0.00646 71526 00616	0.00897 12068 42484
2	−0.00010 60188 22352	−0.00017 13895 98262
3	−0.00000 41406 57716	0.00000 65547 92550
4	−0.00000 02916 95418	−0.00000 03595 19190
5	−0.00000 00365 71574	0.00000 00250 24412
6	−0.00000 00075 81590	−0.00000 00020 74924
7	−0.00000 00019 23009	0.00000 00001 97224
8	−0.00000 00004 20439	−0.00000 00000 20946
9	−0.00000 00000 39372	0.00000 00000 02441
10	0.00000 00000 19007	−0.00000 00000 00308
11	0.00000 00000 10138	0.00000 00000 00042
12	0.00000 00000 01331	−0.00000 00000 00006
13	−0.00000 00000 00677	0.00000 00000 00001
14	−0.00000 00000 00312	
15	0.00000 00000 00012	
16	0.00000 00000 00040	
17	0.00000 00000 00005	
18	−0.00000 00000 00005	
19	−0.00000 00000 00001	
20	0.00000 00000 00001	

A.6 Airy–Fresnel integral

$$Af(x,y) = Ai(x) \cdot 10^{-8} \sum_{m,n} (a_{mn} + ib_{mn}) T_m\left(\frac{x}{2} - 1\right) T_n\left(\frac{y}{2} - 1\right)$$

$0 < x < 4,\ 0 < y < 4$

m	a_{m0}	a_{m1}	a_{m2}	a_{m3}	a_{m4}
1	32 292 358	−30 466 459	5 830 480	1 807 316	−1 797 644
2	−5 441 932	13 204 527	−5 872 056	−39 911	1 385 102
3	1 008 084	−2 867 730	1 777 320	−247 460	−364 701
4	−227 939	654 047	−472 762	110 324	83 050
5	54 366	−153 931	120 284	−35 428	−17 912
6	−13 154	36 947	−30 006	10 148	3 826
7	3 217	−8 899	7 457	−2 671	−785
8	−785	2 168	−1 831	709	189
9	187	−536	443	−181	−39
10	−50	123	−116	39	5
11	11	−33	25	−13	−4
12	−3	8	−7	3	
13	1	−1	2		
14		1			

m	a_{m5}	a_{m6}	a_{m7}	a_{m8}	a_{m9}	a_{m10}	a_{m11}	a_{m12}
1	585 863	−4 476	−82 237	35 694	−3 388	−3 406	1 717	−200
2	−711 929	91 627	78 320	−48 316	8 331	3 567	−2 530	467
3	269 379	−62 160	−22 386	21 293	−5 282	−1 189	1 209	−281
4	−82 368	25 807	4 331	−7 111	2 485	172	−490	153
5	22 678	−8 707	−694	2 113	−805	13	126	−65
6	−5 871	2 568	−56	−598	293	4	−41	13
7	1 517	−700	13	121	−84	16	16	−6
8	−351	206	−13	−45	14	−4	−1	3
9	96	−40	11	10	−8	−1	0	
10	−24	14	1	0	2	−1	−1	
11	4	−4	1	1				
12	−2							

m	a_{m13}	a_{m14}	a_{m15}	a_{m16}	a_{m17}	a_{m18}
1	−135	69	−8	−5	2	0
2	160	−113	18	7	−4	1
3	−56	54	−12	−2	2	−1
4	20	−26	5	1	−1	
5	4	11	−3	−1		
6	−3	−1	2			
7	−1					

m	b_{m0}	b_{m1}	b_{m2}	b_{m3}	b_{m4}
1	−22 268 156	−7 051 703	10 902 445	−4 493 735	544 237
2	6 801 660	−2 200 297	−4 808 321	3 762 314	−1 025 007
3	−1 329 123	962 013	916 082	−1 096 759	439 443
4	294 415	−272 940	−176 206	284 156	−143 030
5	−68 838	70 790	35 376	−70 637	40 899
6	16 486	−17 849	−7 427	17 261	−11 016
7	−3 971	4 464	1 645	−4 194	2 841
8	979	−1 080	−346	1 040	−708
9	−232	278	93	−239	190
10	57	−67	−19	62	−41
11	−15	14	3	−16	11
12	3	−5	−2	3	−3
13	−1	1		−1	

m	b_{m5}	b_{m6}	b_{m7}	b_{m8}	b_{m9}	b_{m10}	b_{m11}	b_{m12}
1	418 834	−272 845	60 624	12 414	−13 373	3 606	358	−576
2	−226 327	296 664	−96 934	−4 878	16 483	−5 895	7	786
3	17 580	−103 713	46 809	−2 543	−6 719	3 066	−193	−344
4	9 237	29 203	−16 880	2 457	1 974	−1 300	193	134
5	−5 573	−7 522	5 300	−974	−515	383	−97	−23
6	2 002	1 732	−1 513	415	151	−132	19	6
7	−637	−453	374	−112	−10	42	−13	−5
8	182	88	−120	25	8	−5	5	
9	−38	−20	23	−14	−3	3		
10	17	9	−5	2	−1	−1		
11	−3		3					

m	b_{m13}	b_{m14}	b_{m15}	b_{m16}	b_{m17}	b_{m18}
1	159	13	−22	6	0	−1
2	−282	−1	34	−11	0	1
3	148	−9	−14	6	−1	
4	−72	4	7	−2		
5	30	−6	−3	1		
6	−5	3		−1		
7	2					
8	−1					

$$Af(x,y)=\frac{Ai(x)}{x+x_0}\cdot 10^{-8}\sum_{m,n}(a_{mn}+ib_{mn})T_m\left(\frac{2x}{3}+1\right)T_n\left(\frac{y}{2}-1\right)$$

where $x_0 = 2.338\ 107\ 410\ 46$ is the first root of $Ai(-x)$;

$-3<x<0,\ 0<y<4$

m	a_{m0}	a_{m1}	a_{m2}	a_{m3}	a_{m4}
1	51 593 844	−90 556 795	57 115 587	−19 664 329	−4 356 304
2	41 015 625	−24 434 578	−28 697 756	31 285 863	−7 769 815
3	−1 350 444	4 942 800	5 735 794	−10 866 464	5 124 398
4	−792	−920 582	−1 502 373	3 433 149	−2 096 852
5	39 554	187 686	455 550	−1 080 432	745 681
6	−18 225	−42 390	−144 783	342 253	−251 055
7	6 783	10 482	46 517	−108 910	82 467
8	−2 344	−2 793	−14 956	34 734	−26 757
9	782	789	4 801	−11 088	8 624
10	−256	−233	−1 539	3 541	−2 769
11	83	71	493	−1 131	887
12	−27	−22	−158	361	−284
13	9	7	50	−115	91
14	−3	−2	−16	37	−29
15	1	1	5	−12	9
16			−2	4	−3
17				−1	1

m	a_{m5}	a_{m6}	a_{m7}	a_{m8}	a_{m9}	a_{m10}
1	10 456 912	−6 346 561	1 284 308	864 058	−807 379	240 050
2	−8 126 619	8 552 824	−2 906 804	−699 313	1 143 446	−436 547
3	1 892 769	−3 628 313	1 732 108	82 879	−534 385	264 273
4	−334 511	1 282 005	−762 063	47 421	201 334	−122 950
5	40 564	−421 238	288 895	−38 425	−67 964	48 959
6	1 203	134 767	−101 371	18 235	21 781	−17 786
7	−3 250	−42 779	34 124	−7 194	−6 839	6 122
8	1 605	13 567	−11 229	2 593	2 138	−2 042
9	−623	−4 308	3 648	−890	−670	669
10	221	1 370	−1 177	297	211	−217
11	−75	−436	378	−97	−67	70
12	25	139	−121	32	21	−22
13	−8	−44	39	−10	−7	7
14	3	14	−12	3	2	−2
15	−1	−5	4	−1	−1	1
16		1	−1			

m	a_{m11}	a_{m12}	a_{m13}	a_{m14}	a_{m15}	a_{m16}	a_{m17}	a_{m18}
1	48 702	−71 089	23 242	2 609	−5 051	1 593	178	−306
2	−34 350	106 085	−41 095	−1 770	7 891	−2 832	−175	497
3	−7 917	−53 587	25 880	−953	−4 284	1 868	−13	−289
4	13 450	21 468	−12 721	1 306	1 838	−969	64	132
5	−8 218	−7 512	5 317	−824	−678	426	−48	−52
6	3 748	2 439	−2 004	391	228	−167	26	19
7	−1 479	−763	708	−160	−73	61	−11	−6
8	537	236	−240	60	23	−21	4	2
9	−186	−73	79	−21	−7	7	−2	−1
10	62	23	−26	7	2	−2	1	
11	−20	−7	8	−2	−1	1		
12	7	2	−3	1				
13	−2	−1	1					
14	1							

m	a_{m19}	a_{m20}	a_{m21}	a_{m22}	a_{m23}	a_{m24}
1	84	14	−16	3	1	−1
2	−151	−19	27	−6	−2	1
3	105	6	−17	4	1	−1
4	−58	1	8	−3		
5	27	−2	−3	1		
6	−11	1	1	−1		
7	4					
8	−1					

m	b_{m0}	b_{m1}	b_{m2}	b_{m3}	b_{m4}
1	−48 634 855	24 593 634	19 624 151	−31 688 306	21 535 627
2	−25 834 375	−37 145 679	26 696 507	9 221 729	−21 834 421
3	2 834 736	5 312 588	−8 987 156	5 782	7 098 729
4	−397 687	−1 291 440	2 682 492	−435 415	−2 129 848
5	66 296	390 609	−817 268	199 647	640 998
6	−12 285	−124 680	254 606	−72 733	−196 385
7	2 445	40 160	−80 316	24 666	61 090
8	−518	−12 928	25 494	−8 124	−19 199
9	117	4 152	−8 117	2 639	6 073
10	−28	−1 331	2 588	−851	−1 928
11	7	426	−826	273	614
12	−2	−136	264	−88	−196
13	1	44	−84	28	62
14		−14	27	−9	−20
15		4	−9	3	6
16		−1	3	−1	−2
17			−1		1

m	b_{m5}	b_{m6}	b_{m7}	b_{m8}	b_{m9}	b_{m10}
1	−6 563 819	−1 924 857	3 166 108	−1 456 878	72 656	288 870
2	12 238 582	−421 590	−3 661 481	2 315 841	−351 421	−364 401
3	−5 776 600	1 133 686	1 349 231	−1 187 146	297 911	143 370
4	2 168 955	−652 432	−409 533	483 436	−160 755	−42 760
5	−744 524	275 400	114 905	−174 738	69 298	10 525
6	246 045	−101 993	−31 644	59 376	−26 399	−2 189
7	−79 936	35 387	8 827	−19 549	9 365	366
8	25 760	−11 854	−2 527	6 335	−3 185	−34
9	−8 267	3 893	744	−2 037	1 056	−8
10	2 648	−1 264	−225	653	−345	7
11	−847	408	69	−209	112	−3
12	271	−131	−22	67	−36	1
13	−87	42	7	−21	12	
14	28	−13	−2	7	−4	
15	−9	4	1	−2	1	
16	3	−1		1		
17	−1					

m	b_{m11}	b_{m12}	b_{m13}	b_{m14}	b_{m15}	b_{m16}	b_{m17}	b_{m18}
1	−161 729	20 134	21 263	−12 850	1 722	1 409	−812	89
2	258 773	−47 259	−28 863	21 081	−3 663	−2 054	1 366	−189
3	−141 820	36 433	12 282	−12 271	2 861	971	−842	159
4	61 601	−20 120	−3 804	5 665	−1 658	−340	411	−99
5	−23 324	9 033	893	−2 252	784	93	−172	49
6	8 156	−3 568	−142	815	−323	−19	65	−21
7	−2 727	1 301	0	−279	122	2	−23	8
8	890	−451	12	92	−43		8	−3
9	−287	151	−7	−30	15		−3	1
10	92	−50	3	10	−5		1	
11	−29	16	−1	−3	2			
12	9	−5		1	−1			
13	−3	2						
14	1	−1						

m	b_{m19}	b_{m20}	b_{m21}	b_{m22}	b_{m23}
1	86	−42	2	5	−2
2	−134	73	−6	−8	3
3	71	−48	6	5	−2
4	−28	25	−4	−2	1
5	9	−11	2	1	−1
6	−3	4	−1		
7	1	−2			
8		1			

A.7 Generalised Fresnel integral

$$G(x, y) = \tfrac{1}{2}F(y) - \frac{1}{2\pi}\operatorname{arctg}\frac{x}{y} - \frac{1}{2\pi}\sum_{m,n=0}^{\infty} 10^{-8}(a_{mn} + ib_{mn})T_{2m+1}\left(\frac{x}{2}\right)T_{2n+1}\left(\frac{y}{2}\right)$$

$-2 < x, y < 2$

m	a_{m0}	a_{m1}	a_{m2}	a_{m3}
0	−153 034 498	49 125 028	25 570 050	−7 012 733
1	20 981 037	4 822 754	−8 621 412	−206 810
2	−1 830 250	−4 912 776	354 514	661 941
3	−769 488	383 891	447 390	−48 798
4	140 932	147 501	−37 196	−23 126
5	11 228	−15 377	−10 333	1 939
6	2 972	−2 332	922	414
7	89	260	137	−34
8	35	23	−12	−5
9		−3	−1	

m	a_{m4}	a_{m5}	a_{m6}	a_{m7}	a_{m8}	a_{m9}	a_{m10}
0	−1 132 750	269 007	24 803	−5 103	−320	59	3
1	544 120	−978	−14 291	127	209	−2	−2
2	−32 173	−26 470	1 075	518	−18	−6	
3	−31 101	2 199	864	−48	−13	1	
4	2 589	990	−77	−20	1		
5	782	−84	−23	2			
6	−64	−19	2				
7	−11	1					
8	1						

m	b_{m0}	b_{m1}	b_{m2}	b_{m3}
0	−4 319 052	−80 836 302	23 884 651	6 000 641
1	−5 916 832	19 734 966	1 317 300	−2 491 517
2	4 735 505	−761 983	−2 229 886	124 301
3	−678 027	−938 458	159 256	136 712
4	−101 199	89 589	73 698	−11 164
5	22 418	19 816	−6 471	−3 320
6	1 072	−2 159	−1 264	277
7	−342	−245	113	46
8	−6	28	13	−4
9	3	2	−1	

m	b_{m4}	b_{m5}	b_{m6}	b_{m7}	b_{m8}	b_{m9}	b_{m10}
0	−1 530 104	−179 635	39 819	2 967	−578	−31	5
1	−16 578	96 088	−607	−1 839	19	21	0
2	147 533	−6 501	−3 987	151	59	−2	−1
3	−11 535	−5 668	350	114	−6	−1	
4	−5 359	486	153	−10	−2		
5	448	146	−13	−3			
6	99	−12	−3				
7	−8	−2					
8	−1						

$$G(x,y)=\frac{1}{2\pi}\exp i(x^2+y^2)\sum_{m,n=0}^{\infty}10^{-8}(a_{mn}+ib_{mn})T_m\left(\frac{2}{3}x-\frac{7}{3}\right)T_n(y-1)$$

$2<x<5,\ 0<y<2$

m	a_{m0}	a_{m1}	a_{m2}	a_{m3}
0	266 264	195 048	−72 676	2 437
1	−377 788	−262 957	115 858	−5 531
2	186 039	118 220	−66 837	5 132
3	−72 700	−40 681	30 557	−3 554
4	24 052	11 303	−11 685	1 938
5	−6 978	−2 551	3 861	−876
6	1 812	441	−1 124	341
7	−426	−42	291	−117
8	91	−7	−67	36
9	−18	5	14	−10
10	3	−2	−2	3
11		1		−1

m	a_{m4}	a_{m5}	a_{m6}	a_{m7}	a_{m8}	a_{m9}	a_{m10}
0	3 304	−677	−43	38	−4	−1	0
1	−5 479	1 221	66	−68	8	2	−1
2	3 338	−893	−26	50	−7	−1	
3	−1 585	540	−2	−30	5	1	
4	606	−275	11	15	−3		
5	−187	121	−10	−7	2		
6	44	−47	7	2	−1		
7	−6	16	−3	−1			
8	−1	−5	1				
9	−1	1	1				
10	1						

m	b_{m0}	b_{m1}	b_{m2}	b_{m3}
0	1 439 526	1 227 208	−228 612	−7 294
1	−1 475 841	−1 177 290	316 237	3 584
2	514 837	370 404	−148 729	3 539
3	−144 425	−89 452	54 517	−3 844
4	34 447	16 920	−16 557	2 167
5	−7 090	−2 261	4 279	−904
6	1 240	75	−945	308
7	−172	75	174	−89
8	13	−31	−24	23
9	2	8	1	−5
10	−1	−2	1	1

m	b_{m4}	b_{m5}	b_{m6}	b_{m7}	b_{m8}	b_{m9}	b_{m10}
0	8 301	−909	−156	54	−2	−2	0
1	−12 786	1 655	242	−95	5	3	−1
2	6 845	−1 181	−122	66	−5	−2	
3	−2 796	669	38	−38	4	1	
4	909	−311	−2	18	−3		
5	−235	122	−6	−7	2		
6	44	−41	4	2	−1		
7	−4	12	−2	−1			
8	−2	−3	1				
9	1	1					

$$G(x, y) = \frac{1}{2\pi} \exp i(x^2 + y^2) \sum_{m,n=0}^{\infty} 10^{-8}(a_{mn} + ib_{mn}) T_m\left(\frac{2}{3}x - \frac{7}{3}\right) T_n\left(\frac{2}{3}y - \frac{7}{3}\right)$$

$2 < x < 5,\ 2 < y < 5$

m	a_{m0}	a_{m1}	a_{m2}	a_{m3}	a_{m4}
0	300 700	−93 180	4 153	2 847	−1 169
1	−377 214	147 800	−14 564	−2 641	1 700
2	158 289	−75 198	12 323	−214	−669
3	−52 560	28 058	−6 269	801	103
4	14 853	−8 347	2 275	−479	42
5	−3 700	2 075	−636	179	−36
6	826	−442	142	−49	14
7	−166	81	−25	10	−4
8	30	−13	3	−1	1
9	−5	1			
10	1				

m	a_{m5}	a_{m6}	a_{m7}	a_{m8}	a_{m9}
0	279	−46	4	1	0
1	−474	87	−9	−1	1
2	267	−63	10	−1	
3	−96	32	−7	0	
4	19	−11	3	−1	
5	2	2	−1		
6	−3				
7	1				

m	b_{m0}	b_{m1}	b_{m2}	b_{m3}	b_{m4}
0	2 421 384	−134 435	−86 876	25 732	−4 551
1	−1 961 091	448 889	30 415	−27 141	6 965
2	522 286	−214 191	18 380	5 563	−2 852
3	−110 946	62 342	−12 362	594	595
4	19 842	−13 078	3 967	−770	10
5	−2 939	2 005	−834	274	−53
6	313	−182	109	−59	20
7	−3	−12	−2	6	−4
8	−10	10	−4	0	1
9	4	−3	1		
10	−1	1			

m	b_{m5}	b_{m6}	b_{m7}	b_{m8}	b_{m9}
0	540	−21	−10	4	−1
1	−1 101	87	10	−6	1
2	687	−100	4	2	−1
3	−256	59	−8		
4	53	−21	5		
5	0	4	−2		
6	−4				
7	1				

A.8 *Ff*-integral

$$Ff(x,y)=\exp(iy^2)\left[\frac{1}{4\pi}(\mathrm{Ci}(r^2)-i\cdot\mathrm{Si}(r^2))-\frac{3}{8}i-\frac{i}{2}\,\mathrm{sign}(r)\cdot F^*(|r|)\right]$$

$$+\exp(ix^2)\cdot 10^{-8}\sum_{m,n=0}(a_{m,n}+ib_{m,n})\left(T_m\left(\frac{-2}{5}x-1\right)T_n(2r)\right)$$

$-0.5<r=x-y<0.5;\ -5<x<0$

m	a_{m0}	a_{m1}	a_{m2}	a_{m3}	a_{m4}
0	5 879 219	3 935 174	−13 611 857	−4 312 545	5 246 915
1	−5 760 416	−1 639 786	−15 373 741	−5 677 831	8 167 958
2	−7 998 818	−7 201 986	1 439 067	−827 606	3 573 978
3	4 273 638	−111 532	3 914 308	918 543	495 262
4	280 420	1 335 142	363 272	362 707	−263 386
5	−286 407	−174 942	−376 128	−48 614	−108 307
6	31 224	−8 473	−12 649	−28 209	3 199
7	−18 072	−9 768	13 331	2 765	5 943
8	11 630	4 012	−3	554	3
9	−3 996	−373	−60	13	−147
10	989	−122	−87	−25	0
11	−150	91	28	2	2
12	−21	−40	−6		
13	28	13	1		
14	−14	−3			
15	5	1			
16	−1				

m	a_{m5}	a_{m6}	a_{m7}	a_{m8}	a_{m9}	a_{m10}
0	1 247 078	−858 952	−164 517	77 717	12 564	−4 481
1	2 018 496	−1 462 993	−285 163	138 007	22 536	−8 151
2	1 025 732	−890 071	−183 718	96 094	16 193	−6 119
3	261 559	−364 765	−84 704	51 547	9 199	−3 763
4	−17 368	−83 130	−25 220	20 489	4 021	−1 867
5	−30 226	126	−3 151	5 512	1 278	−728
6	−4 779	5 671	761	730	255	−212
7	1 132	1 086	359	−82	13	−41
8	272	−110	25	−51	−8	−3
9	−30	−46	−11	−5	−2	1
10	−5	1	−1	1		
11	0	1				

m	a_{m11}	a_{m12}	a_{m13}	a_{m14}	a_{m15}
0	−630	180	22	−5	−1
1	−1 153	332	42	−10	−1
2	−883	262	33	−8	−1
3	−563	175	23	−6	−1
4	−295	99	13	−4	
5	−124	46	7	−2	
6	−41	18	3	−1	
7	−10	5	1		
8	−1	1			

m	b_{m0}	b_{m1}	b_{m2}	b_{m3}	b_{m4}
0	−1 399 496	13 971 554	5 134 704	−9 636 860	−2 618 075
1	4 993 723	6 782 332	4 546 782	−13 589 330	−3 955 663
2	2 293 357	−12 298 799	−2 538 946	−3 629 798	−1 512 324
3	−3 242 788	−3 225 558	−2 026 550	1 023 575	−30 669
4	675 693	2 095 395	161 026	764 688	196 710
5	−92 016	109 289	216 817	13 667	47 291
6	67 716	−96 472	−22 558	−52 470	−8 618
7	−22 439	2 370	−3 683	−844	−3 019
8	1 998	−1 153	−519	1 522	306
9	1 234	1 484	327	6	59
10	−936	−486	−28	−15	−1
11	427	110	−6	−5	−2
12	−147	−17	4	1	
13	38	−1	−2		
14	−6	2	1		
15	0	−1			
16	1				

m	b_{m5}	b_{m6}	b_{m7}	b_{m8}	b_{m9}	b_{m10}	b_{m11}
0	2 314 855	490 316	−275 289	−48 223	19 618	2 948	−935
1	3 807 512	826 943	−480 309	−85 237	35 311	5 349	−1 715
2	2 054 521	486 399	−316 177	−58 476	25 647	3 983	−1 323
3	631 338	184 292	−152 362	−30 485	14 865	2 413	−852
4	37 293	32 849	−49 821	−11 505	6 719	1 169	−454
5	−44 423	−4 821	−8 630	−2 777	2 263	439	−197
6	−12 001	−3 632	483	−234	511	120	−67
7	802	−385	586	89	51	20	−17
8	564	120	82	30	−9	0	−3
9	−7	21	−11	1	−4	−1	
10	−12	−3	−3	−1			

m	b_{m12}	b_{m13}	b_{m14}	b_{m15}
0	−123	32	4	−1
1	−228	59	7	−2
2	−179	48	6	−1
3	−119	33	4	−1
4	−66	19	2	−1
5	−30	10	1	
6	−11	4	1	
7	−3	1		
8	−1			

$$Ff(x,y) = -i\exp(iy^2) + \frac{1}{r}\exp(ix^2)10^{-8}$$

$$\times \sum_{m,n=0} (a_{m,n} + ib_{m,n})\left(T_m\left(\frac{-2}{5}x - 1\right)T_n\left(1 - \frac{1}{r}\right)\right)$$

$-5 < x < 0;\ r = x - y > 0.5$

m	a_{m0}	a_{m1}	a_{m2}	a_{m3}	a_{m4}	a_{m5}
0	398 634	1 628 071	181 531	−103 291	−58 034	−7 680
1	−1 079 493	−2 426 288	−267 033	188 622	103 126	12 662
2	1 001 792	1 425 631	116 125	−148 913	−74 419	−6 781
3	−696 752	−700 344	−12 352	102 755	44 606	1 361
4	391 741	286 198	−29 259	−61 956	−22 236	1 692
5	−183 889	−91 573	32 685	32 525	8 926	−2 508
6	72 296	17 592	−22 544	−14 734	−2 535	2 091
7	−23 076	2 956	12 101	5 628	156	−1 336
8	5 289	−5 085	−5 352	−1 701	402	706
9	−326	3 126	1 961	317	−349	−316
10	−479	−1 388	−571	41	195	120
11	337	487	111	−75	−85	−37
12	−146	−132	3	44	31	8
13	48	23	−15	−19	−9	0
14	−12	1	9	6	2	−1
15	2	−3	−4	−2		1
16		2	1			
17		−1				

m	a_{m6}	a_{m7}	a_{m8}	a_{m9}	a_{m10}	a_{m11}	a_{m12}	a_{m13}	a_{m14}
0	6 608	4 806	1 244	−335	−487	−227	−27	40	35
1	−12 566	−8 878	−2 228	670	922	421	46	−77	−66
2	10 796	7 030	1 595	−653	−779	−338	−26	70	56
3	−8 320	−4 805	−888	593	588	233	4	−59	−43
4	5 706	2 839	339	−484	−397	−138	11	45	30
5	−3 463	−1 437	−26	350	238	68	−18	−32	−18
6	1 849	605	−95	−224	−126	−26	17	20	10
7	−862	−194	106	126	58	5	−13	−11	−5
8	343	31	−77	−63	−22	2	8	6	2
9	−111	16	44	27	7	−3	−5	−3	−1
10	25	−19	−21	−10	−1	3	2	1	
11	0	12	9	3	−1	−1	−1		
12	−4	−6	−3	−1	1	1			
13	3	2	1						
14	−1	−1							

m	a_{m15}	a_{m16}	a_{m17}	a_{m18}	a_{m19}
0	14	0	−4	−3	−1
1	−26	−1	7	5	2
2	21	0	−6	−5	−2
3	−15	1	5	4	1
4	9	−2	−4	−3	−1
5	−4	2	3	2	
6	1	−2	−2	−1	
7		1	1		
8		−1	−1		

m	b_{m0}	b_{m1}	b_{m2}	b_{m3}	b_{m4}	b_{m5}
0	4 942 629	132 555	−415 540	−116 023	13 534	23 343
1	−4 945 749	−15 907	707 134	195 485	27 815	−42 871
2	2 182 409	−220 201	−490 360	−123 815	28 062	33 638
3	−817 486	276 714	289 346	59 673	−25 073	−22 846
4	236 696	−217 101	−146 629	−19 861	19 119	13 500
5	−34 619	131 254	63 139	1 739	−12 439	−6 915
6	−14 661	−65 154	−22 211	3 579	6 954	3 021
7	16 115	27 015	5 603	−3 528	−3 350	−1 075
8	−8 980	−9 205	−345	2 177	1 381	269
9	3 739	2 380	−664	−1 052	−473	−8
10	−1 227	−309	517	418	122	−41
11	300	−113	−255	−136	−15	32
12	−37	109	97	33	−7	−16
13	−12	−52	−29	−4	7	6
14	11	19	6	−2	−3	−2
15	−5	−5	−1	1	1	1
16	2	1		−1		

m	b_{m6}	b_{m7}	b_{m8}	b_{m9}	b_{m10}	b_{m11}	b_{m12}	b_{m13}
0	8 373	−253	−1 863	−1 016	−186	126	128	54
1	−15 035	700	3 522	1 882	328	−246	−242	−100
2	10 962	−1 098	−2 976	−1 502	−221	227	206	81
3	−6 498	1 368	2 247	1 033	101	−194	−156	−56
4	3 044	−1 340	−1 514	−611	−11	151	106	33
5	−993	1 073	906	305	−36	−106	−64	−16
6	64	−725	−479	−122	47	67	34	6
7	211	421	221	33	−39	−38	−16	−1
8	−206	−212	−86	1	26	19	6	−1
9	131	92	26	−9	−14	−8	−2	1
10	−66	−34	−5	7	7	3	0	−1
11	27	10	−1	−4	−3	−1		
12	−10	−2	2	2	1			
13	3	0	−1	−1				

m	b_{m14}	b_{m15}	b_{m16}	b_{m17}	b_{m18}	b_{m19}
0	4	−12	−10	−4	0	1
1	−6	23	18	7	0	−2
2	2	−21	−16	−6	0	2
3	3	18	12	4	0	−2
4	−6	−14	−9	−3	1	1
5	6	10	5	1	−1	−1
6	−6	−6	−3	0	1	1
7	4	3	1			
8	−3	−2	−1			
9	1	1				
10	−1					

$$Ff(x,y)=\frac{1}{r}\exp(ix^2)10^{-8}$$

$$\times\sum_{m,n=0}(a_{m,n}+ib_{m,n})\left(T_m\left(\frac{-2}{5}x-1\right)T_n\left(\frac{-1}{r}-1\right)\right)$$

$-5<x<0;\ r=x-y<-0.5$

m	a_{m0}	a_{m1}	a_{m2}	a_{m3}	a_{m4}	a_{m5}	a_{m6}
0	2 185 120	26 818	−70 070	25 935	−8 160	2 388	−654
1	−3 253 559	364 563	66 125	−40 891	14 688	−4 544	1 287
2	1 849 485	−490 629	22 496	16 936	−9 397	3 525	−1 132
3	−855 604	367 312	−55 831	1 349	3 524	−1 984	790
4	320 783	−199 341	47 257	−7 928	136	660	−397
5	−89 941	83 002	−27 090	7 014	−1 316	48	113
6	11 858	−25 361	11 591	−4 027	1 138	−242	19
7	5 789	4 005	−3 572	1 696	−631	195	−47
8	−5 678	1 287	550	−502	254	−102	34
9	2 921	−1 467	194	65	−69	38	−17
10	−1 111	773	−211	35	6	−9	6
11	321	−292	108	−32	7	0	−1
12	−61	82	−39	16	−5	1	
13	−1	−14	10	−5	2	−1	
14	7	−1	−1	1	−1		
15	−4	2					
16	1	−1					

m	a_{m7}	a_{m8}	a_{m9}	a_{m10}	a_{m11}	a_{m12}	a_{m13}
0	164	−35	5	1	−1	1	0
1	−333	74	−11	−1	2	−1	1
2	326	−83	17	−2	−1	1	
3	−268	80	−21	4	0		
4	169	−61	19	−5	1		
5	−76	35	−14	5	−1		
6	17	−14	7	−3	1		
7	6	2	−2	1	−1		
8	−9	2					
9	6	−2					
10	−3	1					
11	1						

m	b_{m0}	b_{m1}	b_{m2}	b_{m3}	b_{m4}	b_{m5}
0	3 535 180	−887 239	138 573	−22 450	2 919	36
1	−2 522 796	1 294 788	−265 308	49 008	−7 384	378
2	542 561	−632 105	185 961	−44 706	9 051	−1 305
3	90 949	198 705	−93 828	30 220	−8 051	1 769
4	−180 918	−12 656	31 761	−14 941	5 202	−1 513
5	122 210	−33 407	−3 624	4 990	−2 455	921
6	−59 190	28 312	−4 075	−564	770	−407
7	22 516	−14 910	3 822	−621	−63	115
8	−6 573	5 890	−2 054	572	−109	−2
9	1 191	−1 737	807	−300	91	−21
10	105	304	−230	113	−45	15
11	−212	37	36	−30	16	−7
12	114	−60	8	4	−4	2
13	−43	31	−9	2		
14	12	−11	4	−1		
15	−2	3	−1	1		

m	b_{m6}	b_{m7}	b_{m8}	b_{m9}	b_{m10}	b_{m11}	b_{m12}
0	−250	140	−59	22	−8	2	−1
1	372	−244	109	−42	15	−5	1
2	−40	141	−80	34	−13	4	−1
3	−262	−19	38	−22	10	−4	1
4	363	−60	−3	9	−5	2	−1
5	−291	77	−15	0	2	−1	1
6	165	−57	17	−4	0		
7	−67	29	−11	4	−1		
8	16	−11	5	−2	1		
9	1	2	−2	1			
10	−4	1					
11	3	−1					
12	−1						

Index